HANDBOOK OF
CONSTRUCTION
CONTRACTING

Volume 2
Estimating, Bidding, Scheduling

By

Jack P. Jones

Craftsman Book Company
6058 Corte del Cedro / P.O. Box 6500 / Carlsbad / CA 92018

Acknowledgments

The author wishes to express his appreciation to the following companies and organizations for furnishing materials used in the preparation of various portions of this book:

American Plywood Association — Tacoma, Washington

Asphalt Roofing Manufacturers Association — Rockville, Maryland

Brick Institute of America — McLean, Virginia

The Celotex Corporation — Tampa, Florida

Georgia-Pacific Corporation — Atlanta, Georgia

Small Business Administration — Washington, D. C.

The Construction Specifications Institute — Alexandria, Virginia

The Burke Company — San Mateo, California

Symons Corporation — Des Plaines, Illinois

Dennison (National) — Holyoke, Massachusetts

Benjamin Moore & Company — Montvale, New Jersey

Caterpillar Tractor Company — Peoria, Illinois

This book is dedicated to Gregory

Library of Congress Cataloging-in-Publication Data

Jones, Jack Payne, 1928-
　Handbook of construction contracting

　Includes indexes.
　Contents: v. 1. Plans, specs, building -- v. 2.
Estimating, bidding, scheduling.
　1. House construction--Handbooks, manuals, etc.
2. Building--Estimates--Handbooks, manuals, etc.
I. Title.
TH4813.J66　　　1986　　　692'.8　　　86-8925
ISBN 0-934041-11-3 (v. 1)
ISBN 0-934041-13-X (v. 2)

©1986 Craftsman Book Company
Eighth printing 2004

Contents

Chapter 1

Introduction to Estimating 5
Establish Your Own
Construction Estimate File (CEF) 6
Using the Manhour Guides 7
Begin with the Plans 9
Before Estimating Begins 12
Compiling the Estimate 17
Checking the Estimate 20
Setting Your Price 21

Chapter 2

Keeping and Controlling Costs 23
Cost Keeping or Cost Accounting 23
Analysis of Unit Costs 24
The Masterformat 25
Keeping Accurate Labor Costs 29
Getting Cost Data from the Field 33

Chapter 3

Estimating Excavation 36
Surveying the Site 36
Measuring for the Excavation 40
Estimating Excavation 40
General Excavation 41
Trench and Pit Excavation 43
Backfilling 49
Excavating with Power Equipment 50
The Quantity Estimate 57
The Cost Estimate 60
The Unit Cost Estimate 61

Chapter 4

Estimating Concrete 67
Taking Off Concrete Quantities 67
Designing Concrete Forms 68
Designing Wall Forms 73
Floor Forms 79
Estimating Concrete Forms 80
Estimating Slabs 84
Reinforced Concrete 85

Chapter 5

Estimating Masonry 87
Estimating Concrete Block 87
Estimating Brick Masonry 91
Estimating Glass Building Blocks 98
Estimating Stonework 98
How to Estimate Fireplaces 104

Chapter 6

Estimating Rough Carpentry 110
Lumber Grading 110
Lumber Characteristics 113
Expressing Dimensions 113
Taking Off Lumber Estimates 114
Estimating Wall Sheathing 132
Estimating Rafters 135
Estimating Roof Areas 143
Rough Carpentry Labor 145

Chapter 7

Estimating Roof Covering 146
Estimating Surface Area for Complex Roofs . . 147
Estimating Asphalt Roofing 149
Additional Material Estimates 152
How to Estimate Wood Shingles 155

Chapter 8

Estimating Insulation 159
Forms of Insulation 159
Vapor Barriers 160
Estimating Insulation and Labor 161

Chapter 9

Estimating Doors and Windows 165
How to List Doors 165
Estimating Labor for Installing Doors 166
How to List Windows 168
Types of Windows 169

Chapter 10

Estimating Interior Wallboard.............**172**
Estimating Wallboard Materials............173
Labor for Wallboard Installation...........176
Estimating Lath and Plaster...............177
Figuring Surface Areas....................179
Estimating the Costs.....................181
Estimating Ornamental Plaster Work.......185

Chapter 11

Estimating Exterior Finish Carpentry........**186**
Estimating Siding........................186
Hardboard Shakes.......................191
Wood Siding.............................191
Estimating Corner Boards.................195
Estimating Cornices......................196

Chapter 12

Estimating Interior Finish Carpentry........**199**
Estimating Stairs.........................199
Stair Dimensions.........................200
Estimating Cabinets......................202
Estimating Wood Flooring.................204
Estimating Fireplace Mantels..............206
Estimating Molding and Trim..............207
Finish Carpentry Labor...................210

Chapter 13

Estimating Specialty Finishes.............**212**
Ceramic Tile............................212
Resilient Flooring........................213
Paints..................................215

Chapter 14

Scheduling Work Flow....................**221**
Introducing the Schedule..................222
The Critical Path Method.................223
CPM Scheduling.........................229

Chapter 15

Successful Management....................**235**
The Function of Management..............235

Building Your Management Ability..........236
Building a Strong Company................238
Building a Management Team..............239
How to Plan a Job.......................240
Record Keeping.........................244

Chapter 16

Finding the Work to Stay Busy.............**252**
The Basics of Advertising..................253
The Basics of Selling.....................254
The Process of Selling....................255

Chapter 17

Spec Building and Land Development.......**264**
Do the Research Before You Build..........267
Merchandising for the Spec Builder.........268
Land Developing.........................270

Chapter 18

Making a Business Plan...................**273**
What Business Am I In?..................275
Making the Marketing Decisions............276
Organizing to Get the Job Done............278
Is Your Plan Workable?..................284

Chapter 19

Selecting the Legal Structure for Your Firm...**286**
The Sole Proprietorship...................286
The Partnership.........................287
The Corporation.........................288

Chapter 20

Your Business and the SBA................**290**
Kinds of Financial Assistance Available......290
Management Assistance Program...........291
How to Get Help from the SBA............292
SBA Field Offices.......................294

Index...................................**297**

Introduction to Estimating

The best builder in town won't make a dime if his estimates are bad. This chapter will suggest an estimating system that can help you compile estimates that are consistently reliable — and keep your business growing and profitable.

Your skill as an estimator is the key to business success. No construction company is better than its estimates. And in most small construction companies the owner is the chief estimator. No one else can be trusted to make important decisions about selling prices. That's why every successful builder needs good estimating skills: knowledge of construction, an organized approach to compiling costs, the care required to produce valid estimates, and an instinct for situations that can make a job more or less expensive than other similar jobs.

There are four ways to estimate any job:

The Guess Estimate, which is just that, a guess, based on rules of thumb and vague recollections of past experience. Bidding by "guesstimate" is a fine idea if you intend to get into some other line of work in couple of months.

The Area Estimate is based on a cost factor for area alone, applied to the proposed area without consideration for any variables. This method of estimating can provide you with some big surprises, including fabulous profits on some jobs and disastrous losses on others.

The Piece Estimate is by far the most accurate method of determining job cost. You list each and every piece of material as well as the labor necessary to do each step of the job. If you start out with proper plans and a detailed job survey, then take each step systematically and comprehensively, you're pretty certain to come out with an accurate estimate. You also have a material list to use in ordering and coordinating the required materials. This estimating system takes time and will be accurate if you remember to include every cost item in the job.

The Unit Cost Estimate combines the principles of the area estimate and the piece estimate. You use a unit cost for each material, installed, for a given area. If your unit costs are correct, you can prepare an estimate that's nearly as accurate as the piece estimate, but in a lot less time. It's absolutely essential, however, to keep up-to-date unit costs based on current material and labor costs.

If you don't have a unit cost guide based on your own records and experience, I recommend the *National Construction Estimator* and the *Building Cost Manual,* published by Craftsman Book Company. There's an order form at the back of this book. These books are revised every year and are, for the price, probably the best sources available on the subject.

Labor Placing Concrete Block

Work Element	Unit	Man-Hours Per Unit
Concrete block, lightweight		
4" block	100 S.F.	10.50
6" block	100 S.F.	11.70
8" block	100 S.F.	12.80
10" block	100 S.F.	15.00
12" block	100 S.F.	17.90
Concrete block, hollow, standard weight		
4" block	100 S.F.	11.00
6" block	100 S.F.	12.00
8" block	100 S.F.	13.00
10" block	100 S.F.	15.00
12" block	100 S.F.	18.10

Time includes set-up, clean-up, joint striking one side only, cutting, pointing, steel alignment and grout.
Suggested Crew: Small jobs, 1 mason, 1 helper

Construction estimate file (CEF) — concrete block
Figure 1-1

Establish Your Own Construction Estimate File (CEF)

No matter which estimating method you use, there's no substitute for your own manhour and productivity figures, based on your own crew and work methods. You can't have accurate estimates without accurate guidelines for labor. Throughout this volume you'll find examples of labor data and guidelines for filing in your *Construction Estimate File.* I call these *CEF* forms.

To make a basic CEF, you'll need good estimating references. One that I can recommend is *Construction Estimating Reference Data* published by Craftsman Book Company. The address is on the order form in the back of this book. Photocopy the charts that are pertinent to your work and tape them to standard 5" x 8" index cards. Set up a file box for the cards, and you've got a Construction Estimate File with labor guidelines at your fingertips.

Start with the information you'll find in this volume and in your other references, but don't stop there. Change the data if it's not accurate for your conditions. Add cards with information that reflects your experience. Continually update the cards to keep up with the changes in your operation. Your file has to change with the times, just like your business. Of course, it'll take time to keep the file accurate and up-to-date, but it will save you lots more time when you use it to compile fast, accurate estimates. And believe me, it'll save money, too.

Your CEF will not, except in a few instances, show price or cost. Material prices and labor pay rates are constantly changing and vary from location to location. You must use current local prices and pay rates or your estimate won't be valid. Since material prices change on a daily basis, my practice is to obtain current prices after completing the take-off. Pay rates for your tradesmen and laborers are easier to keep up with since they're not subject to daily change.

How many manhours will it take to lay the foundation block? Refer to your CEF for a quick answer. Figure 1-1 shows what your CEF may look like.

How many manhours will it take to frame the floor? Refer to your CEF. (Look at ours, Figure 1-2.)

Labor for Rough Carpentry		
Work Element	Unit	Man-Hours Per Unit
Mudsill, 2″ x 6″		
Bolted	1000 B.F.	21
Shot	1000 B.F.	18
Basement beams (girders)		
2″ x 8″, built-up	1000 B.F.	33
2″ x 10″, built-up	1000 B.F.	25
Basement posts	1000 B.F.	18
Box sills	1000 B.F.	29
Floor joists		
2″ x 6″ to 2″ x 8″	1000 B.F.	16
2″ x 10″ to 2″ x 12″	1000 B.F.	14
Headers, tail joists and trimmers		
2″ x 6″ to 2″ x 8″	1000 B.F.	16
2″ x 10″ to 2″ x 12″	1000 B.F.	14
Bridging 2″ x 3″	50 sets of 2	4
Subflooring, boards		
Straight	1000 B.F.	13
Diagonal	1000 B.F.	15
Subflooring, plywood, 4′ x 8′	1000 S.F.	10
Stud walls, including plates, blocks and bracing		
2″ x 4″	1000 B.F.	22
3″ x 4″	1000 B.F.	21
2″ x 6″	1000 B.F.	20
Ceiling joists		
2″ x 6″ to 2″ x 8″	1000 B.F.	16
2″ x 10″ to 2″ x 12″	1000 B.F.	14
Ceiling backing		
2″ x 6″ to 2″ x 8″	1000 B.F.	15
2″ x 10″ to 2″ x 12″	1000 B.F.	14
Attic floor	1000 B.F.	13
Headers for wall openings		
2″ x 4″	1000 B.F.	25
2″ x 6″	1000 B.F.	20
Gable-end studs	1000 B.F.	20
Fire stop wall blocks	1000 B.F.	20
Corner braces	1000 B.F.	20

Labor For Rough Carpentry (continued)		
Work Element	Unit	Man-Hours Per Unit
Partition plates and shoe	1000 B.F.	20
Partition studs	1000 B.F.	20
Wall backing	1000 B.F.	20
Grounds	1000 L.F.	12
Knee wall plates and studs		
2″ x 4″	1000 B.F.	25
2″ x 6″	1000 B.F.	25
Wall sheathing, boards		
1″ x 6″ diag., includes paper	1000 B.F.	15
1″ x 8″ diag., includes paper	1000 B.F.	14
1″ x 10″ diag., includes paper	1000 B.F.	13
Wall sheathing, plywood		
4′ x 8′ sheets, includes paper	1000 S.F.	11
Wall sheathing, composition		
½″	1000 S.F.	9
¾″	1000 S.F.	10
1″	1000 S.F.	11
Siding, plywood, 4′ x 8′ sheets	1000 S.F.	13
Corner boards	1000 B.F.	40
Common rafters	1000 B.F.	17
Hip, valley, jack rafters	1000 B.F.	29
Roof sheathing, boards		
1″ x 6″, S4S	1000 B.F.	15
1″ x 6″, center match	1000 B.F.	18
1″ x 8″, shiplap	1000 B.F.	17
1″ x 10″, shiplap	1000 B.F.	13
Roof sheathing, plywood, 4′ x 8′ sheets	1000 S.F.	12
Window and door headers	Each	.6
Make and install rough door buck	Each	1.2
Furring concrete or masonry walls	1000 L.F.	46
Wood plaster grounds on masonry	1000 L.F.	38
Attic stairways	Each	10
Basement stairways	Each	7

Time includes layout, all precutting, stacking, repairing and clean-up as required.
Suggested Crew: 2 carpenters and 1 laborer

CEF — rough carpentry
Figure 1-2

How many brick will you need to veneer this house? Your CEF will tell you. (See Figure 1-3.)

How many manhours will it take to lay the brick on this job? Go to your CEF. (See Figure 1-4.)

The CEF will help you lower future costs and make it easier to compare jobs being estimated. If your CEF, for example, shows that it takes 7.5 manhours to brush one coat of paint on 1,000 SF of exterior wood siding, and it actually took 10 hours, you can check on the conditions the workmen faced. You might also want to check on the workmen. Why did it take so long? You've learned something for future use.

In figuring unit costs, bear in mind the season of the year when the work is to be done. A carpenter, for instance will be more productive on a warm, sunny day than when the cold is stiffening his hands and interfering with his movements. Keep a close record on the average output of tradesmen for all seasons of the year and include these results in preparing estimates. Record rainy seasons too.

Using the Manhour Guides

Experienced construction estimators recognize that no two jobs are exactly alike. Labor productivity varies widely from job to job, even if the crew remains the same. Job progress in hot summer months will be different from the progress made in cold winter months. Thus, judgment is an essential element in estimating any construction project. And judgment will be required when using the labor guidelines in this book.

The manhour guides in this volume *are not* based on "ideal" conditions. They assume the kind of conditions most contractors encounter on better planned and managed jobs. The labor productivity indicated in the guides will be accurate to the extent that these conditions apply to the job you're figuring.

Number of Nominal 2⅔" x 4" x 8" Modular Brick in Common Bond per S.F. of Wall									
	Nominal 4" Wall			Nominal 8" Wall			Nominal 12" Wall		
Square Feet Wall Area	No. of Brick	C.F. Mortar		No. of Brick	C.F. Mortar		No. of Brick	C.F. Mortar	
		3/8" Joint	1/2" Joint		3/8" Joint	1/2" Joint		3/8" Joint	1/2" Joint
1	6.75	.07	.08	13.5	.16	21	20.25	.26	.33
10	67.5	.66	.83	135	1.62	2.08	202.5	2.59	3.33
20	135	1.31	1.67	270	3.25	4.17	405.0	5.18	6.67
30	202.5	1.97	2.50	405	4.87	6.25	607.5	7.78	10.01
40	270	2.62	3.34	540	6.50	8.34	810.0	10.37	13.34
50	337.5	3.28	4.17	675	8.12	10.42	1012.5	12.96	16.68
60	405	3.93	5.00	810	9.74	12.51	1215.0	15.55	20.01
70	472.5	4.59	5.84	945	11.37	14.59	1417.5	18.15	23.35
80	540	5.25	6.67	1080	12.99	16.68	1620.0	20.74	26.68
90	607.5	5.90	7.51	1215	14.62	18.76	1822.5	23.33	30.02
100	675	6.56	8.34	1350	16.24	20.85	2025.0	25.92	33.35
200	1350	13.12	16.68	2700	32.48	41.69	4050	51.85	66.71
300	2025	19.67	25.02	4050	48.72	62.54	6075	77.77	100.06
400	2700	26.23	33.36	5400	64.96	83.38	8100	103.70	133.41
500	3375	32.79	41.70	6750	81.20	104.23	10125	129.62	166.76
600	4050	39.35	50.04	8100	97.45	125.08	12150	155.54	200.12
700	4725	45.91	58.38	9450	113.69	145.92	14175	181.47	233.47
800	5400	52.46	66.72	10800	129.93	166.77	16200	207.39	266.82
900	6075	59.02	75.06	12150	146.17	187.61	18225	233.32	300.18
1000	6750	65.58	83.40	13500	162.41	208.46	20250	259.24	333.43

CEF — modular brick
Figure 1-3

Bricks

Mortar Joint	Wall Thickness	Material				Labor 100 Square Feet Wall	
		Brick		Wall Ties	Mortar		
		100 Sq. Ft. Wall	Sq. Ft. Per 1000 Brick	Per 100 Square Ft.	Cubic Ft. Per 100 Sq. Ft.	Mason	Laborer
1/4	4"	698	143	100	4.48		
3/8	4"	655	153	93	6.56	6½ Hours Average	5 Hours Average
1/2	4"	616	162	88	8.34		
5/8	4"	581	172	83	10.52		
3/4	4"	549	182	78	12.60		

NOTE: Mortar includes 20% waste for all head and bed joints.

CEF — brick, general data
Figure 1-4

The guidelines apply only to new construction. Repair, replacement and remodeling usually involve different problems, like difficult access, trying to match materials, working with nonstandard sizes, patching, and control of the construction environment. Your CEF will, however, be a useful guide for repair or remodeling work that's similar to new construction.

The guidelines and estimating data presented in this manual are the result of actual observations compiled, interpreted, and verified by professional estimators. But there's no guarantee that the figures used here will apply to the job *you're* estimating. As a rule, though, the manhour estimates presented in the CEF samples will be accurate within about 20% on most jobs where conditions are similar to the conditions outlined. On most of the remaining jobs the figures will be too high by 20% or more, resulting in more manhours than are actually required. This is intentional, as an estimate slightly too high is better than one too low.

The CEF guidelines, when used with judgment and modified with your experience, will help you develop reliable estimates. But remember that estimating is more art than science. The first construction estimator probably underestimated the time required to hollow out the cave he planned to move into. Procedures have changed a lot since then. But the results haven't necessarily improved very much. On many jobs the difference between high bidder and low bidder will be 20% or more. And all bids may be right! At least right for the contractors that submitted them.

There's room for legitimate disagreement on what the correct cost is, even when complete plans and specifications are available and when all bidders are buying materials from the same suppliers and paying the same labor rates. No cost fits all jobs. Good estimates are custom-made for a particular project and a single contractor through judgment, analysis and experience.

Begin with the Plans

When you receive the plans, specifications and perspective drawings, study them carefully. Try to get a mental picture of the proposed building. Examine the specifications; review the plans, giving special attention to details; make a detailed take-off using the procedures I'm going to recommend; then double-check for omissions. This is the best way to make sure you've considered all the costs in any job.

The working drawings should include all plans, elevation views and sections, as well as any necessary detail drawings showing special construction or finish details. Properly drawn plans or working drawings, made to scale and correctly read, should give you a mental picture of the completed building. Detail drawings and sectional views show in detail how the structure is to be built and the material to be used. If they're not available, refer to the specifications for information on these points.

Over the years, I've developed the habit of *looking* for errors in plans and specifications. I've found that identifying errors is the best way to anticipate problems.

Reading the Floor Plans and Elevations

A *floor plan* is the drawing of a horizontal section cut through the building walls, showing the exact position of windows and doors. *Elevations* represent the sides of buildings and show the heights of doors and windows, the pitch of the roof, and other details which can't be made plain on a floor plan. A *sectional drawing* is a representation of the construction of part of the building, showing what members or parts go into the structure.

No matter how well a plan is worked out, the architect can't always include every minor item necessary for the builder's guidance. In almost every plan there's some point at which you must make use of your technical training and visualize what's needed. This should be one of the purposes of your preliminary examination.

Picture in your mind the kind of house or structure represented. Compare the rooms with others you've built or are familiar with. Study the heights, compare the various dimensions with those you consider ideal for similar purposes. Acquire the habit of visualizing and constructively criticizing every plan that you examine. Have you had problems with a similar plan in the past? What were those problems? How were they solved? Did they affect the estimate or building costs?

The specifications have detailed information and descriptive instructions needed to build the structure. If there's a point you don't understand in the drawings, the answer is probably in the specs. The principal object of the specifications is to define the general conditions under which the work is to be done, and to describe the quality of materials to be used — points not easily shown on a plan. Specifications stipulate the kind and quality of labor and materials desired. They should contain all the written instructions and descriptions you need to do the job. If they don't, take it up with the architect.

The Basement Plan

Study each drawing. Start with the basement plan if there is one. It's the base on which the structure rests. The floor above follows logically both in structure and in development.

The first thing you observe when looking at the basement plan are the foundation lines. The outer line represents the outside of the foundation upon which the house or building will be built. The inner line represents the inside line of the foundation. Figures between these lines indicate the thickness of the foundation walls. Note the various dimensions indicated, adding them to see if the totals check with the overall lengths indicated. If the totals don't agree, there's an error somewhere in

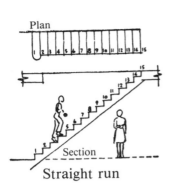

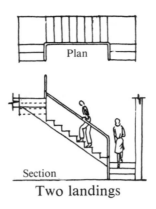

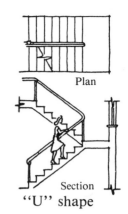

Straight run   Two landings   "U" shape

Typical stair sections
Figure 1-5

the figuring, either on your part or by the draftsman.

You can sometimes correct the error yourself by comparing plans, checking figures, and using the scale where figures are incorrect. If you can't discover the error, or if you just want to make absolutely sure, call the architect for the solution.

Notice the footings. If they're not shown with the foundation walls, find the information on one of the other drawings. Check the relation of the grade to the top of the foundation wall, look at the upper and lower floors to determine the finish ceiling, then consult the specifications to confirm this information. Note any special accommodations such as laundry room, drainage outlets, layouts for heating and cooling equipment, fuel storage, closets, tubs, toilets, and so on. Examine in detail the placement of supporting columns or special piers for porch or stairs.

The elevations will show the height of the foundation above the ground, the style of basement windows, and other particulars. If your study of the basement plan shows that every feature is correct and needs no modification or added data, proceed to the next plan.

The First Floor Plan
As with the basement, study this plan in a general way to get an understanding of the relationships. Then note the details of each room: dimensions, special cabinet work, closets, and so on. If bookcases, china closets, fireplace and similar items are included, look for detail drawings on a larger scale.

Windows, shown by two parallel lines, should be checked for location and style. Are they grouped,

mullion, or triple? You may have to refer to the elevations for this. Then carefully check doors for number, position, and swing. Note the kitchen cabinets and all shelving.

Next examine the stairs. Look at the risers, steps, platforms and any special construction. Form a mental picture of the staircase. Is it located between two walls, or is it open on one side, requiring posts, rails and balusters? Check the basement stairs. Determine whether stairs are single flight with a straight run or have turns and landings. Figure 1-5 shows some typical stair sections.

Are there fireplaces? Note the details. Figure 1-6 is an example. You'll have to figure each fireplace down to the face brick, lining required, hearth, and ashpit — unless you're subbing it out. You'll probably find notes on the plan showing other special

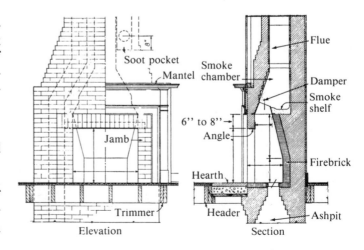

A typical fireplace plan
Figure 1-6

features, such as seating and bookcases worked around the chimney.

Look for and note electric outlets, plumbing, heating and cooling, interior finishes and decorating requirements.

Finally, study any porches on the first floor plan, and look at the elevations for the lines which represent steps, posts and other details. You may find large scale drawings of details of the porches.

The Second Floor Plan

When you look at the second floor plan, you'll find the same outline as the first floor, except where dotted lines indicate a slight variation. Everything you checked on the first-floor plans needs to be checked here also — partitions, doors, windows and other standard features. In addition, you'll probably find full-size bathrooms showing fixture arrangements. The conventional way of showing these seldom varies from one architect to another.

Look for the continuation of the chimney, as well as all flues. Your plan may also show porch roofs and the roof covering to be used.

Note any special construction in the second floor. At the same time, compare all the construction methods indicated with similar work you've done before. This might suggest methods of construction that will save you time and money. Be sure not to overlook the placement of electric outlets, location of plumbing, heating/cooling, and interior finishes.

This extensive examination of the plans should include the roof, cornices, ridges, vents, gutters, and so on. The plans, attached notes, and specs should contain everything necessary for you to correctly erect the building, from basement to roof. If not, aside from certain details which are assumed in all construction, consult the architect who prepared the plans. Note any omissions or errors you discover.

The Elevations

Now study the front, rear and two side elevations. From the elevations, you'll get all necessary information about the heights, widths and style of all doors and windows, the shape and slope of the roof, the style of cornice, porches, balustrades and outside trim. Conventional drawings or notations detail the materials used for wall and roof coverings. Together with the floor plans and specifications, these elevations give you a true mental picture of the building. You can read the dimensions

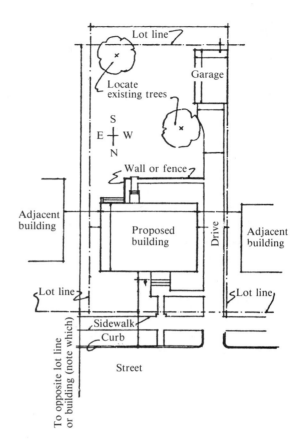

Plot plan
Figure 1-7

directly or scale them off where they're not given.

The plans may include large scale drawings to show the style and construction of such things as cornices, porches or bay windows. Additional scale sections through walls or rooms provide the sizes of joists, rafters, or moldings.

The Plot Plan

With large building plans, the architect often includes a *plot plan*. This shows the relative position of the house on the lot and the relationship of outbuildings. It also shows any wells, gas and water mains and connections, septic tank or sewer connections, sidewalks, driveways and fences. All of these features are essential to the estimator. Figure 1-7 shows a typical plot plan.

The preliminary reading of plans should give you a picture of the entire building structure and the relationship of all the parts. You acquire a general knowledge of all the requirements, perhaps come across mistakes, make slight corrections that are self-evident, and end up in a better position to intelligently take off material quantities.

There's one more step before you begin listing materials. Unless you're already familiar with it, visit the site.

Examine the Site

On your visit to the site, determine the soil type, the levels of the lot, the distance from sources of material supply, and accessibility for material delivery.

Preparation of the site is always necessary before building can begin. You want to learn what preparation is necessary and see what space is available for receiving and storing materials. A thorough estimator is seldom taken by surprise.

Learn what you can about the subsoil and the presence of possible stone ledges or water below the surface. Special timbering or other costly preparations may be required. Drill test holes where you're in doubt. Ground water can be a serious problem. The contract may specify a waterproof job. That's usually O.K., but in some areas "waterproof" means you'll have to build the structure like a ship.

You may also find on your visit to the lot that the building will go on a hillside, or some distance from existing roads. There may not be water available, or the site may be densely covered with trees and underbrush. In the city, you may be so hemmed in by buildings or fences that there's no space to handle materials. Also, keep in mind that operation of heavy equipment close to a building can cause damage due to vibration. These things are certain to affect your costs and final bid. So don't just visit the site. *Study* it. A patient and complete examination can save you a bundle.

Before Estimating Begins

Don't rush the estimating. Prospective home owners and others are always in a hurry to know *how much it's going to cost.* I've always advised such people that no one can know what a structure will cost until every stick and nail is figured. Then I might offer a wide range of possible square foot prices, based on similar houses I've built.

Leave the estimating of plumbing, electrical work, heating, ventilating and air conditioning, and extensive excavating to people with thorough knowledge of these trades, especially if the job under consideration is a large one. In securing bids

on such work, select people you feel you can trust. And get subs who have established reputations in their respective fields. You'll find it's worth the little extra you may have to pay.

Your Estimating Workbook

Develop an orderly, systematic method of approaching your estimating problems. Make up your sheets in the order in which you'll need them on the job. Preparing the site is usually the first consideration, followed by excavation, foundation, floor framing, wall framing, and so on, as indicated by the sequence of estimating chapters in this book. Your estimating workbook might be a loose-leaf notebook with 8½" x 11" pages. You'll need a number of sheets for each category, but all sheets should bear the same job or estimate number in the upper right-hand corner. You might want to include the title of the building, the owner's and architect's names, the date, and any other necessary identifying information.

In taking off quantities, list the materials in the order in which they'll be used in the building. This habit of estimating according to a definite routine reduces the chances of overlooking items. Keep each section of the estimate separate as far as possible. In this way, should any errors occur or prices change, you can make changes in one section without affecting other sections. Take off each item separately, but you may group small items so that each group or class can be found easily. The work of pricing small items is simplified by fewer operations or multiplications.

Once you've completed the estimating on the worksheets, transfer the figures to a summary sheet like the one in Figure 1-8. Now you've got all the figures in one place for easy reference. The summary sheet also acts as a checklist that can remind you of items you've forgotten.

Uniform Classification of Operations

Decide at the beginning what terms you'll apply to the different building components, and stick with them. It's easiest to follow a standard numbering system which assigns a number for every major expense. I recommend the Masterformat, which we'll talk about in detail in the next chapter. Each cost category has a number which you use both in estimating and in accounting for project costs. You may not need a system this detailed. But it's important to find and use a standard list which meets the needs of your operation.

ESTIMATE SUMMARY

Owner _____ Date _____ , 19____

Owner's Address _____ Telephone _____

Job Address _____ Lot _____ Blk. _____ Tract _____

GENERAL CONDITIONS	LABOR	OTHER	TOTAL		LABOR	OTHER	TOTAL
Supervision:				SUB TOTAL FORWARD			
Superintendent				Temporary Utilities (Cont.)			
General Foreman				Light			
Foreman				Water			
Master Mechanic				Heat			
Engineer				Fuel			
Timekeeper				Temporary Toilets			
Assistant Timekeepers				Surveys			
Payroll Clerk				Photographs — Damage And Progress			
Material Checkers				Lost Time, Weather Etc.			
Watchman				Removing Utilities			
Waterboy				Cutting And Adjusting For			
Others				Sub Contractors			
Permits:				Repairing Damage			
Blasting				Patching After Sub Contractors			
Building				Cleanup, General			
Sidewalk Bridge				Clean Windows			
Street Obstruction				Clean Floors			
Sunday Work				Cleanup After Sub Contractors			
Temporary				Removing Debris From Job			
Wrecking				Signs			
Other				Pumping			
Bonds:				Protection Of Construction			
Completion — Bond				Protection Of Adjoining Land			
Maintenance — Bond				And Buildings			
Street Encroachment — Bond				Barricades			
Street Repair — Bond				Temporary Fences			
Insurance:							
Workmen's Compensation				SPECIAL CONDITIONS			
Builder's Risk Fire Insurance				Plant And Equipment. (See Checklist)			
Completed Operations Public				Acoustical			
Liability Insurance				Air Conditioning			
Equipment Floater Insurance				Architectural Concrete			
Public Liability Insurance				Architectural Terra Cotta			
Truck And Automobile Insurance				Bins			
Licenses:				Cabinetwork:			
Local Business License				Book Cases			
State Contractor's License				Cases			
Taxes:				Coat Closets			
Excise — Taxes				Counters			
Payroll Taxes				Displays			
Sales Taxes				Linen Closets			
Field Office:				Metal Cabinets			
Owner's				Special Cabinets			
Job				Stationary Cabinets			
Maintenance Of Office				Store Fixtures			
Telephone:				Telephone Booths			
Owner's				Wardrobes			
Job				Wood Cabinets			
Job Office Supplies				Other			
Shanties:				Cellular — Steel Floors			
Storage				Cofferdams			
Tool				Concrete:			
Watchman's				Admixtures			
Transportation Of Equipment				Columns			
Delivery Charges				Curbs And Gutters			
Travel Expense				Flatwork			
Temporary Utilities:				Floors			
Power				Footings			

Estimate summary
Figure 1-8

	LABOR	MATERIAL	TOTAL	
SUB TOTAL FORWARD				
Concrete (Cont.)				
Foundations.				
Frost Protection.				
Pumpcrete.				
Sidewalks.				
Slabs.				
Vacuum Concrete.				
Walls.				
Walks.				
Other.				
Concrete Curing Compounds.				
Contingencies.				
Conveyors.				
Corrugated Steel:				
Roofing.				
Siding.				
Culverts:				
Concrete Box.				
Concrete Pipe.				
Corrugated Metal.				
Metal Arch.				
Doors:				
Exterior Wood:				
Panel.				
Flush.				
Dutch.				
French.				
Exterior Metal:				
Revolving.				
Kalamein.				
Overhead.				
Garage.				
Screen.				
Tin Clad.				
Industrial.				
Other.				
Glass.				
Plastic.				
Interior Metal:				
Shop.				
Office.				
Interior Wood:				
Panel.				
Flush.				
Sliding.				
Frames:				
Metal.				
Wood.				
Electric Fixtures.				
Electric Wiring:				
Building Wiring.				
Service.				
Power Wiring.				
Motors.				
Electric Signs.				
Other.				
Elevators.				
Excavation:				
Clearing And Grubbing.				
Removing Obstructions.				
General Excavation.				
Blasting.				
Shoring.				
Structural Excavation.				
Trench Excavation.				
Backfill.				
Rough Grading.				
Fine Grading.				
Rock.				
Fences And Railing.				
Fire Alarm.				
Fireproofing.				

	LABOR	MATERIAL	TOTAL	
SUB TOTAL FORWARD				
Flooring:				
Asphalt Tile And Linoleum.				
Composition.				
Cork.				
Flagstone.				
Hardwood.				
Marble.				
Rubber.				
Tile.				
Slate.				
Terrazzo.				
Formwork:				
Arches.				
Beams				
Beam And Slab Floors.				
Bridge Piers.				
Caps.				
Columns.				
Fiber Tubes.				
Flat Slabs.				
Floor Pans.				
Footings.				
Foundation Walls.				
Girders.				
Metal Pans.				
Movable Forms.				
Other.				
Foundations:				
Wall Footings.				
Piers.				
Spread Footings.				
Piles, Bearing:				
Wood.				
Steel.				
Concrete.				
Piles, Sheet:				
Wood.				
Steel.				
Concrete.				
Caissons.				
Frost Protection.				
Hardboards.				
Hardware:				
Rough.				
Finish.				
Heating.				
Incinerator.				
Insulation:				
House.				
Cold Storage.				
Piping.				
Rigid.				
Flexible.				
Foil.				
Other.				
Lath:				
Metal:				
Ceilings.				
Exterior — Walls.				
Interior — Walls.				
Soffits.				
Partitions.				
Beads.				
Plaster Boards:				
Walls.				
Ceilings.				
Arches.				
Lift Slab.				
Loading Dock.				
Lumber Construction:				
Heavy:				
Beams.				

Estimate summary
Figure 1-8 (continued)

	LABOR	MATERIAL	TOTAL	
SUB TOTAL FORWARD				
Lumber Construction Heavy (Cont.)				
Blocks.				
Braces.				
Built-Up Beams.				
Caps.				
Centering.				
Columns.				
Falsework.				
Girders.				
Girts.				
Joists.				
Lagging.				
Laminated Members.				
Plank And Laminated Floors.				
Planking.				
Plates.				
Posts.				
Sheeting.				
Stringers.				
Timber Purlins.				
Trusses.				
Trussed Beams.				
Wales.				
Wedges.				
Light:				
Beams.				
Blocking.				
Bracing.				
Bridging.				
Columns.				
Cripples.				
Dormers.				
Fascia.				
Furring.				
Girders.				
Half Timber Work.				
Headers.				
Hips.				
Jacks.				
Joists, Floor, Roof & Ceiling.				
Outlooks.				
Pier Pads.				
Plates.				
Plywood, Flooring, Sheathing & Roofing.				
Posts.				
Rafters.				
Ribbons.				
Ridges.				
Roof Trusses.				
Sheathing — Roof, Wall.				
Sills.				
Steps.				
Studs.				
Subfloor.				
Trimmers.				
Valleys.				
Manholes.				
Marble.				
Masonry:				
Ashlar.				
Common Brick.				
Fase Brick.				
Concrete Blocks.				
Precast Concrete Panels.				
Clay Tile.				
Dimension Stone.				
Rubble Stone.				
Flue.				
Metal Work:				
Art Metal Work.				
Base.				

	LABOR	MATERIAL	TOTAL	
SUB TOTAL FORWARD				
Metalwork Cont.)				
Casings.				
Chair Rail.				
Column Guards.				
Cornices.				
Elevator Entrances.				
Fire Escapes.				
Freight Doors.				
Grillwork.				
Information Boards.				
Linen Chutes.				
Lintels.				
Mail Chutes.				
Metal Doors And Frames.				
Platforms.				
Railings.				
Shop Front.				
Shutters.				
Stairs.				
Treads.				
Trap Doors.				
Transoms.				
Wheel Guards.				
Wainscoating.				
Other.				
Millwork And Finish:				
Base.				
Built-Ins.				
Casings.				
Caulking.				
Corner Boards.				
Cornice.				
Doors.				
Frames.				
Jambs.				
Mantels.				
Molding.				
Paneling.				
Sash.				
Screens.				
Shelving.				
Siding.				
Sills.				
Stairs.				
Stops.				
Storm Doors.				
Trim.				
Windows.				
Wood Carving.				
Other.				
Painting And Decorating:				
Aluminum Paint.				
Doors And Windows.				
Finishing.				
Floors.				
Lettering.				
Masonry And Concrete.				
Metal.				
Paperhanging.				
Plaster.				
Roofs.				
Shingle Stain.				
Stucco.				
Wood.				
Pavements:				
Asphalt.				
Base.				
Block.				
Brick.				
Concrete.				
Wood.				
Other.				

Estimate summary
Figure 1-8 (continued)

	LABOR	MATERIAL	TOTAL		LABOR	MATERIAL	TOTAL
SUB TOTAL FORWARD				**SUB TOTAL FORWARD**			
Pipelines:				Structural Steel (Cont.)			
Cast Iron Pipelines.				Painting Steel And Removing Rust.			
Concrete Pipelines.				Rivets.			
Corrugated Pipelines.				Steel Stacks.			
Steel Pipelines.				Welding.			
Trenches.				Engineering And Shop Details.			
Vitrified Tile Pipelines.				Inspection.			
Plaster:				Freight.			
Bases.				Unloading.			
Coves.				Tanks.			
Cement:				Test Holes.			
Exterior.				Thresholds.			
Interior.				Tile.			
Interior.				Tilt-Up Concrete.			
Keene's Cement.				Trench Shoring.			
Models.				Vaults And Vault Doors.			
Ornamental.				Wallboard.			
Perlite.				Waterproofing.			
Special Finish.				Weatherstrips.			
Plumbing:				Well Drilling.			
Interior.				Windows:			
Exterior.				Frames:			
Prestressed Concrete.				Wood.			
Railroad Work.				Steel.			
Reinforcing Steel:				Other.			
Bars.				Sash:			
Mesh.				Wood.			
Spirals.				Steel.			
Stirrups.				Aluminum.			
Retaining Walls.							
Roofing:				Sub Total:			
Asbestos.				Contingency.			
Asphalt Shingles.				Profit.			
Built-Up.							
Concrete.				Total:			
Copper.							
Corrugated:							
Aluminum.							
Asbestos.							
Steel.							
Gravel.				**PLANT AND EQUIPMENT CHECK LIST**			
Gypsum — Poured And Plank.				Air Compressors.		Pumpcrete Equipment.	
Slate.				Asphalt Plants.		Pumps.	
Shingles.				Asphalt Tools.		Radio Units.	
Steel.				Backhoes.		Rollers And Compactors:	
Tile.				Batch Plants.		Grid Rollers.	
Tin.				Bulldozers.		Rollers And Compactors:	
Sandblasting.				Cable.		Power Rammers.	
Service Lines.				Clamshells.		Rubber-Tired Rollers.	
Sheet Metal.				Concrete Buckets.		Sheepfoot Rollers	
Shutters.				Concrete Chutes.		Steel Rollers.	
Skylights.				Concrete Hoppers.		Vibrator Compactors.	
Sound Deadening.				Concrete Mixers.		Salamanders.	
Stacks.				Conveyors.		Saws, Power.	
Stalls.				Cranes:		Scaffolds.	
Structural Steel:				Crawler.		Scrapers.	
Anchors For Structural Steel.				Truck.		Shop Equipment.	
Bases Of Steel Or Iron.				Diving Apparatus.		Shores.	
Beams, Purlins And Girts.				Distributors.		Small Tools.	
Bearing Plates And Shoes.				Draglines.		Surveyor's Instruments.	
Brackets.				Drilling Equipment.		Tractor Shovels.	
Columns Of Steel, Iron Or Pipe.				Electric Generators.		Tractors:	
Crain Rails And Stops.				Elevating Graders.		Crawler.	
Door Frames.				Hoist Equipment.		Rubber Tired.	
Expansion Joints.				Hooks.		Shovels.	
Floor Plates.				Hoses.		Trailers.	
Girders.				Jack Hammers.		Travel-Loaders.	
Grillage Beams.				Jacks.		Trenchers.	
Hangers Of Structural Steel.				Lift-And-Carry Cranes.		Trucks.	
Lintels.				Lift Trucks.		Vibrators.	
Margnees — Structural Frame.				Light Plants.		Wagons.	
Monorail Beams.				Motor Graders.		Water Trucks.	
				Night Flares.		Welding Units.	
				Power Shovels.		Wellpoints.	

Estimate summary
Figure 1-8 (continued)

Some system of classification helps ensure that all items or categories are listed and can be compared quickly with work you've done in the past.

Compiling the Estimate

To begin compiling the estimate, organize all the plan sheets in front of you. Spread out the several elevations and sections. Keep all the floor plans on the table before you, placing the foundation or basement plan on the top, the first floor plan next, the second floor plan below that, and so on through the series. We will assume, for the purpose of our discussion, that the plans are for a two-story brick residence with a basement.

Now open the specifications.

Assume that the site contains two trees and an assortment of brush to be cleared. For a tree about 16 inches in diameter we estimate it would take two men using a chain saw three to four manhours to fell the tree, remove the branches and cut the trunk into short lengths. We further estimate that it will take 10 manhours to remove the stump, using picks, shovels and axes.

So, to remove both trees and stumps we estimate a total of 26 to 28 manhours. Clearing of brush will depend on the thickness and size. Light clearing with axes and brush hooks is estimated at 100 square yards in 2.5 manhours.

Now consider the survey and the erection of batterboards and lines. Possibly two carpenters and one laborer can erect the markers, while a surveyor and his helper go over the ground. Figure how much time for each person will be required for labor, plus the surveyor's time and that of his helper unless the survey work is on a flat fee basis. Include the cost of necessary lumber, nails and lines. This, and the clearing expense, gives you the cost of preparation of the site.

Follow this system for each item on your estimating worksheet. The chapters that follow give detailed information on estimating each phase of the construction project. Regardless of the type of work involved, the estimating method is the same: Analyze the work to be done by considering the time required, the materials needed, and the equipment necessary to perform each operation.

A good working knowledge of the type of work involved is essential. By that, I don't mean to imply you must know how to lay brick in order to estimate the cost of brick veneer on a house. You don't. But you have to be able to figure it out,

which is fairly simple if you follow the guidelines available in your CEF. The most important thing is not to leave anything out. Not even the nails required to install the batterboards.

It's usually best to leave the estimating of plumbing and electric work to plumbing and electrical subcontractors. Plumbing can be estimated on the basis of a certain unit cost per fixture, including piping, waste and vent stock. Electrical wiring can also be figured per outlet, fixture or switch. Commercial or industrial work requires specialized knowledge and requires expertise in the field to compile estimates.

The Contractor's "Labor Burden"

Even after all the labor, material and equipment costs have been computed and added, you haven't found your cost for the job. The "labor burden" will add between 18% and 28% to the labor cost. For every dollar of payroll, the contractor must pay an additional 18 to 28 cents in taxes and insurance to government agencies and insurance carriers.

Most states levy an unemployment insurance tax on employers. This tax is based on the total payroll for each calendar quarter. The actual tax percentage is usually based on the employer's history of unemployment claims and may be from 1% to 4% or more.

The federal government also levies an unemployment insurance tax based on payroll (FUTA). This tax runs about 0.7%. And the federal government collects Social Security and Medicare taxes (FICA). These two taxes come to about 7% of payroll, and are collected from the employer at specified intervals.

States generally require employers to maintain worker's compensation insurance to cover their employees in the event of job-related injuries. The cost of the insurance is a percentage of payroll and is based on the type of work each employee performs. Most light construction trades have a rate between 5% and 15%. The cost varies from one period to the next, depending on the history of injuries for the previous period.

Most contractors also carry liability insurance to protect them in the event of an accident. Liability insurance usually costs about 1% to 2% of payroll. The higher the liability limits, the higher the cost.

Here is the approximate total cost for taxes and insurance:

State Unemployment Insurance	4.0%
FICA and Medicare	7.0%
FUTA	0.7%
Worker's Compensation Insurance	5.0 to 15.0%
Liability Insurance	2.0%
Total	18.7% to 28.7%

Direct Overhead

You must bear many costs that aren't associated with any particular phase of construction, but that are incurred as a result of taking on the job. These costs are usually called *direct overhead.* Think of them as administrative costs. Here are some items that are usually included as direct overhead: fire insurance, surety bonds, building permit, sidewalk permit, telephone, water, electricity, sewer and water connection fee, timekeeper, watchman, temporary office, repairs to adjoining property, and job toilets.

You can probably think of many more direct overhead items. Some contractors include in direct overhead the cost of supervision and other non-productive labor such as the cost of estimating the job. The time you spend on each job should be charged against that job. These are very real costs and must be included somewhere in the estimate. Since they are incurred as a result of taking each particular job, they can properly be included under direct overhead.

Indirect Overhead

After everything else is figured, you're left with certain expense items necessary in conducting your business, but not chargeable against any particular job. They include office rent, telephone and utilities, office staff, small tools, office insurance, printing, service car, postage, and advertising.

Some builders take a monthly total of this overhead as the cost of doing a certain amount of building work. Then they reduce this figure to a weekly expense. If you are doing two jobs of about the same size during a given week, each would bear one-half of the indirect overhead cost for that week. Multiply by the duration of the project and you have the indirect overhead cost for that project.

Some contractors figure the indirect overhead cost as a cost per productive manhour, or a percentage of the total job cost. Other contractors have reduced the indirect overhead to a cost per square foot of floor area. Any one of these systems is good if it works for you. Keep a record of your indirect overhead and develop some method of dividing this cost among your jobs. And don't forget to include this important item in your bid. The combined cost of direct and indirect overhead will be 10% or more of the total job cost for most light construction work. This 10% can make the difference between a profit and a loss on almost every job you have.

Contingency and Escalation

I always add a small amount to my bid to allow for *contingencies* — the unexpected problems that always seem to come up during the course of construction. Work seldom goes faster than planned, but you'll frequently come up against problems that make the work go slower. The right amount to add for contingency depends on the contractor and the job. However, for most light construction, I recommend about 2%.

For you remodelers, one of the most common estimating errors is failing to provide for the possibility of encountering hidden problems. Contingencies are the essence of remodeling work. Be alert to all of the problems you might face in any job you're estimating. While no standard fits all jobs, a 5% to 10% allowance should cover contingencies for most remodeling jobs, *if you have done a thorough examination of the job.* Remember, contingency reserves can buffer unforeseen costs to some extent, but they aren't a cure for sloppy estimating.

At current labor costs, the time consumed in correcting for just one unforeseen condition could make the difference between profit and loss. What are some of these contingencies? Here are a few:

• Underground service lines that have to be removed and replaced.

• Heat ducts or electrical service lines in walls that have to be relocated.

• Carpeting that may have to be removed and reinstalled to protect it during construction.

• Trees and shrubs that have to be removed for access to the building.

• A lack of available skilled labor. This means additional costs because it takes unskilled labor more time to perform the task.

Mineed careful.

• The weather, which can impose delays on even the most perfectly conceived jobs. The weather alone can cause increased labor costs of one-third or more.

Escalation is the increase in costs of labor, materials, and equipment between the time the bid is submitted and the time work is actually done and paid for. Even though you're sure of the price of 1,000 board feet of framing lumber when you submit your bid, you may not know what you'll pay for that lumber when you actually go to purchase it from the yard.

Lumber, plywood, and a number of other building materials change in price quite rapidly. If you can't secure firm quotes for materials to be delivered in the future, allow some amount for price increases, or specifically *exclude* price increases from your bid.

Profit
Profit is the contractor's return on his investment. You should be able to pay yourself a wage for the work you perform and, in addition, pay yourself a return on the money you have invested in the business. If you have $50,000 invested, you should receive a return on your investment of $4,000 to $6,000 per year (8% to 12% of investment), in addition to a reasonable wage. If you don't, you're better off putting your money into a bank account (where there's no risk) and going to work for somebody else, where the headaches also are somebody else's. Think of this profit as interest on the money invested in equipment, office, inventory, work in progress and everything else associated with running a construction business.

How much should you include in your estimate for profit? You'll hear many conflicting figures. Some estimators say a 20% profit is a good target figure and they try to end up with a "profit" of 20% of the total contract price after all bills are paid. Maybe some builders do operate efficiently enough to achieve a 20% profit, but they're the exception.

The contractor who talks about a 20% profit most likely means that after he's paid for his labor, material and equipment, he has 20% left over for himself. This 20% is really his wage, and though it may be substantial, it isn't a profit in the true sense. A profit is what remains after *all* costs are considered.

The builder should include the cost of his own

work under *direct overhead*. What, then, is a realistic profit in the true sense? Dun & Bradstreet compiled figures on general contractors for many years. They reported the average net profit after taxes for all contractors sampled was consistently between 1.2% and 1.5% of gross receipts.

The sample included many contractors who reported losses or became insolvent. A 1.5% profit is a fairly slim margin. Not many contractors, especially contractors on residential projects, include so small a profit in their bid.

On extremely large projects such as highways, power plants or dams, the contractor may allow only 1% or less for profit, especially if the compensation is based on the contractor's actual cost rather than a fixed bid.

Residential construction, especially remodeling and repair work, traditionally carries a higher profit margin because the jobs are much smaller than other types of work and the risk of significant cost overruns is larger. Probably 8% to 10% profit is a reasonable expectation on most jobs, with small jobs or remodeling work running to as much as 25%.

It's not all that simple, however. There's more to "profit" than just how much profit you would like to earn. Sharp competition will reduce the amount of profit you can figure into your estimate. If you include too much profit in your bids, you'll lose desirable jobs to lower bidders. If you've developed a specialty *and* you can do a particular type of work better than other contractors in the area *and* you have enough work to keep you busy, *then* you may increase your profit by 1% or 2%, or even more.

Spec builders have been very successful at maintaining high profit margins when real estate values are increasing. When work is less plentiful, however, many contractors take work at little or no profit in order to keep their best crews busy (and themselves in the construction business).

In practice, there's no single profit figure that will fit all situations. For most residential work, an 8% to 10% profit is a very nice expectation. If you have all the work you want and are asked to bid on more, maybe a 15% profit isn't excessive.

Your profit should give you a good annual return on the money you've invested in the business (after you've taken a reasonable wage for yourself). You should earn a profit equal to 8% to 12% of the "tangible net worth" of your business. The tangible net worth is the value of all the assets

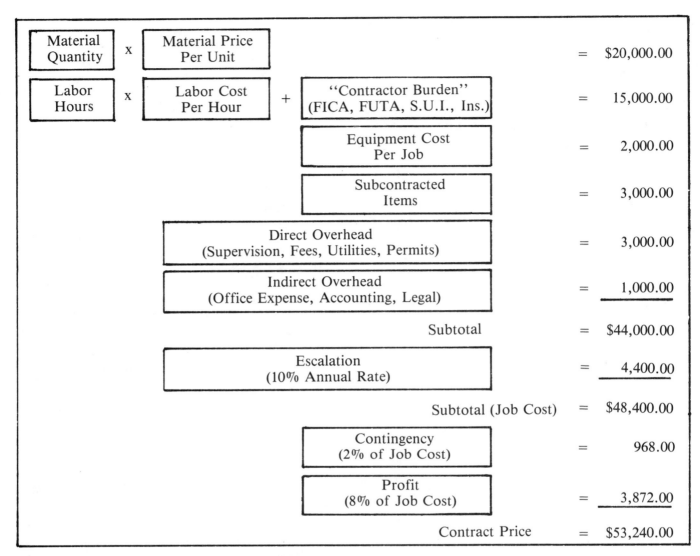

Typical bid including all costs
Figure 1-9

of your business less the liabilities (anything your business owes) and less any intangible items such as goodwill, patents or copyrights.

The small contractor who has only a vehicle, some tools and a few hundred dollars in working capital may have a tangible net worth of less than $15,000. He'll probably plow any profit back into the business to buy additional equipment and increase his working capital. Still, he should include a profit in each job which will give him a return at the end of the year of between 8% and 12% of his $15,000 tangible net worth.

Look at Figure 1-9. It shows a typical bid with direct and indirect overhead, contingency and escalation, and profit included in the bid price.

Checking the Estimate
The object of preparing an estimate is to determine the *cost* of a building or structure. You can't give a bid for furnishing the materials and labor for a job unless you know in detail how much material of each kind is required, as well as the extent and kind of labor needed to install it.

Contractor's estimates are usually made either by the contractor himself (if his business is small) or by a trusted employee who's an experienced estimator. Since the bid is based on the estimate, you can see how serious it is to have errors in the estimate.

If you bid too low because an item was omitted from the estimate, the chances are that you'll be awarded the contract and probably lose money in the process. If the error, on the other hand, makes your bid too high, a competitor with a more accurate estimate will come in at a lower price and get the job. You'll lose business which might have proved profitable. But everyone knows that the estimator who has never made a mistake has never estimated a job. In some cases you may go over the calculations again and again and still not find the error because you've become "blind" to the defect.

Following a fixed *system* in preparing estimates helps eliminate the possibility of errors. Such a system should permit a quick and easy method of self-checking. A methodical estimator always double-checks his work.

A practical method of detecting errors is to have a second person go over your figures, or have a second person prepare completely new calculations according to his own method and then compare the results. There are many ways errors may enter into an estimate. Some of the more common ones are:

Math errors: Errors in arithmetic such as addition, subtraction, multiplication, division, and decimal points in the wrong place.

Copying errors: Mistakes in copying items from one paper to another.

Omissions: Leaving out items of materials, labor, equipment, or overhead. Check these items against a reminder list.

Labor estimate errors: Errors in estimating the length of time required to do a certain piece of work.

Doing the take-off methodically according to a system, double-checking all your work, using a checklist and having someone else check your work are all protections you should use on every estimate.

Estimating Tips

Estimating requires precision and care. Most successful estimators consciously or unconsciously follow certain practices which tend to reduce the chance of error or miscalculation. Follow these tips to make your estimates more accurate:

1) List the materials in the order in which they appear in the building. For example, don't figure the hauling of the spoil before the loosening of the soil. Don't list the finished material on a house before the framework.

2) Every estimate *must* be checked over by one or more people before it's accepted as final. Therefore, the estimate must be written in detail, showing every item. The checker must be able to see at a glance how the estimator arrived at the answers. Where it's practical, draw sketches on the estimate sheet showing how the dimensions were obtained. For example, in figuring the area of a floor, draw a small sketch giving the dimensions.

3) Wherever practical, use the unit cost method in your estimates. There's less likelihood of error since you're dealing with fewer figures.

4) Check your figures. Don't use a figure that you suspect. If it doesn't "feel" right, it's probably wrong.

5) Don't price each subtotal if the same item occurs in different places. Make a complete total before figuring cost.

6) Use only standard abbreviations.

7) Use decimals instead of fractions. Wherever possible, refer to a table of decimal equivalents. You'll find one in Chapter 3, Figure 3-25. This will make your work much easier.

8) Be careful to distinguish between linear measure, square measure and cubic measure. For example, *never* use the symbols ' and '' for abbreviating square feet or square inches. Twenty-five square feet should never be written 25 sq'. Always write 25 sq. ft., 25 SF or 25 square feet. You might think I'm being picky here over a trivial point. But just mistake square feet for linear feet once, and you won't think it's so trivial.

9) Write as plainly as possible. You don't want to lose money because you or the person checking your work mistook a one for a seven, or inches for feet.

Setting Your Price

Once you've arrived at the total cost for the job, you have to consider at what price you'll offer it to the customer. Here's what I recommend:

When you've come up with a figure that will provide a normal profit based on a definite return on your investment, this price is your price. Stay with it, even if you lose the job. If you can't make a profit on the job, thank your competitor for taking it. You may soon have one competitor less.

Even after the estimate is complete, the bid is submitted and the job is done, those hard-won estimate figures have a continuing value. The real test of any estimate is an item-by-item comparison with actual construction costs. When there's a sizable variation, it's important to know just what caused the discrepancy.

If a particular job condition was responsible for the result, you probably don't need to change your unit cost factors. On the other hand, if the unit cost used in estimating wasn't a realistic one, then it's time to update the unit cost factor in your cost file.

This review of each job estimate is all-important. It's the best means of keeping your estimating process current in the face of changing conditions.

In the next chapter, we'll talk about cost keeping and cost accounting.

Chapter **2**

Keeping and Controlling Costs

Congratulations! You've compiled an estimate, submitted a successful bid, and won the job. Now comes the hard part: doing the work *and making a profit*. To do that, you've got to make actual costs match estimated costs. Notice the two parts of that important task: First, be able to identify the actual cost of the job. Second, you have to keep that cost under control.

There are actually *two profits* possible on every job. The first is the profit figured in the bid. The second is the extra profit you can make by reducing actual costs below estimated costs. That takes careful purchasing, efficient management, and good supervision. This "extra profit" *may* be the result of some good luck. But more often it's the product of a good cost recording system.

This chapter is intended to show what a cost keeping system can do and suggest the system that's right for your business.

If you have little or no payroll and personally watch every part of every job, you may feel like you're in complete control of all job costs. You're probably right. But when a business grows, no one can follow every penny that's spent. Successful builders recognize two facts: First, efficiency is essential to success in the construction business. Second, cost control is the essence of an efficient operation.

Good judgment in handling employees and subcontractors is an asset. But that by itself doesn't guarantee good cost control because cost control doesn't stem from observation alone. Construction is a complex manufacturing process. You can't be everywhere on every job, observing all the workers at the same time, judging productivity, and making sure the pace of work is matching estimated production rates. That's why most builders use some simple system of reports or summaries that show where time and materials are being used correctly and where money is being wasted.

Cost Keeping or Cost Accounting

Cost keeping is a system for recording the cost of each unit of production or work. A good cost-keeping system:

1) Lets you compare unit costs from job to job so you know what's going right and what's going wrong.

2) Provides data that can help with future cost estimates.

To get this information, you'll need a record of anything that has a major impact on costs: weather, tradesmen employed, hours of work, overtime, rates of pay, special conditions, materials used, and the labor performed.

The purpose is to find the actual cost for each unit of work done: cost of labor, materials, and overhead. By analyzing these *unit costs* and comparing them with other jobs and standards, you'll discover:

- Waste of material
- Waste of supplies
- Inefficiency in workers or supervisors
- Time lost due to faulty scheduling
- Poor coordination between trades
- Excess labor costs or padded payrolls

Most construction work goes to the lowest bidder. The contractor who has the lowest costs is always a step closer to getting the award. He may make a profit on work that others could handle only at a loss. It isn't very likely that you'll become the lowest cost contractor by hiring tradesmen at lower rates or buying cheaper materials. But you can cut costs by operating more efficiently, reducing waste, and specializing in the type of work you do best. That's what good cost keeping can do for you.

Here's an important point to remember: No cost-keeping system will tell you what your costs *should* be; it will simply tell you what your costs actually *are* for each job. You have to decide where savings are possible and do what's required to reduce wasted time and materials.

Keeping cost records isn't the same as accounting. Cost keeping may rely on some of the same records used in bookkeeping. But don't confuse the two. Cost keeping sees the project through the eyes of an engineer: in unit costs and quantities of materials. The accountant is concerned with finance, dealing with cash balances and the flow of assets. No matter how indifferent you are to accounting, be alert to the importance of good record keeping.

The record-keeping system you use may be simple or complex. In a smaller company, the simpler, the better. This chapter will outline a more detailed system appropriate for a larger company handling 6 or 8 jobs at once. If your company is smaller than that, the system you use need not be as detailed.

Essentials of a Cost System
Regardless of the size of your organization, the basic principles of cost accounting are the same. The difference in the systems used in large organizations and small ones is the details recorded. We don't have the space in this volume to develop a full cost-keeping system with all its

forms. You probably don't need that much information, anyhow. But I hope to plant the seed of a good basic cost-keeping system in your mind.

Any cost-keeping system you use should meet the following standards. It must be:

1) Reliable
2) Simple
3) Inexpensive
4) Allow quick analysis
5) Flexible enough to cover all your jobs

The term *cost* means money spent for any business purpose. In the construction industry, costs come in three forms, labor, materials (including equipment), and general expense or overhead. Any good cost system has to consider every project expense, from preparing the estimate to the completion of the project — for labor, material and overhead.

Analysis of Unit Costs
Simply dividing all costs into the labor, material and overhead categories isn't enough. It won't help us figure future costs, or help control costs on a current project. Your system has to subdivide costs into more detailed accounts which will show costs per unit installed. Of course, we need to know the total cost of labor, material and overhead. But even more we need to know the cost of concrete work per cubic yard, masonry per 100 block, carpentry per 1,000 board feet, plastering per square yard, painting per square foot, and so on.

But even that level of detail isn't enough. We want to know the cost of labor, material and overhead for each item. If we have a total of $15,400 charged for concrete work on a certain job, we want to know how much was spent for forms, reinforcing, and concrete placing. Knowing the total cost, even knowing it very precisely and with a high degree of certainty, isn't enough. We want a complete breakdown of expenses by each work item.

Suppose you're the foreman on a cattle ranch. The owner sends you across the river to buy more livestock. You come back having spent $50,000. The owner asks the price. Your reply, "The whole bunch cost $50,000." He'd probably bellow out, "Specifics, man! Break it down. What'd that ugly-looking bull cost? How much a head for those heifers? What's the story on the old cows?" Without the details it's hard to know if you struck a good deal or got taken. It's even harder to know what the bulls, heifers and cows have to sell for to turn a profit.

Like a cattle rancher, every contractor needs a record-keeping system that *breaks down the costs into enough detail so that comparisons are possible.*

In the case of concrete work, we'll divide it into labor, material and other expenses, of course. But we also want to know the cost of concrete work for foundations, slabs, sidewalk, and so on. For brick work, we want to know the labor cost for walls, porch piers and chimneys.

If you had a full-time accountant on every job, breaking out the manhours and materials used would be no problem. Your cost accountant would just make an entry in the job log, "Brick work, labor building walls" or "Brick work, material on porch piers." But making narrative entries like that would take a lot of time — and require more time back at the office for reading and interpreting. No contractor can afford to do that on residential and small commercial jobs. But there's an easier way. As long as these are some manageable number of work items that appear regularly on most of your jobs, it's much easier and faster to use a code for each work classification. Most cost-keeping systems use a simple number code system that's ideal for builders because it's both simple and flexible. It's called the *Masterformat* and can be used to identify nearly every work item in most jobs.

The Masterformat
The Construction Specification Institute (C.S.I.) developed the Masterformat. The specifications on most jobs follow the order of the Masterformat, and estimating is much easier if the take-off follows the same order as the specifications. Figure 2-1 shows the broadscope section titles of the Masterformat. You can get a copy of the entire Masterformat from:

The Construction Specifications Institute
601 Madison Street
Alexandria, VA   22314

In the Masterformat, each subclassification has its own assigned number. The system of account codes divides any construction project into about 10,000 components, and classifies these under 16 major divisions. But complete as this is, it won't be enough for your purposes. Here's why. The Masterformat doesn't identify labor, material and expense. That part is up to you. Fortunately, it's easy to append another digit to indicate which of the three categories applies: 1 for labor, 2 for material, and 3 for general expense.

Let's say we're using the number 04211 for common brick. Each operation is represented by a five-digit number. By adding a sixth digit, either a 1, 2 or 3, we identify the whole operation as either labor, material or expense. So we would immediately recognize the number 04211-1 as a charge for labor against the classification, common brick. Number 04211-2 would indicate material used in masonry, while number 04211-3 would represent an expense against masonry (possibly scaffolding or other equipment).

If your operation is small and you don't do a wide variety of work, consider setting up your own classification system instead of following the Masterformat. Concentrate on detailed breakdowns of the kind of work you do regularly. There are only three important requirements: First, classify your operations in a way that makes sense to you. Second, have the codes written down so they're available to everyone who needs them. And third, use them for estimating, cost keeping and bookkeeping. The best system in the world is no good if it's not followed.

Forms for Cost Keeping
You'll find a wide selection of estimating, cost keeping and bookkeeping forms at any well-stocked office supply store. Take time to compare the available forms and pick the ones best suited to your operations.

I like to use a daily journal to record receipts, disbursements and activity. It works well for a small company and gives you a bird's-eye view of all your functions. Use a 12-column analysis pad for your daily journal. See Figure 2-2.

From the daily journal, you can develop total equipment, material and labor costs for each job and for each trade. That's essential for estimating the expected profit or loss on the job as work progresses.

Let's say we estimated $2,200 for brick, $1,200 for labor and $620 for expenses, or a total of $4,020 for a particular job. The daily journal shows charges of $3,100 when only two-thirds of the brickwork is finished. It's obvious that costs are running high for this job. Unless someone gets the lead out, you're going to lose your shirt.

BIDDING REQUIREMENTS, CONTRACT FORMS, AND CONDITIONS OF THE CONTRACT

00010	PRE-BID INFORMATION
00100	INSTRUCTIONS TO BIDDERS
00200	INFORMATION AVAILABLE TO BIDDERS
00300	BID FORMS
00400	SUPPLEMENTS TO BID FORMS
00500	AGREEMENT FORMS
00600	BONDS AND CERTIFICATES
00700	GENERAL CONDITIONS
00800	SUPPLEMENTARY CONDITIONS
00850	DRAWINGS AND SCHEDULES
00900	ADDENDA AND MODIFICATIONS

Note: Since the items listed above are not specification sections, they are referred to as "Documents" in lieu of "Sections" in the Master List of Section Titles, Numbers, and Broadscope Explanations.

SPECIFICATIONS

DIVISION 1—GENERAL REQUIREMENTS

01010	SUMMARY OF WORK
01020	ALLOWANCES
01025	MEASUREMENT AND PAYMENT
01030	ALTERNATES/ALTERNATIVES
01040	COORDINATION
01050	FIELD ENGINEERING
01060	REGULATORY REQUIREMENTS
01070	ABBREVIATIONS AND SYMBOLS
01080	IDENTIFICATION SYSTEMS
01090	REFERENCE STANDARDS
01100	SPECIAL PROJECT PROCEDURES
01200	PROJECT MEETINGS
01300	SUBMITTALS
01400	QUALITY CONTROL
01500	CONSTRUCTION FACILITIES AND TEMPORARY CONTROLS
01600	MATERIAL AND EQUIPMENT
01650	STARTING OF SYSTEMS/COMMISSIONING
01700	CONTRACT CLOSEOUT
01800	MAINTENANCE

DIVISION 2—SITEWORK

02010	SUBSURFACE INVESTIGATION
02050	DEMOLITION
02100	SITE PREPARATION
02140	DEWATERING
02150	SHORING AND UNDERPINNING
02160	EXCAVATION SUPPORT SYSTEMS
02170	COFFERDAMS
02200	EARTHWORK
02300	TUNNELING
02350	PILES AND CAISSONS
02450	RAILROAD WORK
02480	MARINE WORK
02500	PAVING AND SURFACING
02600	PIPED UTILITY MATERIALS
02660	WATER DISTRIBUTION
02680	FUEL DISTRIBUTION
02700	SEWERAGE AND DRAINAGE
02760	RESTORATION OF UNDERGROUND PIPELINES
02770	PONDS AND RESERVOIRS
02780	POWER AND COMMUNICATIONS
02800	SITE IMPROVEMENTS
02900	LANDSCAPING

DIVISION 3—CONCRETE

03100	CONCRETE FORMWORK
03200	CONCRETE REINFORCEMENT
03250	CONCRETE ACCESSORIES
03300	CAST-IN-PLACE CONCRETE
03370	CONCRETE CURING
03400	PRECAST CONCRETE
03500	CEMENTITIOUS DECKS
03600	GROUT
03700	CONCRETE RESTORATION AND CLEANING
03800	MASS CONCRETE

DIVISION 4—MASONRY

04100	MORTAR
04150	MASONRY ACCESSORIES
04200	UNIT MASONRY
04400	STONE
04500	MASONRY RESTORATION AND CLEANING
04550	REFRACTORIES
04600	CORROSION RESISTANT MASONRY

DIVISION 5—METALS

05010	METAL MATERIALS
05030	METAL FINISHES
05050	METAL FASTENING
05100	STRUCTURAL METAL FRAMING
05200	METAL JOISTS
05300	METAL DECKING
05400	COLD-FORMED METAL FRAMING
05500	METAL FABRICATIONS
05580	SHEET METAL FABRICATIONS
05700	ORNAMENTAL METAL
05800	EXPANSION CONTROL
05900	HYDRAULIC STRUCTURES

DIVISION 6—WOOD AND PLASTICS

06050	FASTENERS AND ADHESIVES
06100	ROUGH CARPENTRY
06130	HEAVY TIMBER CONSTRUCTION
06150	WOOD-METAL SYSTEMS
06170	PREFABRICATED STRUCTURAL WOOD
06200	FINISH CARPENTRY
06300	WOOD TREATMENT
06400	ARCHITECTURAL WOODWORK
06500	PREFABRICATED STRUCTURAL PLASTICS
06600	PLASTIC FABRICATIONS

DIVISION 7—THERMAL AND MOISTURE PROTECTION

07100	WATERPROOFING
07150	DAMPPROOFING
07190	VAPOR AND AIR RETARDERS
07200	INSULATION
07250	FIREPROOFING
07300	SHINGLES AND ROOFING TILES
07400	PREFORMED ROOFING AND CLADDING/SIDING
07500	MEMBRANE ROOFING
07570	TRAFFIC TOPPING
07600	FLASHING AND SHEET METAL
07700	ROOF SPECIALTIES AND ACCESSORIES
07800	SKYLIGHTS
07900	JOINT SEALERS

DIVISION 8—DOORS AND WINDOWS

08100	METAL DOORS AND FRAMES
08200	WOOD AND PLASTIC DOORS
08250	DOOR OPENING ASSEMBLIES
08300	SPECIAL DOORS
08400	ENTRANCES AND STOREFRONTS
08500	METAL WINDOWS
08600	WOOD AND PLASTIC WINDOWS
08650	SPECIAL WINDOWS
08700	HARDWARE
08800	GLAZING
08900	GLAZED CURTAIN WALLS

DIVISION 9—FINISHES

09100	METAL SUPPORT SYSTEMS
09200	LATH AND PLASTER
09230	AGGREGATE COATINGS
09250	GYPSUM BOARD
09300	TILE
09400	TERRAZZO
09500	ACOUSTICAL TREATMENT
09540	SPECIAL SURFACES
09550	WOOD FLOORING
09600	STONE FLOORING
09630	UNIT MASONRY FLOORING
09650	RESILIENT FLOORING
09680	CARPET
09700	SPECIAL FLOORING
09780	FLOOR TREATMENT
09800	SPECIAL COATINGS
09900	PAINTING
09950	WALL COVERINGS

Masterformat
Figure 2-1

DIVISION 10—SPECIALTIES

10100	CHALKBOARDS AND TACKBOARDS
10150	COMPARTMENTS AND CUBICLES
10200	LOUVERS AND VENTS
10240	GRILLES AND SCREENS
10250	SERVICE WALL SYSTEMS
10260	WALL AND CORNER GUARDS
10270	ACCESS FLOORING
10280	SPECIALTY MODULES
10290	PEST CONTROL
10300	FIREPLACES AND STOVES
10340	PREFABRICATED EXTERIOR SPECIALTIES
10350	FLAGPOLES
10400	IDENTIFYING DEVICES
10450	PEDESTRIAN CONTROL DEVICES
10500	LOCKERS
10520	FIRE PROTECTION SPECIALTIES
10530	PROTECTIVE COVERS
10550	POSTAL SPECIALTIES
10600	PARTITIONS
10650	OPERABLE PARTITIONS
10670	STORAGE SHELVING
10700	EXTERIOR SUN CONTROL DEVICES
10750	TELEPHONE SPECIALTIES
10800	TOILET AND BATH ACCESSORIES
10880	SCALES
10900	WARDROBE AND CLOSET SPECIALTIES

DIVISION 11—EQUIPMENT

11010	MAINTENANCE EQUIPMENT
11020	SECURITY AND VAULT EQUIPMENT
11030	TELLER AND SERVICE EQUIPMENT
11040	ECCLESIASTICAL EQUIPMENT
11050	LIBRARY EQUIPMENT
11060	THEATER AND STAGE EQUIPMENT
11070	INSTRUMENTAL EQUIPMENT
11080	REGISTRATION EQUIPMENT
11090	CHECKROOM EQUIPMENT
11100	MERCANTILE EQUIPMENT
11110	COMMERCIAL LAUNDRY AND DRY CLEANING EQUIPMENT
11120	VENDING EQUIPMENT
11130	AUDIO-VISUAL EQUIPMENT
11140	SERVICE STATION EQUIPMENT
11150	PARKING CONTROL EQUIPMENT
11160	LOADING DOCK EQUIPMENT
11170	SOLID WASTE HANDLING EQUIPMENT
11190	DETENTION EQUIPMENT
11200	WATER SUPPLY AND TREATMENT EQUIPMENT
11280	HYDRAULIC GATES AND VALVES
11300	FLUID WASTE TREATMENT AND DISPOSAL EQUIPMENT
11400	FOOD SERVICE EQUIPMENT
11450	RESIDENTIAL EQUIPMENT
11460	UNIT KITCHENS
11470	DARKROOM EQUIPMENT
11480	ATHLETIC, RECREATIONAL AND THERAPEUTIC EQUIPMENT
11500	INDUSTRIAL AND PROCESS EQUIPMENT
11600	LABORATORY EQUIPMENT
11650	PLANETARIUM EQUIPMENT
11660	OBSERVATORY EQUIPMENT
11700	MEDICAL EQUIPMENT
11780	MORTUARY EQUIPMENT
11850	NAVIGATION EQUIPMENT

DIVISION 12—FURNISHINGS

12050	FABRICS
12100	ARTWORK
12300	MANUFACTURED CASEWORK
12500	WINDOW TREATMENT
12600	FURNITURE AND ACCESSORIES
12670	RUGS AND MATS
12700	MULTIPLE SEATING
12800	INTERIOR PLANTS AND PLANTERS

DIVISION 13—SPECIAL CONSTRUCTION

13010	AIR SUPPORTED STRUCTURES
13020	INTEGRATED ASSEMBLIES
13030	SPECIAL PURPOSE ROOMS
13080	SOUND, VIBRATION, AND SEISMIC CONTROL
13090	RADIATION PROTECTION
13100	NUCLEAR REACTORS
13120	PRE-ENGINEERED STRUCTURES
13150	POOLS
13160	ICE RINKS
13170	KENNELS AND ANIMAL SHELTERS
13180	SITE CONSTRUCTED INCINERATORS
13200	LIQUID AND GAS STORAGE TANKS
13220	FILTER UNDERDRAINS AND MEDIA
13230	DIGESTION TANK COVERS AND APPURTENANCES
13240	OXYGENATION SYSTEMS
13260	SLUDGE CONDITIONING SYSTEMS
13300	UTILITY CONTROL SYSTEMS
13400	INDUSTRIAL AND PROCESS CONTROL SYSTEMS
13500	RECORDING INSTRUMENTATION
13550	TRANSPORTATION CONTROL INSTRUMENTATION
13600	SOLAR ENERGY SYSTEMS
13700	WIND ENERGY SYSTEMS
13800	BUILDING AUTOMATION SYSTEMS
13900	FIRE SUPPRESSION AND SUPERVISORY SYSTEMS

DIVISION 14—CONVEYING SYSTEMS

14100	DUMBWAITERS
14200	ELEVATORS
14300	MOVING STAIRS AND WALKS
14400	LIFTS
14500	MATERIAL HANDLING SYSTEMS
14600	HOISTS AND CRANES
14700	TURNTABLES
14800	SCAFFOLDING
14900	TRANSPORTATION SYSTEMS

DIVISION 15—MECHANICAL

15050	BASIC MECHANICAL MATERIALS AND METHODS
15250	MECHANICAL INSULATION
15300	FIRE PROTECTION
15400	PLUMBING
15500	HEATING, VENTILATING, AND AIR CONDITIONING (HVAC)
15550	HEAT GENERATION
15650	REFRIGERATION
15750	HEAT TRANSFER
15850	AIR HANDLING
15880	AIR DISTRIBUTION
15950	CONTROLS
15990	TESTING, ADJUSTING, AND BALANCING

DIVISION 16—ELECTRICAL

16050	BASIC ELECTRICAL MATERIALS AND METHODS
16200	POWER GENERATION
16300	HIGH VOLTAGE DISTRIBUTION (Above 600-Volt)
16400	SERVICE AND DISTRIBUTION (600-Volt and Below)
16500	LIGHTING
16600	SPECIAL SYSTEMS
16700	COMMUNICATIONS
16850	ELECTRIC RESISTANCE HEATING
16900	CONTROLS
16950	TESTING

Masterformat
Figure 2-1 (continued)

DISBURSEMENTS

DAILY JOURNAL   DATE: 11/6/xx

DATE	RECEIPTS (Brought Forward)	JOB NO.	INCOME	DISBURSEMENT	MATERIAL	LABOR	INSURANCE	INTEREST	OFFICE	FEES	TOOLS VEHICLES	NOTES
(Brought Forward)	876400		876400	248900	130000	76500			8700	13500	21000	
11/6/xx	River Lumber Co	A-33		596031	596031							Po Check No.132
	J. Smith	A-33				12000						8 Carpenter Hrs
	B. Brown	A-33				6200						8 Carpenter Hrs
	C. Crown	A-33				7100						8 Carpenter Hrs
	A. Adams	B-801				13100						8 Mason Hrs
	H. Hardy	B-801				5500						8 Laborer Hrs
	J. Johnson	A-48				13000						9 Painter Hrs
	Gulf S.S.	A-36 A-48		2000							2000	Gas
	City Hall	A-35		65000						65000		Perm It
	Z. Holt	B-800	350000	350000								Final Payment
	Metro Insurance	A-33		35000			35000					Builders Risk
	Lee Heating	A-33		220000		220000 (sub)						HVAC Sub
	H&H Rentals	A-33		13000							13000	Reut Floor Sander Forms
	Office Supply			3300					3300			Forms
	C. Hill Co.	A-33		58000		58000 (sub)						Hang Sheetrock
	DAILY TOTAL		350000	992331	596031	344600	35000		3300	65000	15000	
	MONTHLY TOTAL		1226400	1241231	726031	421100	35000		12000	78500	36000	

Daily journal
Figure 2-2

From bookkeeping records and the daily labor reports, you should have all the data needed to compute accurate unit costs. All you need to know is the manhours spent, the amount of material and the overhead expense, and the units completed, whether it's excavating, foundation walls, brickwork or carpentry. If your records show the amount of time needed for completion, it's easy to develop a unit cost that covers all labor, material and expense.

Here's an example. Suppose your records show these costs for the erection of a small wall:

```
Code
04211   : 5,000 common brick at $142.00 per M   =   $710.00
04100   : 15 bags mason mix (mortar) @ $4.00    =     60.00
041002  : 3 tons sand @ $10.00                  =     30.00
042111  : Bricklayer:  35 hours @ $20.00 hour   =    700.00
042111  : Helper:  36 hours @ $15.00 hour       =    540.00
                                                   $2,040.00
                              Overhead - 10%         204.00
                                                   $2,244.00
```

You can add your profit at this point and arrive at a cost for building this wall. Or, if you want to use the total cost as a basis for future estimates, divide $2,240 by 5,000. Your unit cost is $0.45 per brick or $45.00 per 100 brick. Computing unit costs like this is simple if you use a sensible coding system that's coordinated with your bookkeeping system. Some coding system like this is the basis for nearly all cost keeping.

Keeping Accurate Labor Costs

The key to good cost keeping is accurate tracking of labor time. I recommend the following system. It will appeal to many smaller builders. You can use it separately as a condensed and practical system by itself, or as a part of a more complete cost-keeping system as outlined above. It's adaptable for all sizes of projects, from small jobs with only four or five workers, to the largest jobs, employing a hundred or more.

All labor cost keeping begins with a record of how much time was spent on each work item. Simple, isn't it? Let's see how this works in practice.

Weekly time sheet— We'll start with a weekly time sheet and distribute the employee's time into the types of work done each day. Figure 2-3 shows a typical weekly labor record.

The weekly time sheet is self-explanatory. The foreman, or whoever is supervising the job, makes notations on time cards each day. In this example there are three bricklayers, three carpenters and four laborers working on the job. We note that Friday, for example, shows 24 hours bricklayer's time, 22 hours carpenter's time and 29 hours helper's time — a total of 75 hours.

Payroll summary— Every contractor keeps a weekly or monthly payroll record showing each employee's name, trade, hours worked each day, total hours worked each payroll period, and rate of pay. This information is needed to calculate the deductions made from each employee's pay and compute the net amount paid each employee in each payroll period. Even if your payroll is prepared by a computer service bureau, you'll need a form like Figure 2-4, payroll summary.

Labor distribution record— The third form you'll need is the labor distribution record. I use a 12-column analysis form. Each day after work, enter the time worked by each man and the work completed. Figure 2-5 shows labor distribution records for three types of work done during a typical week. This provides a record for carpenter, bricklayer and laborer time, separate from the weekly time sheet we talked about earlier.

Use Figure 2-5 as a daily, weekly, or semi-monthly report. In the main column, Class of Work, list the different kinds of work done. Use the first column for the code for the various trades; the second column, occupation, and in the third column, total hours worked by each trade for the class of work listed in the main column.

There's space to insert the hourly rate of the different trades, the total amount paid each, and the total cost of each class of work for the payroll period. The remaining columns of this sheet have space for figuring the amount of each kind of work completed during the payroll period, the amount completed to date, the unit labor costs on each, and notes.

If you're using this form as a daily report, insert and total the labor hours for the various trades each day. The totals here must balance with the weekly time sheets.

Builders who follow this method prefer to keep labor records in *manhours per unit* rather than cost per unit. Manhours per unit can be used at any wage scale and are easier to modify. You can, of course, show both manhours and labor cost per unit, as in Figure 2-5.

Page 1 of 1 pages

Weekly Time Sheet

For period ending 9-20-XX

KNIGHT GARAGE job

	Name	Exemptions	Days SEPT 20: 15 M	16 T	17 W	18 T	19 F	20 S	Rate	Hours worked: Reg.	Over-time	Total earnings	
1	K. MISPAGEL F/MASON		8	8	-	8	8	4	25^{00}	36		900	00
2	J. MARINO - MASON		8	-	-	8	8	4	22^{00}	28		616	00
3	V. RICE - MASON		8	8	RAINED	8	8	4	22^{00}	36		792	00
4	B. GROTE - CARPENTER		8	8		8	8	4	23^{00}	36		828	00
5	G. WILSON - "		8	8		8	6		23^{00}	30		690	00
6	D. HUGES "		8	8		8	8	4	20^{00}	36		720	00
7	C. KASSLE - LABORER		9	9		9	9	4.5	12^{00}	40	0.5	489	00
8	W. JOHNSON - "		8	8	-	8	8	4	12^{00}	36		432	00
9	J. JACOBS - "		-	8	-	8	8	4	12^{00}	28		336	00
10	G. RICE - "		8	8	-	8	4	4	10^{00}	32		320	00
11													
12													
13													
14													
15													
16													
17													
18													
19													
20	TOTAL		73	73		81	75	36.5		338	0.5	6123	00

Daily Log

Monday LAYED BLOCK - EXTERIOR WALLS - BUILT ROOF TRUSSES

Tuesday " "

Wednesday RAINED

Thursday LAYED BLOCK - INTERIOR WALLS - INSTALLED ROOF TRUSSES

Friday " "

Saturday LAYED FACE BRICK - FRONT - FRAMED OFFICE PARTITIONS

Weekly time sheet
Figure 2-3

PAYROLL AND SUMMARY

CLOCK NO. **34**
DATE OF BIRTH **11/20/50**
SEX **M** SINGLE MARRIED ✓

WAGES AND HOURS

S.S. ACCT. NO. **XXX-XX-XXXX** NAME **JOHN A. SMITH**
OCCUPATION **LABORER** ADDRESS **100 S. MAIN**
HRS. FULL TIME WEEK **40** NUMBER OF WITHHOLDING EXEMPTIONS **3**

SOCIAL SECURITY   WITHHOLDING TAX

	PERIOD OR WK. ENDING	TOTAL HRS.	REGULAR RATE OF PAY	EARNINGS AT REG. RATE FOR TOTAL HRS.	EXTRA FOR OVERTIME	TOTAL EARNINGS	DEDUCTIONS F.I.C.A / WITHHOLD / STATE	NET PAY	CHECK NO.	
	FORWARD									
1	1/4	40	1300	52000		52000	378 6600 3700	37982	XX	1
2	1/11	40	1300	52000		52000	378 6600 3700	37982	XX	2
3	1/18	24	1300	31200		31200	2230 3744 1341	23885	XX	3
4	1/25	28	1300	3640		3640	2602 4622 1565	27611	XX	4
5	2/1	44	1300	52000	7800	59800	425 7594 2571	47530	XX	5
6	2/8	40	1300	52000		52000	378 6600 3700	37982	XX	6
7	2/15	40	1300	52000		52000	378 6600 3700	37981	XX	7
8	2/22	8	1300	10400		10400	743 1320 449	7890	XX	8
9	3/1	40	1300	52000		52000	378 6600 3700	37982	XX	9
10	3/8	48	1300	52000	15800	67800	833 8565 4100	50082	XX	10
11	3/15	38	1300	49400		49400	358 6273 2134	37421	XX	11
12	3/22	40	1300	52000		52000	378 6600 3700	37982	XX	12
13	3/29	40	1300	52000		52000	378 6600 3700	37982	XX	13
	TOTAL 1ST QUARTER	470		595400	23400	618800	4491 78338 36930	458123		
14	4/5	16	1300	20800		20800	1487 2641 894	15718	XX	14
15	4/12	24	1300	31200		31200	2230 3744 1341	23885	XX	15
16	4/19	48	1300	52000	15800	67800	833 8565 4100	50082	XX	16
17	4/26	40	1300	52000		52000	378 6600 3700	37982	XX	17
18	5/3	40	1300	52000		52000	378 6600 3700	37982	XX	18
19	5/10									19
20	5/17	40	1300	52000		52000	378 6600 3700	37982	XX	20
21	5/24	48	1300	52000	15800	67800	833 8565 4100	50082	XX	21
22	5/31	24	1300	31200		31200	2230 3744 1341	23885	XX	22
23	6/7	40	1300	52000		52000	378 6600 3700	37982	XX	23
24	6/14	40	1300	52000		52000	378 6600 3700	37982	XX	24
25	6/21	38	1300	49400		49400	358 6273 2134	37421	XX	25
26	6/28	40	1300	52000		52000	378 6600 3700	37982	XX	26
	TOTAL 2ND QUARTER	488		578600	31600	579800	4453 73123	429615		
	TOTAL TO DATE	708		1144000	54600	1196600	8944 151461 74118	887138		

AMOUNT TAXABLE — F.I.C.A

Payroll summary
Figure 2-4

LABOR DISTRIBUTION RECORD JOB NO. A-33 WEEKENDING SEPT. 20

CLASS OF WORK	CODE	OCCUPATION	TOTAL HOURS	HR/RATE	TOTAL PAY	WORK IN PLACE / QUANTITY	UNIT	UNIT COST	DAILY INSPECTION	NOTES
SETTING FORMS	03/061	CARPENTER	100	23.00	2300.00	1236 LIN FT	LIN FT	$1.97/1.93 LF		DRIZZLE RAIN 2 HOURS
		LABORER	8	12.00	96.00					
				TOTAL	$ 2446.00					
MIXING AND PLACING CONCRETE	03300	MASON	8	22.00	176.00	12 CY	CY	38.66 / 0.38 CY		MIXER DOWN 30 MIN, BELT BROKE
		LABORER	24	12.00	288.00					
				TOTAL	464.00					
COMMON BRICK	04911	F/MASON	4	25.00	100.00	7000	1000 BK	374.00 / 53.8 BK		HAD GOOD WEATHER FOR LAYING BK
		MASON	56	22.00	1239.00					
		LABORER	24.5	12.00	82.00					
				TOTAL	2708.00					
FACE BRICK	04181	F/MASON	24	25.00	6000.00	3000	1000 BK	35.00 / 53.6 BK		
		MASON	8	22.00	1760.00					
		LABORER	24	12.00	2880.00					
				TOTAL	10640.00					

**Labor distribution record
Figure 2-5**

The CEF figures in the chapters to follow are presented in *manhours per unit* to make your estimating easier.

Getting Cost Data from the Field

I'll emphasize again the importance of recording cost data on your jobs. The most reliable cost data I'll ever find is the cost data I develop myself. No matter how thorough and efficient your office procedures may be, if the data from the field is hogwash, your cost records will be worse than useless. Daily reports should be made out by supervisors who understand your cost system and know how important it is to keep accurate records.

A foreman who knows his crews have been unproductive may yield to the temptation to exaggerate a little here and there. Then you'll be basing your estimates and bids on inaccurate records — with some potentially disastrous results.

Individual tradesmen have little incentive to fill out reports properly. It's better if no one has to report on their own work. Emphasize to your foreman that these daily reports become the basis for future estimates — and can make or break the company. I've found that most foremen will report accurately the number of employees working on each part of the job, the rate of pay, the number of hours worked by each, and the amount of work completed — if it can be measured simply. But don't require too much detail. Maybe you don't need a precise breakdown on manhours to install plates and sills. Maybe just manhours for wall framing will be enough.

The Foreman's Daily Report

The foreman's daily report, shown in Figure 2-6, is useful for time keeping, as a progress report, and to compile the daily journal, the labor distribution record, and other reports.

This form, to be really useful to the builder, should show not only the class of work performed, but the class of material used in the work. Including this information is simple when the foreman is familiar with the coding system you're using.

The main columns of this report show the class of work that was done. The first column indicates the code number for the class of labor. Column two covers the code number of material used. The

description can be more detailed if necessary. The three column divisions under "Hours" should show regular time, overtime and a total of the two.

This report helps the timekeeper distribute labor by operation on the time sheets. It's the basic record for finding unit labor costs. Here you show the number of units completed for each employee. Add the price of materials actually used and the overhead to find the total unit cost. Now you can quote a price for laying brick, framing walls, or any other work.

Although we've been calling this the foreman's daily report, many builders place responsibility for completing this form on the superintendent. There are good reasons for this, especially if you are the superintendent. It simplifies the task if one person in your company completes this form. That makes it more likely that categories are consistent and coding is uniform.

There's no hard and fast rule as to who should complete the daily reports. You'll have to consider the ability of the workers, the size of the job, and the size of your business.

Summary of Estimate

We've talked about a number of useful reports for the builder who needs a practical, accurate system for cost keeping. The last form I'll recommend is the Summary of Estimate, Figure 2-7. This sheet lists 45 classes of work, and provides space for recording the estimated labor and material costs or the subcontract cost and the actual costs for each class.

You can use a form like this to compare estimated and actual costs after each job is completed. This provides a continuous check on the accuracy of the unit costs you're using.

You may not need all the cost-keeping forms I've outlined in this chapter. Use as much or as little as appropriate. Adapt the parts that will work in your company. Find a system that fits your requirements now and is flexible enough to grow along with you. If you decide to automate, the same account codes can be used in a computerized record-keeping system.

Experiment to find what works for you. It's worth the time and effort spent. Your cost-keeping system will pay for itself many times over, and keep you on the road to a profitable operation.

FOREMAN'S DAILY REPORT JOB NO. A-33

DATE	CLASS OF WORK	CODE NO. LABOR	CODE NO. MATERIAL	QUANITY WORK DONE	DESCRIPTION OF WORK	NO. OF EMPLOYEES	HOURS REG	HOURS O.T.	HOURS TOTAL	RATE	TOTAL
1/20/XX	MASONRY	042111	04211	12000	BUILDING EXT. WALL (CB)	12	96	48	144	2200	369600
	MASONRY	042111	04212	4000	BUILDING ENTRYWAY (FB)	4	32	8	40	2200	96800
	CARPENTRY	061001	06100	3200 SF	NAILED ON ROOF DECKING	4	64	0	64	2100	134400
	CARPENTRY	061002	06100	3200 SF	APPLIED FELT	2	16	0	16	2000	32000
	CARPENTRY	073001	07300	33 SQ.	INSTALLED SHINGLES	3	51	0	51	2000	102000
MATERIALS USED											
12600	COMMON BRICK		04211								
4200	FACE BRICK		04212								
75	SACKS CEMENT-LIME MORTAR		04100								
7	TONS SAND		041002								
20	LBS. WALL TIES		04150								
36	PCS 4x8x½ PLYWOOD		06100								
35	SQ. 3TAB SHINGLES		07300								
9	ROLLS #15 FELT		06100								
50	LBS 7/16 ROOFING NAILS		06050								
35	LBS 6d BOX NAILS		06050								

Foreman's daily report
Figure 2-6

RESIDENTIAL ESTIMATE BREAKDOWN

Owner _____   Contractor _____

Prop. Address _____   Address _____

Estimate By _____   Tel. No. _____

Legal Desc. _____   Net Bldg. Fund _____ Loan No. _____

No.	Item	Quantity	Unit	Material	Labor	Subcontract	Total	Actual Cost
1	Excavation							
2	Trenching							
3	Concrete Rough							
4	Concrete Finish							
5	Asphalt Paving							
6	Rough Lumber							
7	Finish Lumber							
8	Door Frames							
9	Windows and Glass							
10	Fireplace							
11	Masonry							
12	Roof							
13	Plumbing							
14	Heating							
15	Electric Wiring							
16	Electric Fixtures							
17	Rough Carpentry							
18	Finish Carpentry							
19	Lath and Plaster							
20	Drywall							
21	Garage Door							
22	Doors							
23	Painting							
24	Cabinets							
25	Hardwood Floors							
26	Ceramic Tile							
27	Formica							
28	Hardware							
29	Linoleum							
30	Asphalt Tile							
31	Range and Oven							
32	Insulation							
33	Shower Door							
34	Temp. Facilities							
35	Miscellaneous							
36	Clean Up							
37	Carpet							
38	Gutters and Downspouts							
39	Septic Tank							
40	Sewer Connection							
41	Overhead							
42	Supervision							
43	Water Meter							
44	Bldg. Permit							
45	Plan and Specs							
46								
47	Total Cost							

Summary of estimate
Figure 2-7

Chapter 3

Estimating Excavation

Now that we've covered the estimating basics, we'll talk about each specific area of construction estimating, literally from the ground up. We'll start with estimating excavation. And estimating excavation begins with the building survey.

The owner or architect will usually furnish a survey of the property showing the lot lines. If not, you'll have to have the survey done yourself.

Surveying the Site

The purpose of a building survey is to determine the exact location of the building site and the position of the building on that site. The first step is to accurately locate and mark the lot lines. If you put up a building that extends onto someone else's property, expect to go to court — and expect to lose.

Survey Markers

In most urban areas you'll find fixed marks called *monuments* or *markers* which were placed at street intersections by municipal surveyors. You'll measure the property lines from these markers. It's quite common, however, for these markers to shift out of place due to freezing and thawing of the ground, rain and washouts, or your friendly neighborhood kids. If you have any doubt about the accuracy of the markers, call in a professional surveyor to verify the location and measurements of the property.

In cities and nearby suburbs, a granite monument is often used as a marker. They're usually from 4'' to 8'' square, depending on the location and use, and extend from 4'' to 18'' above ground. On the top of a marker is the *topographical spot* or measuring point. This is sometimes a hole drilled in the stone, a copper plug or two crossed V's cut into the stone, intersecting at the spot.

If a monument is moved because the area is paved, the topographical spot will be transferred to the pavement. The "topo cut" usually consists of a small hole drilled in the pavement with four long right-angle wings and four short diagonal wings extending from it. Figure 3-1 shows this mark, along with three common variations.

Lot markers may be made of wood, usually 2'' square, with iron pins, or a tack or crossed grooves marking the spot.

A careful builder will make sure that lot lines have been established and clearly marked by a qualified surveyor. Few builders are competent surveyors or have the equipment necessary to lay out lines between markers. And the cost of a mistake is simply too high. Before you begin construction on any site, insist on a plot showing the boundaries. Check to make certain that the corners are staked, and be sure that the plot shows the exact location of the building on the lot.

After you've located the four corners of the

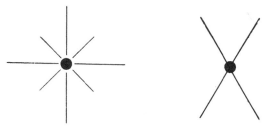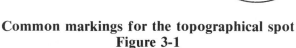

Common markings for the topographical spot
Figure 3-1

building site, you can mark the location of the building according to the foundation plans. It's good practice to allow a couple of inches on the inside of lot lines for variations of construction when the building is located adjacent to the line.

Determining Elevations

The surveyor usually refers elevations to a system of known reference points called *bench marks*. The bench marks generally refer to some definite level surface, such as mean sea level. This surface is known as the *datum*, or *datum plane*. If mean sea level hasn't been established, some fixed object such as the water table or other permanent point of an adjoining building is used as a datum. The height of a point above the datum is called its *elevation.*

You can measure elevations above and below the datum with a contractor's level and a leveling rod. Just measure the distance above or below the line of vision of the telescope. Use this method to find the depth of excavations and the height of joists or sills.

Here's an example that shows how to use the datum level. An architect's drawings of a building show that the elevation of the top of the footing, referring to a city bench mark, is 92, and the elevation of the top of the wall is 97.5. How high is the foundation wall?

The number 92 means that the top of the footing is 92 feet above a datum plane, and 97.5 means that the top of the wall is 97½ feet above the same datum. The difference, 5½ feet, or 5'6'', is the height of the foundation wall.

In actual practice, the datum used for the drawings usually refers to the elevation of the first floor, or the established grade. This practice eliminates a lot of interpolation when referring to "city datum." So in the above example, the city bench

mark of 92 feet would be shown as 0.00', and the elevation of the top of the wall would be plus 5.5'.

Staking Out the Building

After you've surveyed the site, it's time to stake out the building. You'll drive stakes and boards called *batterboards* into the ground to mark off the dimensions of the building. Then you can attach guidelines to the batterboards and use them as measuring bases in laying out the foundations. Figures 3-2 and 3-3 show two of the common types of batterboards.

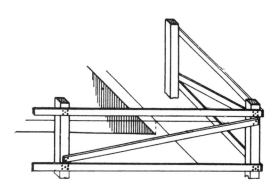

Typical batterboards
Figure 3-2

For large permanent structures, this work will be done by an engineer or surveyor. However, some general contractors prefer to stake out the building or have it done by their masonry subcontractor.

Be sure that the measurements and elevations are accurate and the corners true. In most cases, it's best to first lay out a suitable rectangle. Look at Figure 3-4. Points A, B, C and D are the corner points of the rectangle used in the illustration. The lines of the rectangle form base lines from which

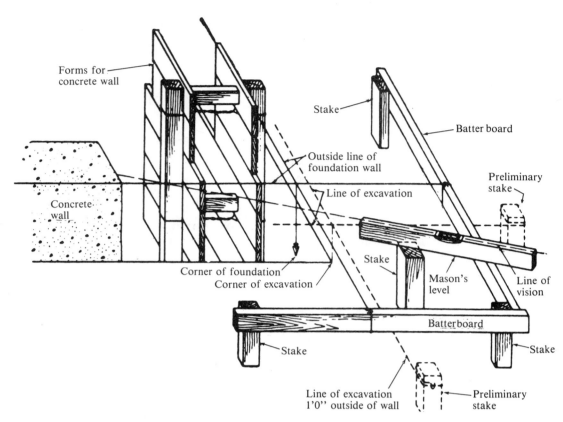

Building lines
Figure 3-3

you take measurements when locating irregularities in the shape of the excavation. The more irregular the outline, the more you need the rectangular base lines.

Squaring the Corners
To make sure the corners are true, measure the diagonals of the base rectangle. If they're not exactly the same, the corners aren't square. Use a steel tape to get an accurate measurement. Measure from each corner of the base rectangle to the corner diagonally opposite it. The diagonals are shown by the lines marked A-C and B-D in Figure 3-4.

Another common method of squaring corners is known as the "6, 8 and 10 rule." Measure off 6 feet one way from a corner of the rectangle and 8 feet the other way. If the corner is square, the measurement across the corner between the two points will be exactly 10 feet. If it isn't exactly 10 feet, adjust the side or sides that are out of line until it *is* 10 feet. Now the corners are square. You

can also use any convenient multiples of these figures, such as 12, 16 and 20.

After you've accurately located a corner of the base rectangle, set up the batterboards. Drive three stout stakes (2 x 4 or 4 x 4) into the ground, 3 feet to 6 feet back from the proposed excavation lines. Nail horizontal pieces (1 x 6 or 2 x 4) to the stakes. Brace them as shown in Figure 3-2.

Attaching the Guidelines
Now you can begin stretching the guidelines on the batterboards. The first lines are the *excavation lines,* the outlines of the general excavation which guide the workmen as they excavate. If forms are needed on the outside of the foundation walls below grade, make an allowance of at least 2 feet for working space. This will bring the excavation lines considerably outside of the actual wall lines, as shown in Figure 3-3. You'll also need to make allowance for the offset of the footings and for the drain tile, if it's used.

The *footing line,* which is the line of the outside

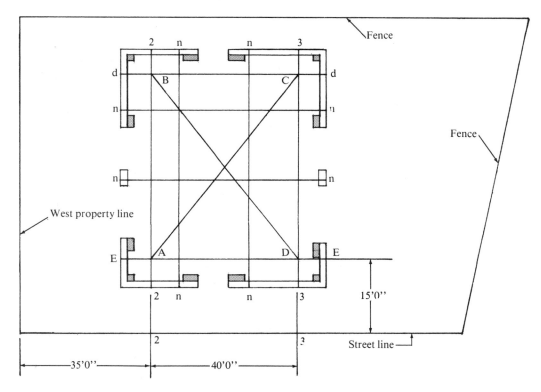

Building layout using a rectangle
Figure 3-4

of the footing, is usually omitted for small structures. The *neat lines* or *face lines* are the lines of the outside of the foundation wall. Where outside forms are used for concrete walls, these lines serve as a guide in lining up the forms. In masonry construction these lines are often called *ashlar lines*. It isn't usually necessary to stretch a line for the inside of the foundation wall or for the inside lines of the footing. These walls and footings are usually made to some easily measured thickness and you can use blocks to separate the concrete forms.

Let's use the building shown in Figure 3-4 to do a step-by-step building stakeout. In the figure, the front of the building is parallel to and 35'0" from the west property line.

Step one— Locate the approximate corners A, B, C, and D with a tape so you can set the stakes for the batterboards. Then nail boards horizontally to the stakes as shown in Figures 3-2 and 3-3. If there's no elevation marked on the plan, you may be told to set the top of the foundation a certain distance above the sidewalk or above the surface of some part of the lot. If there's a reference to a city

datum plane on the plan, level from the nearest bench mark and set the batterboards at the grade given.

Step two— Measure 35'0" from the western boundary line and get the two points marked (2) in the rectangle in Figure 3-4. Then measure 15'0" from the street line and get the two points marked (E).

Step three— Find the points marked (3) by measuring 40'0" from the points marked (2).

Step four— Measure the diagonals AC and BD of the reference rectangle as a check. They must be the same length. If one is longer than the other, correct the error.

Step five— Locate the points marked (n) and stretch lines to mark the irregularities of the excavation by measuring the required distances with a tape.

It's often necessary to set a new building with its front on line with an existing building. If the ex-

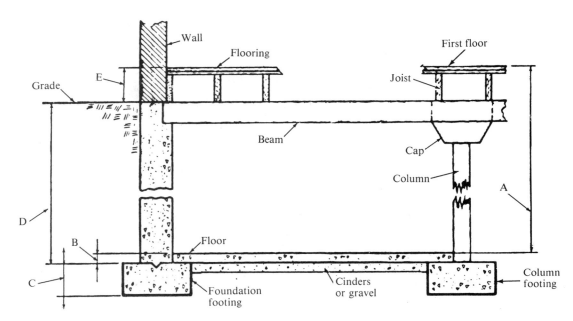

Typical sectional drawing
Figure 3-5

isting building isn't parallel to the street line, establish points (2) and (3) on a line parallel to and a definite distance from the existing building. Then proceed as above.

Measuring for the Excavation
Establish a bench mark to give a point of permanent reference during the construction of the building. It's best to set the construction bench mark about 4'0'' above the first floor or finish grade. A nail in a nearby tree or a mark on an adjoining building makes a good bench mark.

Figure 3-5 is a sectional drawing of a typical residence, showing the first floor, basement, foundation and footing, basement girder and column, and the basement floor. Architects usually show a dimension on their blueprints to indicate the distance from the top of the basement floor to the top of the first floor. This is dimension A in Figure 3-5. Using this dimension, you can easily calculate the depth of the excavation. To find this depth, carefully examine the dimensions marked A, B, C, D, and E. Dimension E shows the height of the first floor above the grade or ground level. Subtracting dimension E from dimension A gives the depth below ground level of the surface of the basement floor. To this add dimension B, the thickness of the basement floor. The total is the dimension marked D, the distance between the

ground level and the top of the footing. Excavate to a depth equal to dimension D, plus the thickness allowed for the cinder or gravel base under the floor slab.

Estimating Excavation
Before starting an estimate for any kind of construction, visit the building site and make notes about what you find. Also make a list of the costs that you'll include in the estimate.

When compiling the estimate, follow the sequence of operations as they *actually take place* in the field:

1) Clear the site
2) Strip and store topsoil
3) General excavation
4) Trench and pit excavation
5) Backfill
6) Excavation and fill for new site grades

We'll cover these steps one at a time, discussing general estimating information for each step and how to determine volume measurements to use in compiling the actual estimates.

Clearing the Site
In clearing the site, you'll have to remove standing buildings, paved areas, fences, poles, freestanding

Volume of Material in Piles

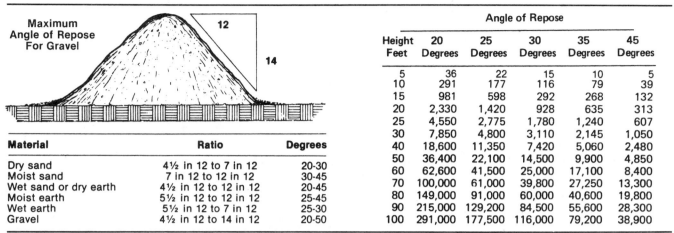

Material	Ratio	Degrees
Dry sand	4½ in 12 to 7 in 12	20-30
Moist sand	7 in 12 to 12 in 12	30-45
Wet sand or dry earth	4½ in 12 to 12 in 12	20-45
Moist earth	5½ in 12 to 12 in 12	25-45
Wet earth	5½ in 12 to 7 in 12	25-30
Gravel	4½ in 12 to 14 in 12	20-50

	Angle of Repose				
Height Feet	20 Degrees	25 Degrees	30 Degrees	35 Degrees	45 Degrees
5	36	22	15	10	5
10	291	177	116	79	39
15	981	598	292	268	132
20	2,330	1,420	928	635	313
25	4,550	2,775	1,780	1,240	607
30	7,850	4,800	3,110	2,145	1,050
40	18,600	11,350	7,420	5,060	2,480
50	36,400	22,100	14,500	9,900	4,850
60	62,600	41,500	25,000	17,100	8,400
70	100,000	61,000	39,800	27,250	13,300
80	149,000	91,000	60,000	40,600	19,800
90	215,000	129,200	84,500	55,600	28,300
100	291,000	177,500	116,000	79,200	38,900

Cubic yards assuming the pile is freestanding with a circular base.

Volume of material and angle of repose
Figure 3-6

walls and any other obstacles above the existing grades. Some of this work is actually wrecking. When there's a lot involved, it's usually best to get a bid from a wrecking contractor instead of trying to estimate this specialty yourself.

Stripping and Storing the Topsoil
Most specifications call for stripping the topsoil to a depth equal to the total thickness of the topsoil. Find this depth by taking borings or by digging a test hole to determine the actual ground condition. For a building of any size, always take soil borings or test holes to determine the value of the soil before beginning actual construction. You'll usually figure the stripping of the topsoil to about 5 feet outside the actual building lines.

To find the volume of topsoil to be stripped, multiply the length of the building plus 10'0'' by the width plus 10'0'' by the depth of the topsoil.

Where you need to preserve the sod for starting a lawn after the final grading, cut it into strips, roll the strips with the grass on the inside and stack it in some convenient place. It's best to cut the sod back 8 to 10 feet beyond the building lines because grass within that distance will be destroyed during construction.

General Excavation
When you figure mass excavating of large areas, never skimp on the size of the cut required. In general, you need about 2 feet outside the wall footings for working space to install concrete forms. You'll also have to slant the sides to about a 45-degree slope in sandy soil to keep the banks from sliding. The stiffer or more stable the excavated material, the less slope is required.

Figure 3-6 gives the angle of repose for various soils. This is the angle, measured from the horizontal, at which a pile of earth will stand. It will vary according to the amount of moisture in the soil. You'll have to use judgment in determining the angle to be used in a particular excavation project. The angles of repose given are approximate and will vary. Vibration by construction equipment can cause banks to slide into open cuts. Leave a little margin of error in your estimate. It's usually wise to allow a greater angle of repose than absolutely necessary.

Calculating the volume of excavation— Find the plan sheet that shows the sections through the basement. Estimate how much larger the excavation must be than the building size to allow for projecting footings plus working room. Generally you'll need to allow about 2 feet from the outside of the walls to give enough work room. Next, get the area of the building from the foundation plan. Then look at the elevations for the natural grades and the depth of the basement. Take the depth to the bottom of the concrete floor or gravel, if any, for the

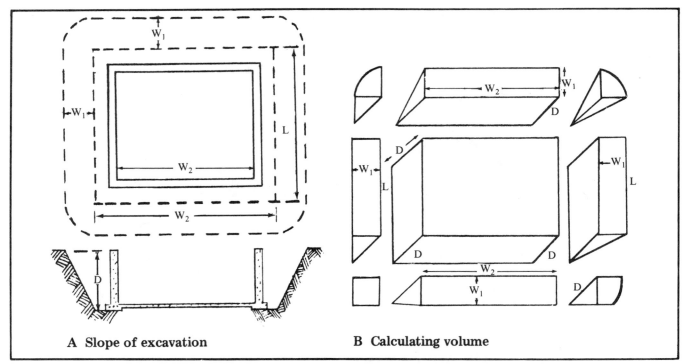

A Slope of excavation   **B Calculating volume**

Calculating volume of excavation
Figure 3-7

general basement level and put it down on the estimate sheet under the heading of General Excavation.

In finding the area covered by the building, take off the larger or main portions first; then take off the smaller portions, additions, or wings. Don't attempt to use "overall" dimensions on odd-shaped buildings unless you first make sure that all of the parts actually do have to be excavated.

Here's how to calculate the volume of excavation when sloping banks are considered. Look at Figure 3-7. In this figure the banks are sloped. First calculate the volume of the central box, then add the wedge-shaped pieces and sum up. Here's the formula:

$$\text{Volume} = LW_2D + LW_1D + W_1W_2D + \frac{\pi DW_1^2}{3}$$

| Vol. of center | Vol. of left & right wedges | Vol. of top and bottom wedges | Vol. of cone pieces |

For example, assume you're digging an excavation that's 100'0" long, 50'0" wide, and 10'0"

deep. First add 2'0" to each side, making the dimensions 104'0" by 54'0". Then add for the appropriate slope. Visualize this slope as being in the shape of a right angle along the side of the building, with the base at the top of the excavation and the hypotenuse lying along the 45-degree slope. Since only a right triangle with two equal sides has a 45-degree hypotenuse, the width of the excavation at the top will equal the depth, 10 feet. So you'll have two equal triangles, one on each side of the building. Together they form a square, 10'0" by 10'0". Now add the extra cut for the slope to your 104'0" by 54'0", getting you the dimensions 114'0" by 54'0". This is the actual size of the hole to be dug. If four sides must be sloped, then the actual size of the hole is 114'0" by 64'0".

Calculating volume in sloping ground— Where the natural ground slopes, we have a slight change in the procedure. In this case, the volume calculation must reflect the difference in the depths D1 and D2 in Figure 3-8. Add depths D1 and D2, then divide the sum in half. This is the average depth of the excavation. Multiply the length by the width times the average depth to find the average volume. Here's the formula:

$$\text{Volume} = \frac{LW(D_1 + D_2)}{2}$$

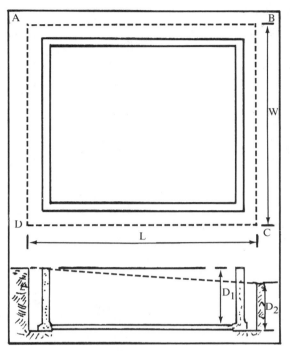

Calculating volume in sloping ground
Figure 3-8

When the ground surface slopes in two directions, the corners — points A, B, C and D of Figure 3-8 — are all unequal. In this case, add the individual heights of the four corners and divide by four. Here's the formula for finding the volume in this case:

$$\text{Volume} = \frac{LW(A + B + C + D)}{4}$$

Notice in Figure 3-8 that the excavating line extends beyond the foundation wall. You must provide room for the footing extension beyond the wall line, and allow sufficient space for the carpenters to set and strip the foundation formwork.

If there is some deeper part, such as an equipment room floor, find the area it covers and multiply by the depth below that you've already figured. If it's large enough to excavate by the same methods, list it with the general excavation. But if it's small, list it with trench excavation.

Calculating volume of irregular shapes— When you're dealing with foundations of irregular shape, break them down into simple sections with regular shapes, as shown in Figure 3-9. Then figure the volume for each subsection, numbered 1, 2 and 3 in the figure, and add them together to get the total volume.

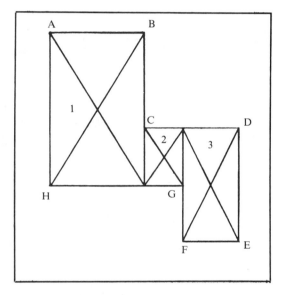

Calculating volume for irregular shapes
Figure 3-9

Pumping and shoring— If any shoring or pumping is needed, analyze the work to be done, estimating how many days you'll need for pumping, and how much labor and material are needed to do the shoring. Think it out and put down the cost while you have the plans in front of you. That way it won't be overlooked in the final estimate.

An earthwork contractor usually regards a basement excavation with a bulldozer as a day's work, regardless of the size of the basement. So his price will be a lump-sum bid. You should estimate it the same way. Consider the cost of excavating a basement with a dozer to be a day's rental on the equipment and a day's work for the crew. But compute the yardage anyway, for comparison with future jobs.

Trench and Pit Excavation
Trench or pit excavations extend below or outside

Depth of trench	Width of trench										
	12"	14"	16"	18"	20"	22"	24"	30"	36"	42"	48"
12"	3.7	4.3	4.9	5.6	6.2	6.2	7.4	9.3	11.1	13.0	14.8
14"	4.2	5.0	5.8	6.5	7.2	7.9	8.6	10.8	12.9	15.1	17.3
16"	4.9	5.8	6.6	7.4	8.2	9.0	9.9	12.4	14.8	17.4	19.8
18"	5.6	6.5	7.4	8.3	9.3	10.2	11.1	13.9	16.7	19.4	22.2
20"	6.2	7.2	8.2	9.3	10.3	11.3	12.3	15.4	18.5	21.5	24.6
22"	6.8	7.9	9.0	10.2	11.3	12.4	13.6	17.0	20.4	23.6	27.2
2'-0"	7.4	8.6	9.9	11.1	12.3	13.6	14.8	18.5	22.2	26.0	29.6
2'-6"	9.4	10.9	12.5	13.9	15.4	17.0	18.5	23.2	27.8	32.4	37.0
3'-0"	11.2	13.0	14.9	16.7	18.5	20.4	22.2	27.8	33.3	38.9	44.5
3'-6"	13.0	15.1	17.3	19.4	21.6	23.8	25.9	32.4	38.9	45.4	52.0
4'-0"	14.8	17.2	19.7	22.2	24.7	27.2	29.6	37.0	44.5	52.0	59.2
4'-6"	16.6	19.3	22.1	24.8	27.6	30.6	33.3	41.6	50.0	58.4	66.7
5'-0"	18.5	21.5	24.6	27.7	30.8	34.0	37.0	46.3	55.5	64.9	74.1
5'-6"	20.4	23.8	27.2	30.5	34.0	37.3	40.7	51.0	61.1	71.3	81.6
6'-0"	22.2	25.8	29.6	33.3	37.1	40.7	44.4	55.5	66.7	77.9	89.0
6'-6"	24.0	27.8	31.8	36.1	40.0	44.2	48.1	60.2	72.2	84.2	96.5
7'-0"	25.9	30.1	34.4	38.9	43.1	47.6	51.9	64.8	77.8	90.8	103.8
7'-6"	27.8	32.3	36.9	41.6	46.2	51.0	55.6	69.5	83.4	97.3	111.4
8'-0"	29.6	34.4	39.2	44.5	49.4	54.5	59.2	74.1	88.9	102.0	118.6
8'-6"	31.5	36.6	41.8	47.8	53.0	58.5	63.0	78.7	94.5	110.0	126.0
9'-0"	33.3	38.7	44.1	50.0	55.5	61.2	66.7	83.5	100.0	117.0	133.3
9'-6"	35.2	40.9	46.7	52.7	58.5	64.5	70.5	88.0	105.7	123.2	140.9
10'-0"	37.0	43.0	49.1	55.5	61.5	68.0	74.0	92.7	111.0	129.8	148.1

For shallow trenches (up to 2' deep), use the width of the footing.

For trenches 2' to 4' deep, use the width of the footing or 1'6", whichever is greater.

For trenches 4' to 6' deep, use a width of at least 2'.

Cubic yards per 100 LF of trench
Figure 3-10

of the lines of the general excavation. Estimate the excavation for pits and trenches at their actual size plus 1 foot outside for working room. For walls that will be waterproofed, figure 2 feet all around instead of 1.

Where shoring is required, it may be necessary to make the excavation 2 feet or more larger all around to provide room for placing and bracing the shoring. To estimate costs of shoring and bracing accurately, list the shoring for each class of excavation separately, and include the depth of the excavation.

If you can't get a backhoe in to dig the footing trenches, they'll have to be dug by hand. To figure the volume for this handwork, multiply the *width* of the trench by the *depth*, both in inches. Divide the result by 144 to get the cross-sectional areas in square feet. Multiply this by the *perimeter* of the building in feet, and divide the result by 27. This gives you the volume of the trench in cubic yards.

Multiply the volume by the cubic yard cost of hand digging to get the footing trench cost.

Figure 3-10 is a chart giving the cubic yards per 100 feet of trench. Use it to speed up your calculations. Your CEF for excavating a trench or pier by hand might be like the example shown in Figure 3-11.

For trenches more than 2 feet deep and less than 3 feet wide, figure them as 3 feet wide. It'll require almost this width to give room for digging. Even if the trench is dug narrower, there can be cave-ins. It's safer to figure them as 3 feet wide.

Some drawings may not give the depth of excavation of piers and footings. If you come across one of these, take it up with the architect. In most cases you can assume the depth of footings at below the frost line, but this depends on soil conditions. Check to make sure.

Labor for Trench or Pier Excavation — By Hand

Work element	Unit	Manhours per unit
Up to 2 feet deep		
Normal soil	CY	.9
Sand or gravel	CY	1.2
Medium clay	CY	1.1
Heavy clay	CY	1.4
Loose rock	CY	2.2
Over 2 feet to 6 feet deep		
Normal soil	CY	1.3
Sand or gravel	CY	1.5
Medium clay	CY	1.4
Heavy clay	CY	1.8
Loose rock	CY	2.6
Fine grade excavation bottom	10SY	.9
Trim excavation banks	10SY	.5
Hand work around obstructions	CY	3.0
Backfilling, no compaction	CY	.9
Compacting backfill to 95% of maximum density		
Hand tamping	CY	.5

Excavation includes only piling soil adjacent to the trench or pit. Add for loading on trucks or spreading soil.

CEF — Excavating trench or pier by hand
Figure 3-11

Figure 3-12 shows a typical building plan, including a section through the foundation. Figure the excavation one foot beyond the outside wall, since the soil is clay and will stand straight. Here's how you'd figure the depth of excavation for this building: For the general excavation, use the bottom of the cinders, or 4'9" from grade. Dig the trenches just 3½" deeper, or 5'1/2" below grade.

Calculating the volume of a pit— To calculate pit volume, divide the plan view of the pit into a system of squares and rectangles of 20 to 50 feet on a side, depending on the unevenness of the ground. Then find the difference in elevation between the original ground surface and the grade of the finished excavation for all the corners of the rectangles or squares.

Look at the example shown in Figure 3-13. The area has been divided into squares and rectangles, labeled A, B, C and so on. To estimate the volume of the pit excavation, determine the top elevation from the contours shown for each labeled point. Then subtract the finish grade, 106.0', from each of these elevations, and write the difference at each intersection point. The next step is to average the figures for the four corners and multiply by the area. That gives the volume for each square. Then add the volume of each square to find the total volume of the excavation in the pit.

Let's work out the first square. We'll call this Block 1. There are four corners, A, B, I and J. Note that the elevation line 118 nears corner A at about the 118.2 point. Corner B is 117.2, corner I is 117.8 and corner J is 118.8 feet. So, it looks like this:

```
Corners:
A = 118.2'  —  106.0'  =  12.2'
B = 117.2'  —  106.0'  =  11.2'
I = 117.8'  —  106.0'  =  11.8'
J = 118.8'  —  106.0'  =  12.8'
                            ———
                  Total     48'
```

To get the average depth of the four corners, divide 48 feet by 4 for a total of 12 feet.

Block 1 is 10 feet wide and 10 feet long. To get the area of excavation required, multiply the width of the block by the length: 10 x 10 equals 100 square feet. To get the volume of the excavation, multiply the 100 square feet by the average depth (12 feet) of the four corners: 100 SF x 12' is 1,200 cubic feet. To convert cubic feet to cubic yards, divide 1,200 cubic feet by 27. There's 44.44 cubic yards to be excavated from Block 1.

Using the same procedure, find the cubic yards for each of the blocks and then add the results. You'll find that the total volume of excavation required for Figure 3-13 is 410.5 cubic yards.

You can use the factors in Figure 3-14 to find the cubic yards of dirt per unit of depth when you know the number of square feet of surface area. Just multiply the factor by the square feet to find the volume.

To estimate excavation for piers and abutments, multiply the outside dimensions by the depth. No extra allowances are required unless room is necessary to work in. Concrete footings usually don't require forms. But check soil conditions and the specifications. When the piers are constructed of block or brick, allow extra room on each side for the bricklayer to work.

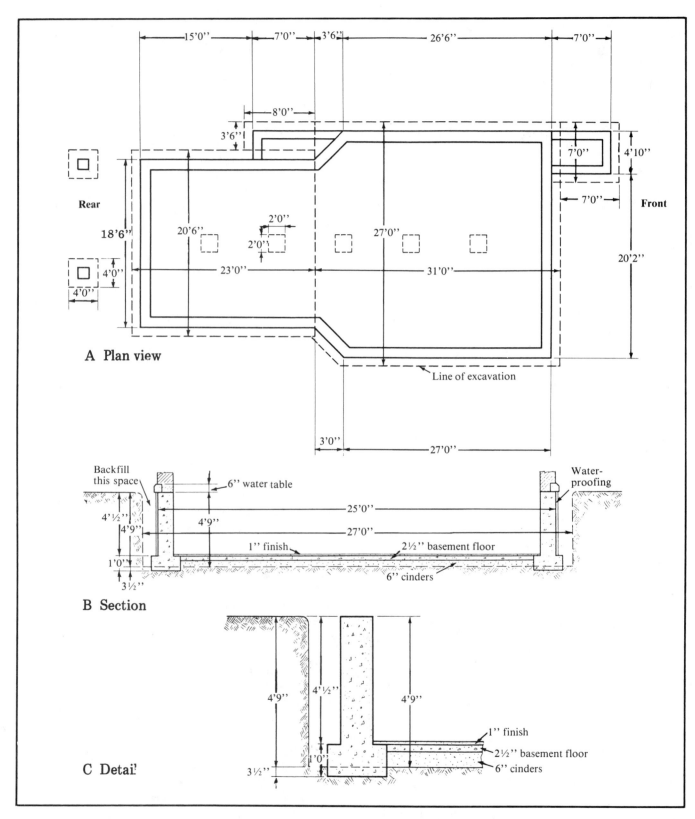

Typical plan including trench and pit excavation
Figure 3-12

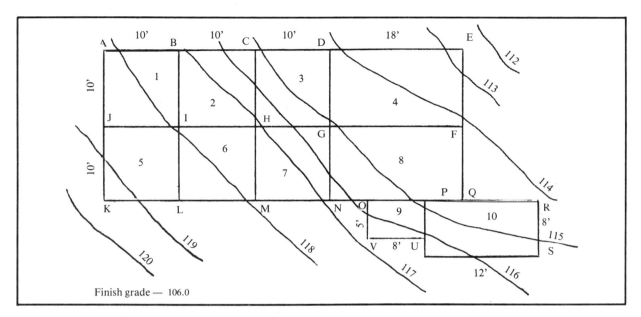

Calculating volume of pit excavation
Figure 3-13

Depth	Cubic yards per square foot	Depth	Cubic yards per square foot
2''	.006	4' 6''	.167
4''	.012	5' 0''	.185
6''	.018	5' 6''	.204
8''	.025	6' 0''	.222
10''	.031	6' 6''	.241
1' 0''	.037	7' 0''	.259
1' 6''	.056	7' 6''	.278
2' 0''	.074	8' 0''	.296
2' 6''	.093	8' 6''	.314
3' 0''	.111	9' 0''	.332
3' 6''	.130	9' 6''	.350
4' 0''	.148	10' 0''	.369

Excavation factors
Figure 3-14

Always figure pits and pier holes more than two feet deep as having an area of at least 12 to 16 square feet, even though they don't require that much area. For example, if a footing 2 feet by 2 feet is to be carried down 5 feet below grade, estimate it as 12 square feet of area times the depth. It requires a space larger than 2 feet by 2 feet to dig a hole 5 feet deep.

Estimating excavation by hand— The amount of trench a man can excavate varies with the kind of dirt, the height to which it is lifted, the extent of digging required and the weather conditions. A pick is usually used if loosening is needed, and lifting is done with a round-pointed, long-handled shovel. As a rule, it takes 150 to 200 shovels of dirt to excavate a cubic yard under normal conditions.

I'm including some tables you can use in estimating excavation under average conditions. But *keep your own records* of the actual time (labor hours) and the cost of different jobs. This will help you to build up tables of your own that will be reliable for your company. Figure 3-15 gives typical hourly rates for hand excavation. The deeper the trench or pit, the more the dirt must be handled. The dirt must be rehandled once for depths of 6 to 10 feet, twice for depths of 11 to 15 feet. For every two or three diggers, you'll need one man for rehandling the dirt. He'll probably stand on a stage or platform that can be moved as required.

Piers generally require more labor for excavating than trenches. They have to be kept straight on four sides and there's less space in which to work. For this reason use the larger number in Figure 3-15, where different values are given for sandy or heavy soil.

Here's the simplest way of estimating trenching

	Sandy or loamy soil	Heavy soil or clay
General excavation	.67	1.0
Shoveling dug earth	1.0	2.0
Loosen with pick	2.0	4.0
Shovel foundation pit to 6 feet	1.0	2.0
Shovel loose earth to truck	1.0	1.0
Hand backfilling	.4	.67
Spread loose earth	.15	.25
Fill 3 CF wheelbarrow and haul 100 feet	.67	1.0

Spreading material by hand:	
Loose rock, dumped in place	.4
Loose rock, one throw and spread	.8
Loose screenings, dumped in place	1.0

Trench by hand:

	Trench depth in feet			
Soil type	3	5	8	10
Light	.7	.75	.85	.9
Medium	.85	.9	.9	1.0
Heavy	1.1	1.0	1.25	1.3
Hard pan	1.3	1.4	1.6	1.7

**Labor hours per cubic yard for hand excavation
Figure 3-15**

costs by hand labor:

1) Compute the quantity of each kind of soil in cubic yards.

2) Estimate the hours required for a man to excavate one cubic yard of each kind of soil.

3) Compute the total cost of each kind of soil.

4) Total the costs.

Example: Estimate the time to excavate and backfill a footing trench 5'6'' deep, 3'0'' wide and 150'0'' long, in ordinary soil. About half of the excavated dirt will be thrown back from the edge of the trench by a worker other than the digger in order to leave a clear space about 3 feet wide be-

tween the trench and the piled dirt. One laborer can excavate a section of the trench about 15 feet long, so this trench will require 10 laborers to complete its length. Other laborers will throw the dirt back from the edge of the trench once the diggers get down about 3 feet. One foreman will also be required to supervise the job.

Solution: Volume = 5.5' x 3' x 150' = 2,475 CF = 92 CY in round figures.

Loosening earth with pick, 92 x 3 hours per CY	276 hours
(3 is the average value of 2 and 4)	
Shoveling earth from trench, 92 x 1.00 hours per CY	92 hours
Shoveling dirt back from edge of trench 46 x 1.0 hours per CY	46 hours
Backfilling, 92 x 0.67 hours per CY	62 hours
Total hours for laborers	476 hours

Total laborers required 10 + 2 (on bank) = 12
Length of job or foreman's time, 476 ÷ 12 = 40 hours in round figures.

We used time estimates in hours per cubic yard for loosening, shoveling and backfilling from Figure 3-15.

You'll probably still use wheelbarrows on small jobs to move excavated dirt for distances of up to 100 feet. The ordinary wheelbarrow will hold about three cubic feet of loose dirt. A laborer should be able to haul it 100 feet and return in about 2½ minutes if the path in which it runs is fairly smooth and hard. Filling the wheelbarrow with loose dirt will take about 2½ minutes. So it takes about five minutes to load and haul three cubic feet of dirt up to 100 feet. This will amount to about one hour for each cubic yard (bank measure).

Estimating caissons— Caissons are used to support heavy structures where the soil near grade won't support the load. Caissons carry the support down to solid earth that can carry the load without settling. The size and depth of caissons vary with the loads to be carried and how far it is to solid ground. In Chicago and New York City, caissons are usually carried down to rock, which may be 100 feet or more below the surface.

Drilled caissons are more economical than hand-dug caissons for several reasons. First, drilling caissons allows for smaller shaft diameters. The

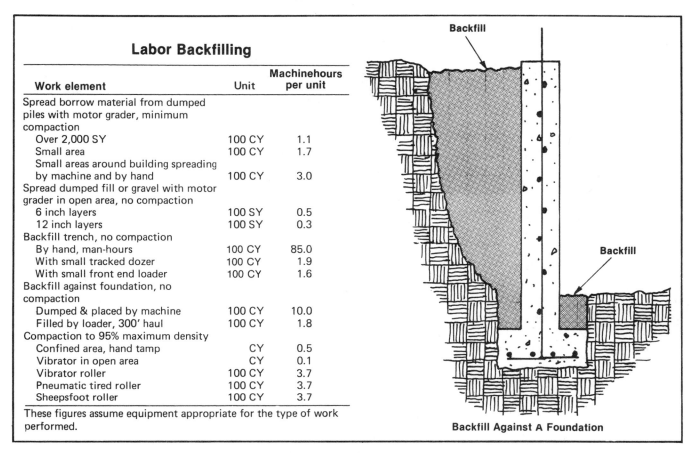

Labor Backfilling

Work element	Unit	Machinehours per unit
Spread borrow material from dumped piles with motor grader, minimum compaction		
Over 2,000 SY	100 CY	1.1
Small area	100 CY	1.7
Small areas around building spreading by machine and by hand	100 CY	3.0
Spread dumped fill or gravel with motor grader in open area, no compaction		
6 inch layers	100 SY	0.5
12 inch layers	100 SY	0.3
Backfill trench, no compaction		
By hand, man-hours	100 CY	85.0
With small tracked dozer	100 CY	1.9
With small front end loader	100 CY	1.6
Backfill against foundation, no compaction		
Dumped & placed by machine	100 CY	10.0
Filled by loader, 300′ haul	100 CY	1.8
Compaction to 95% maximum density		
Confined area, hand tamp	CY	0.5
Vibrator in open area	CY	0.1
Vibrator roller	100 CY	3.7
Pneumatic tired roller	100 CY	3.7
Sheepsfoot roller	100 CY	3.7

These figures assume equipment appropriate for the type of work performed.

Backfill Against A Foundation

CEF — Backfilling labor
Figure 3-16

smallest hand-dug caisson is 4 feet in diameter. This often means that more concrete has to be poured than is required for the load placed upon it. Second, it's much faster to drill caissons than to dig them by hand. This reduces construction time, and that always saves money.

Many firms specialize in drilling caissons. I'd advise you to get figures from these companies any time you have caissons to estimate.

Backfilling
Backfilling is replacing the dirt that's been excavated. Estimate the backfill by taking the total volume of the excavation, less the displacement or total volume of the structure below grade. The difference is the volume of dirt to be handled for backfill.

You may be required to use something besides the originally excavated material for backfill, but usually exterior backfilling may be done with the original material. Interior backfill usually has to be sand, stone, loamy clay or other material suitable for compaction. Some jobs require compaction using mechanical tampers to get the degree of compaction desired. Figure 3-16 is an example of a CEF on labor for backfilling and compaction.

If you've allowed space around the building in excavating, you have to backfill that space between the outside of the wall and the bank of earth. Look back to Figure 3-7. You use the same methods to calculate volume of backfill as you did to calculate volume of excavation.

Figuring backfilling for ordinary basement walls where you've left one or two feet around the building is a simple matter if the edges of the excavation are reasonably straight. Multiply the distance around the excavation by the depth of the excavation. The answer will be in cubic feet. Divide by 27 to convert the answer to cubic yards. As a rule of thumb, assume that for ordinary buildings with 12-inch walls, the backfilling will just about equal the amount of concrete or brick used (in terms of cubic feet) in the foundation.

It's important that you don't do any backfilling until the owner or architect has inspected the outside of the foundation walls and given his permission to start backfilling. This is particularly true where the foundation is made of concrete blocks to be plastered below grade. If there are any gaps in the plastering, water will get into the basement.

Backfilling also includes refilling trenches dug for laying pipes. Excavating and backfilling for sewers and drains is usually done by the plumber.

Establishing New Site Grades
New site grades are always required to prevent pooling of runoff and to provide slopes for paved and landscaped areas. Cuts and fills for new site grades vary from simple to complex. In simple cases, you can estimate the amount of cut or fill required by using the grid method. Just superimpose a system of grids, similar to graph paper, on the survey of the plot plan which shows the elevations of the lot. Lay out the grid in 50 squares on the plot. In the center of each square write the average cut or fill as shown by the nearest elevation. Total the averaged cuts and fills shown in the grids to get the total cubic yards of cut and fill. This is essentially the same procedure we used for Figure 3-13.

After you've established the site grade, the excavating job is finished except for dressing up the area. This may include only enough grading to receive topsoil. Rough grading will include all the areas of cut and fill and the lawn area.

Specifications usually call for spreading 6 inches of topsoil over all the future lawn areas. Calculate topsoil separately, as it's more expensive than ordinary fill. To figure topsoil on your estimate sheet, multiply the rough grading area by the depth of the topsoil required.

Excavating with Power Equipment
The kind of equipment you'll use for excavating depends on the size of the job. Some contractors buy every labor-saving device they can. Others rent what's needed and do some work by hand. Hand excavation is the least efficient and the most expensive on larger jobs. But it will be the only choice if there's no room for power equipment to maneuver.

Bulldozers are most effective for clearing land. They can remove trees and shrubs quickly and easily. A bulldozer or tractor shovel is also appropriate for shallow excavations, where the excavated dirt will be spread over the lot. A crawler-loader is good for excavating basements and loading directly into trucks.

In large excavations, a power shovel or a dragline crane is the most economical equipment to use. The power shovel will excavate and load the trucks as fast as they move in and out of the hole. This minimizes delays and makes the job move much faster. The cost of power shovel excavating will vary with the size of the job, size of the shovel, and the speed that the trucks can move into position. A shovel seldom, if ever, works to capacity on mass excavation.

Modern power excavation equipment has amazing productive capacity. But don't assume that you're going to get the manufacturer's full rated capacity from any machine. In building excavation, delays due to cramped quarters, traffic, congestion of trucks waiting to get in or out of the hole, delays at the dump, and bad weather are almost inevitable. Never figure 100% for either shovels or trucks.

We'll consider the most popular pieces of excavation equipment one at a time, beginning with the bulldozer.

Bulldozers
The bulldozer is probably the most versatile of all power excavating equipment. After using it to clear the site, you can often use it for excavating too. It isn't suitable for most deep excavations, but for shallow cuts and for skinning and backfilling, it's unexcelled.

While there are several sizes of bulldozers, the Caterpillar D4 track-type tractor is the one most used for residential construction. It's shown in Figure 3-17. It rents for up to $200 per day, not including operator, and will move about 200 cubic yards in a day. A track-type bulldozer has to be transported to the job on a trailer or a truck. It will excavate for the basement of a small house in a little over half a day. For a large house, allow a full day. A bulldozer scrapes and pushes the earth ahead of it and out of the excavation. The operator can leave the dirt in piles where desired, either close by or at a distance. But always pile the topsoil separate from the loam or sand. Since it's the first removed, deposit the topsoil farther away. Plan the work so the bulldozer pushes the dirt out of the excavation over two opposite sides of the excavation, usually in the front and rear, instead of over all four sides. Then it can cut a vertical bank on two sides of the hole and leave the other two banks

Caterpillar D4 tractor
Figure 3-17

sloped. Deposit most of the soil at the rear. This requires the least amount of excavation and makes excavating and backfilling more economical.

A bulldozer is economical in excavating down to a depth of about five feet. After that, a power shovel or backhoe is more efficient. At depths over five feet, the ramp or slope needed for the dozer to get out of the excavation becomes too large.

Figure 3-18 is a bulldozer production table showing typical production figures for four different models of bulldozers. These figures are based on slot dozing in good weather with well-maintained equipment. We're assuming a well-organized and supervised job that results in about 50 work-productive minutes each hour. This gives an efficiency factor of 83%, which has been applied to the production figures in the table. Of course, our figures won't be as accurate as the production rates you keep for your own crew and operation. As shown in the table, a D6 should move 75 cubic yards of clay 150 feet in one hour, moving about 2.5 cubic yards with each run.

Power Shovels

The power shovel is good for digging basement ex-

cavations. I recommend a machine with a 1/2 or 3/4 cubic yard bucket. Anything much larger or smaller will cut down efficiency. The shovel may rent for $250 to $320 or more per day and should dig about 220 cubic yards in eight hours. It'll usual-

| Model | Dozer Capacity | Type of material | Load factor | Distance moved | | | | | |
				50 ft.	100 ft.	150 ft.	200 ft.	300 ft.	400 ft.
D8	5 CY	Loose vol.	1	375	208	150	118	83	62
		Clay	.72	270	150	108	85	60	45
		Loam	.80	300	167	126	95	67	50
		Gravel	.80	337	188	135	107	75	56
		Hardpan	.66	250	140	100	80	56	42
		Rock, well blasted	.60	225	125	90	70	50	37
		Sand	.90	337	188	135	107	75	50
		Sandstone Shale, or soft rock	.60	225	125	90	70	50	37
		Traprock	.66	250	140	160	80	56	42
D7	4.1 CY	Loose vol.	1	325	179	147	116	82	62
		Clay	.72	230	130	105	80	60	44
		Loam	.80	260	140	120	90	70	50
		Gravel	.90	290	150	130	100	75	55
		Hardpan	.66	215	120	95	75	55	40
		Rock, well blasted	.60	200	110	90	70	50	35
		Sand	.90	290	150	130	100	75	55
		Sandstone Shale, or soft rock	.60	200	110	90	70	50	35
		Traprock	.66	215	120	95	75	75	40
D6	2.6 CY	Loose vol.	1	237	128	106	84	61	--
		Clay	.72	170	90	75	60	45	--
		Loam	.80	190	100	85	70	50	--
		Gravel	.90	210	115	95	75	55	--
		Hardpan	.66	155	85	70	55	40	--
		Rock, well blasted	.60	140	80	65	50	35	--
		Sand	.90	210	115	95	75	55	--
		Sandstone Shale, or soft rock	.60	140	80	65	50	35	--
		Traprock	.66	155	85	70	55	40	--
D4	1.9 CY	Loose vol.	1	153	87	60	46	32	--
		Clay	.72	110	65	45	33	23	--
		Loam	.80	120	70	50	37	25	--
		Gravel	.90	140	80	55	41	29	--
		Hardpan	.66	100	60	40	30	21	--
		Rock, well blasted	.60	90	50	45	28	19	--
		Sand	.90	140	80	55	41	29	--
		Sandstone Shale, or soft rock	.60	90	50	35	27	19	--
		Traprock	.66	100	60	40	30	21	--

Note: Production is computed for slot dozing. If dozing is done without slots, blade capacity should be reduced approximately 25%.

Bulldozer production table
based on 50-minute hour (83% efficiency)
Figure 3-18

	Production in cubic yards for 60 minute hour - 90° swing						Rock, well blasted
Dipper capacity in CY	Sand and gravel		Common earth		Dense clay		
	Optimum depth of cut in feet	Cubic yards	Optimum depth of cut in feet	Cubic yards	Optimum depth of cut in feet	Cubic yards	Cubic yards
⅜	3.8	80	4.5	65	6.0	45	35
½	4.6	105	5.7	90	7.0	70	55
¾	5.3	150	6.8	125	8.0	105	75
1	6.0	180	7.8	160	9.0	135	115
1¼	6.5	220	8.5	190	9.8	170	145
1½	7.0	260	9.2	225	10.3	200	170
1¾	7.4	290	9.7	260	11.5	230	200
2	7.8	320	10.2	295	12.2	265	230
2½	8.4	400	11.2	360	13.3	320	280
3	9.0	380	12.0	450	14.0	400	350
4	10.5	630	13.5	575	15.5	550	450

Modify the above figures for:
1) Efficiency. Usually figure a 50 minute hour or 83% efficiency.
2) Depth of cut and angle of swing. Use the percentages below.

	Factor Table						
	Angle of swing						
Depth of cut in percent of optimum	45°	60°	75°	90°	120°	150°	180°
40	0.93	0.89	0.85	0.80	0.72	0.65	0.59
60	1.10	1.03	0.96	0.91	0.81	0.73	0.66
80	1.22	1.12	1.04	0.98	0.86	0.77	0.69
100	1.26	1.16	1.07	1.00	0.88	0.79	0.70
120	1.20	1.11	1.03	0.97	0.86	0.77	0.70
140	1.12	1.04	0.97	0.91	0.81	0.73	0.66
160	1.03	0.96	0.90	0.85	0.75	0.67	0.62

Power shovel production table
Figure 3-19

ly dig the basement for an average house in about a half day.

In soil with many large stones, a power shovel is almost undoubtedly your best bet. If you're building on the side of a hill, you're very likely to come up against rock near the bottom. Rock near the surface of the ground is almost always disintegrated and can be removed easily with a power shovel. If the excavated material is to be loaded into trucks or if there's not enough room for maneuvering a bulldozer or piling dirt, a power shovel is a good choice.

If it were possible to use a 3/4 cubic yard shovel to capacity (figuring a 100% dipper load each time), it would excavate 135 cubic yards an hour. But the dipper isn't always filled and there are bound to be delays in loading, moving the shovel from place to place, cleaning up and bad weather. The average on building work will run about 50% to 60% of the full capacity of the shovel.

1. Loading from bank or pile, 30 feet. Travel and 90° turn, loose cubic yards loaded per hour, based on 60 minute hour and 15 seconds per truck for spotting.

Bucket size	Loading to truck size			
CY	4 CY	6 CY	8 CY	10 CY
1¼	90	93	--	--
1¾	131	135	140	--
2½	--	207	220	221

2. Excavation work, load to truck

Bucket size	Haul distance			
CY	50 ft.	100 ft.	150 ft.	200 ft.
1¼	50 CY	33 CY	25 CY	20 CY
1¾	77 CY	50 CY	38 CY	30 CY
2½	110 CY	77 CY	57 CY	45 CY

Above table based on:
6 ft. excavation depth
60 minute hour
Loose yardage - apply load factor for bank measure
Fixed time - 0.6 minute

30 ft. allowance for ramp and maneuvering to truck
Speed - 1st gear forward
Return - 2nd gear forward
For 2nd gear haul decrease production 10% at 50 feet
For 2nd gear haul decrease production 20% at 200 feet

High-lift tractor production table
Figure 3-20

The ownership and operation cost of power shovels includes interest on the investment, insurance, maintenance and repairs, fuel, lubrication, depreciation, supplies, and labor. Make allowance for all of these items in figuring costs.

To estimate the total cost of excavating with a power shovel, include the cost of transporting the shovel to the job, the labor cost of setting up the shovel, the equipment and labor costs of excavating the material, and the cost of removing the shovel when the job is finished. You'll need one or two helpers in addition to the shovel operator, depending on the size of the shovel and the job. A foreman, as a rule, is necessary to supervise the excavating and hauling, so a part of his salary should be charged to the excavation.

Figure 3-19 is a production table for excavating with a power shovel. Note that the production times in this table are gross figures. You'll have to apply the 83% efficiency factor (or another modifier as appropriate) to make them accurate.

Figure 3-20 gives figures for excavating with a high lift tractor.

Backhoes
Backhoes come in many sizes and shapes, but the most widely used is a 1/2-cubic yard machine that digs an 18-inch wide trench. It can be used for footing trenches, water, sewer or drainage ditches. Some models will make a 10-foot deep cut. Average productivity is about 100 cubic yards a day. The machine may rent for about $110 to $140 per day.

The backhoe is a highly versatile piece of equipment that's productive in basement excavation as well as trenching. The backhoe will produce a vertical cut in suitable soil, and has the advantage of being able to work above the level of the excavation. Both rubber tire and crawler types are available with bucket sizes from 3/8 cubic yard to 2 cubic yards. Some of the larger models can cut trenches 25 feet deep, though for most units 12 to 15 feet is the maximum operating depth. You can interchange the buckets from narrower to broader units to suit the type of work and soil conditions. In shale or harder soil, you may need a narrower bucket to overcome greater soil resistance.

Productivity is determined by the bucket size and the turning radius required for dumping. If trimming and leveling are required, as in basement work, trenches and footings, allow extra time. Figure 3-21 is a production table for backhoes.

Output per hour

Bucket size	Sand or gravel	Compact gravel or clay
⅜ CY	20 CY	15 CY
½ CY	25 CY	20 CY
¾ CY	45 CY	40 CY
1 CY	60 CY	50 CY

Note: For cuts beyond 75% of the deepest reach or for swings up to 120°, reduce these figures by 20%.

Backhoe production table
Figure 3-21

In general, the process of digging a basement with a hoe is to dig a trench around the four sides of the basement and scoop out the center as you go. The set-ups for different jobs may vary because of nearby houses and the available space for disposal of materials.

If you have a backhoe, you need a simple, low-cost way of moving the unit from job to job. Your best bet is a single-purpose trailer which can be loaded or unloaded in 15 to 20 minutes. The cost of moving the backhoe from job to job will vary with the travel time.

Excavating with Draglines
Draglines are seldom used for residential and small commercial construction. But there are some cir-

cumstances where only a dragline can do the work. A dragline is a crane rigged with a bucket, so it can stand some distance away from the excavation, on firm ground.

The dragline is limited in three important respects, however. It may be unable to operate where there are overhead restrictions or surrounding structures; time required to swing the boom from excavation to dump will reduce productivity; and precise unloading is difficult. Loading trucks from a dragline bucket will result in a lot of spillage. However, where the dirt is very loose or wet, where long reach is required and where dumping must be done at some distance from the hole, a dragline may be your best choice.

Drag Loaders and Trucks
If the dirt must be moved a short distance away from the excavation site, consider using a drag scraper, "Carryall," or similar self-loading machine. But when the dirt must be hauled to some disposal site, trucks usually do the hauling. The number of trucks or laborers you'll need depends on the distance to be hauled.

Carefully calculate the time required for the round trip, from the time the truck is loaded until it's back empty and ready for another load. Use enough trucks so the excavation equipment doesn't have to wait for trucks to fill — but not so many that the trucks have to wait for a load. A truck that's standing still isn't earning money. Larger excavations warrant the use of heavy equipment, such as power shovels or draglines, which can fill trucks very quickly.

Figure 3-22 shows the round trip time for trucks for various distances and speeds. Before you use it, look it over carefully to make sure your job has conditions similar to the ones the figure is based on. It uses four minutes at the shovel and three minutes at the dump for a total fixed time of seven minutes. The operating factor is based on a 50-minute hour.

Hauling Excavated Dirt
The best method of disposing of excavated material depends on the distance it must be hauled. If the dirt is used on the premises for grading, you don't need trucks. The slowest, but the most simple method is shoveling the dirt back away from the site. Where the dirt must be moved a short distance, such as 50 to 100 feet, use wheelbarrows or a bulldozer.

Haul distance		Time - minutes		
Miles one way	Travel round trip	Fixed time	Round-trip time	Trips per hour
Average speed - 10 M.P.H.				
1	12	7	19	2.63
2	24	7	31	1.62
3	36	7	43	1.16
4	48	7	55	0.91
Average speed - 15 M.P.H.				
1	8	7	15	3.33
2	16	7	23	2.17
3	24	7	31	1.62
4	32	7	39	1.28
Average speed - 20 M.P.H.				
1	6	7	13	3.85
2	12	7	19	2.63
3	18	7	25	2.00
4	24	7	31	1.62
Average speed - 30 M.P.H.				
1	4	7	11	4.55
2	8	7	15	3.33
3	12	7	19	2.63
4	16	7	23	2.18

Above table based on:
1) Loading time 4 minutes
2) Dumping time 3 minutes
   Fixed time 7 minutes
3) 1½ CY loader
4) 6-8 CY trucks
Note: Truck capacity x load factor = pay load

Truck hauling capacity
Figure 3-22

When the excavated dirt is loaded into trucks or tractor-pulled equipment, the hauling cost per cubic yard will depend on the capacity of the hauling unit and length of the haul. Loading a truck with a power shovel takes only a few minutes. But it's difficult for a fully loaded truck to drive out of a deep hole. Often the truck can't be fully loaded.

In large structures having basements and sub-basements, the excavation may extend 40 to 50 feet below grade. An extra truck or dozer may be needed to push the loaded truck out of the hole. Take this into consideration when pricing the excavating job.

Struck and heaped capacity— The capacity of hauling units is usually expressed in tons or cubic yards. When stated in cubic yards, you need to know if it's *struck* or *heaped*. The struck yardage is the volume the unit will hold when it's filled level with the sides; the heaped yardage is the volume when dirt is piled above the sides. The struck capacity is fixed, since it depends on the length, width and depth of the unit. The heaped yardage will depend on the depth of the dirt above the sides and the area of the bed. Some manufacturers specify the heaped capacity based on a 1 to 1 slope of the dirt above the sides.

Find the actual capacity by taking the average of several loads. Heavy trucking units for this work are made to carry 3, 5, 7 or more cubic yards. You have to know the capacity of the truck you're using before you can make an accurate estimate.

All of the quantities in the tables in this book are based on *bank* measure, which means cubic yards removed from the bank rather than cubic yards in a truck. There's a big difference between the two. Dirt *swells* when it's dug or loosened. Figure 3-23 shows the percent of swell for the types of soil commonly encountered in construction. It also gives the formula for finding the volume of material in piles and the average number of cubic yards in stockpiles of various heights. You'll notice that Figure 3-23 includes a column headed "load factor." A load factor of 72% means that 28% of the load is air space. It's the average loaded density for each kind of material.

To convert loose measure to bank measure, multiply the loose measure by a shrinkage factor. For ordinary dirt with a swell of 25%, the loose volume is 1.25 times the bank volume. So the shrinkage factor is 1 divided by 1.25, or 0.8. For example, a truck hauling 10 cubic yards loose measure will have a bank measure load of 8 cubic yards (10 times 0.8 equals 8).

In most areas there are contractors who make a specialty of excavating. They're familiar with the kind of soil in different parts of town. These contractors may quote a price per cubic yard for all of the general excavation, or a lump sum for the entire job.

Backfilling
Backfilling around basement walls or in trenches may be done by hand shoveling, by scrapers, or by dozers. Hand work goes faster in light soil and when no tamping or compacting is required.

| | Approximate material characteristics* | | | |
Material	Pounds per cubic yard-bank	Percent of swell	Load factor	Pounds per cubic yard-loose
Clay, natural bed	2960	40	.72	2130
Clay & gravel, dry	2290	40	.72	1940
Clay & gravel, wet	2620	40	.72	2220
Clay, natural bed				
Anthracite	2700	35	.74	1600
Bituminous	2160	35	.74	1600
Earth, loam, dry	2620	25	.80	2100
Earth, loam, wet	3380	25	.80	2700
Gravel, ¼" - 2" dry	3180	12	.89	2840
Gravel, ¼" - 2" wet	3790	12	.89	3380
Gypsum	4720	74	.57	2700
Iron ore, magnetite	5520	33	.75	4680
Iron ore, pyrite	5120	33	.75	4340
Iron ore, hematite	4900	33	.75	4150
Limestone	4400	67	.60	2620
Sand, dry, loose	2690	12	.89	2400
Sand, wet, packed	3490	12	.89	3120
Sandstone	4300	54	.65	2550
Trap rock	4420	65	.61	2590

*The weight and load factor will vary with such factors as grain size, moisture content, degree of compaction, etc. A test must be made to determine an exact material characteristic.

Material characteristics
Figure 3-23

Figure 3-24 shows productivity rates for shovels in different soils. Use it to calculate what it would cost per cubic yard to backfill around basement walls for ordinary soil. It shows that a man with a shovel can backfill 16 cubic yards per eight-hour day. Here's the calculation for the total cost and the unit cost:

	Cost
Labor: 1 man, 8 hours @ $10.00	$ 80.00
Payroll taxes and insurance (18% of payroll)	14.40
Overhead and profit (20% of labor cost)	16.00
Cost of 16 cubic yards	$110.40
Unit price per cubic yard	$6.90

Of course, you'll have to substitute your own hourly labor rate.

The most efficient way to backfill is with a bulldozer. But don't backfill until the first floor is framed, since the frame braces the new walls against any pressure or shock from the backfilling. When the masonry of the foundation walls has been plastered, dampproofed, and inspected, and

Kind of soil	Cubic yards per hour	Hours per cubic yard	Shovel and tamp hours per cubic yard
Sand	2½ to 4	.25 to .4	.45 to 1.25
Ordinary	2 to 2¼	.4 to .5	.55 to 1.45
Clay or heavy	1 to 2	.5 to 1	.65 to 1.65

Labor table for backfilling by hand
Figure 3-24

Inch	0"	1"	2"	3"	4"	5"	6"	7"	8"	9"	10"	11"
0	0	.0833	.1667	.2500	.3333	.4167	.5000	.5833	.6667	.7500	.8333	.9167
1/32	.0026	.0859	.1693	.2526	.3359	.4193	.5026	.5859	.6693	.7526	.8359	.9193
1/16	.0052	.0885	.1719	.2552	.3385	.4219	.5052	.5885	.6719	.7552	.8385	.9219
3/32	.0078	.0911	.1745	.2578	.3411	.4245	.5078	.5911	.6745	.7578	.8411	.9245
1/8	.0104	.0937	.1771	.2604	.3437	.4271	.5104	.5937	.6771	.7604	.8437	.9271
5/32	.0130	.0964	.1797	.2630	.3464	.4297	.5130	.5964	.6797	.7630	.8464	.9297
3/16	.0156	.0990	.1823	.2656	.3490	.4323	.5156	.5990	.6823	.7656	.8490	.9323
7/32	.0182	.1016	.1849	.2682	.3516	.4349	.5182	.6016	.6849	.7682	.8516	.9349
1/4	.0208	.1042	.1875	.2708	.3542	.4375	.5208	.6042	.6875	.7708	.8542	.9375
9/32	.0234	.1068	.1901	.2734	.3568	.4401	.5234	.6068	.6901	.7734	.8568	.9401
5/16	.0260	.1094	.1927	.2760	.3594	.4427	.5260	.6094	.6927	.7760	.8594	.9427
11/32	.0286	.1120	.1953	.2786	.3620	.4453	.5286	.6120	.6953	.7786	.8620	.9453
3/8	.0312	.1146	.1979	.2812	.3646	.4479	.5312	.6146	.6979	.7812	.8646	.9479
13/32	.0339	.1172	.2005	.2839	.3672	.4505	.5339	.6172	.7005	.7839	.8672	.9505
7/16	.0365	.1198	.2031	.2865	.3698	.4531	.5365	.6198	.7031	.7865	.8698	.9531
15/32	.0391	.1224	.2057	.2891	.3724	.4557	.5391	.6224	.7057	.7891	.8724	.9557
1/2	.0417	.1250	2083	.2917	.3750	.4583	.5417	.6250	.7083	.7917	.8750	.9583
17/32	.0443	.1276	.2109	.2943	.3776	.4609	.5443	.6276	.7109	.7943	.8776	.9609
9/16	.0469	.1302	.2135	.2969	.3802	.4635	.5469	.6302	.7135	.7969	.8802	.9635
19/32	.0495	.1328	.2161	.2995	.3828	.4661	.5495	.6328	.7161	.7995	.8828	.9661
5/8	.0521	.1354	.2188	.3021	.3854	.4688	.5521	.6354	.7188	.8021	.8854	.9688
21/32	.0547	.1380	.2214	.3047	.3880	.4714	.5547	.6380	.7214	.8047	.8380	.9714
11/16	.0573	.1406	.2240	.3073	.3906	.4740	.5573	.6406	.7240	.8073	.8906	.9740
23/32	.0599	.1432	.2266	.3099	.3932	.4766	.5599	.6432	.7266	.8099	.8932	.9766
3/4	.0625	.1458	.2292	.3125	.3958	.4792	.5625	.6458	.7292	.8125	.8958	.9792
25/32	.0651	.1484	.2318	.3151	.3984	.4818	.5651	.6484	.7318	.8152	.8984	.9818
13/16	.0677	.1510	.2344	.3177	.4010	.4844	.5677	.6510	.7344	.8177	.9010	.9844
27/32	.0703	.1536	.2370	.3203	.4036	.4870	.5703	.6536	.7370	.8203	.9036	.9870
7/8	.0729	.1562	.2396	.3229	.4062	.4896	.5729	.6562	.7396	.8229	.9062	.9896
29/32	.0755	.1589	.2422	.3255	.4089	.4922	.5755	.6589	.7422	.8255	.9089	.9922
15/16	.0781	.1615	.2448	.3281	.4115	.4948	.5781	.6615	.7448	.8281	.9115	.9948
31/32	.0807	.1641	.2474	.3307	.4141	.4974	.5807	.6641	.7474	.8307	.9141	.9974
1	---	---	---	---	---	---	---	---	---	---	---	1.0000

Decimal conversion table
Figure 3-25

when the masonry has set up, then you can backfill.

About half of the excavated dirt is generally needed for backfilling; the remaining half, including the topsoil, is available for grading the grounds. The loose dirt can be backfilled or rough graded at the rate of 80 cubic yards per hour. About two hours are required for an ordinary small house. The time required for fine grading with topsoil will usually run about the same as for backfilling, including the necessary manual labor.

The Quantity Estimate

Preparing a list of materials to be estimated is called doing the *take-off*. You "take off" the various materials from the plans and specifications, then list them by kind, number, size or volume on *quantity sheets*. If the job is a large one requiring many quantity sheets, you'll probably carry the totals from the quantity sheets to sum-

Quantity Estimate Sheet

Building: Two-story house

Material: Quantities of excavation

Date: _____

Notes or details: General excavation 1'0'' beyond foundation line.

Contractor: _____

Description	Length	Width	Depth	Cubic feet	Cubic yards
General Excavation:					
Front part	31'	27'	4.75'	3975.75	
Rear part	23'	20.5'	4.75'	2239.63	
Front extension	7'	7'	4.75'	232.75	
Side	8'	3.5'	4.75'	133.00	
Total CF general excavation				6581.13	
Total CY general excavation					244
Trenches for footings:					
Front	25.0'				
One side 27'0'' + 22'0'' + 4'6''	53.5'				
One side 26'6'' + 12'0'' +15'0''	53.5'				
Rear	18.5'				
Total length	150.5'	1.83'	.29'	79.9	
Total CY trenches					3
Piers					
Interior piers: 5	2'	2'	.29'	5.8	
Porch piers: 2	4'	4'	4.0'	128.0	
Total CF piers				133.8	
Total CY piers					5

Sample quantity estimate sheet
Figure 3-26

mary sheets. These totals are used to prepare the final estimates.

In the case of excavation, the quantity estimate would include estimates of cubic yards of general excavation, trench excavation, pier excavation and backfill and square feet of shoring, if required.

We'll do a sample estimate based on Figure 3-12, calculating these quantities:

1) Cubic yards of general excavation
2) Cubic yards of trench excavation
3) Cubic yards of pier excavation

To simplify this estimate, we'll omit the backfill. Since it's easier to compute decimals than fractions, we'll use decimal feet. Use Figure 3-25 to convert fractions of an inch to decimals.

Cost Estimate Sheet

Building: Two-story house

Material: Cost of excavation   **Date:** _____

Notes or details: Clay to be hauled 4 miles

Contractor: _____

Description	Cubic yards	Hours	Rate	Cost
General excavation and loading with tractor shovel	244	6.1	$50.00 ($20 opr)	$ 305.00
Cost of loosening and loading trenches (hand labor)	3	6	$10.00	60.00
Cost of loosening and loading piers (hand labor)	5	11.25	$10.00	112.50
Hauling 244 + 3 + 5 = 252 CY + 20% for expansion = 252 + 50.4 = 302.4 CY to be hauled	302.4	39.65	$26.00 ($14 opr)	$1030.90
Total				$1508.40

Total labor cost ($122.00, 60.00, 112.50, 555.10) = $849.60

Payroll taxes and insurance (18% of labor cost)	$ 152.93
10% overhead	$ 150.84
Subtotal	$1812.17
10% profit (11% of subtotal cost)	$ 199.34
Contract price	$2011.51

Sample cost estimate sheet
Figure 3-27

We'll divide the general excavation into four parts. First, the front part, which is 31'0" x 27'0", including the extra 1'0" on each side. Then the rear part, which is 20'6" x 23'0". We'll convert 20'6" to 20.5' to list it on the quantity sheet. The front area and side areas are also listed separately. The depth to the bottom of the gravel is 4'9" (4.75').

We've listed these dimensions on Figure 3-26, the quantity estimate sheet for this job.

To estimate the trenches, first find the total number of linear feet. We've taken the outside measurement of the building. The exact length of the trench is along the centerline of the wall, but we'll use the outside dimensions since they're the ones given. This gives us a little extra, but it's close enough for excavation work. Note that the trenches are only figured 0.29 feet (3½'') deep. The trenches are only 3½'' below the bottom of the general excavation, so that's all that has to be dug out for the trench.

The five interior piers are the same depth. The two porch piers are figured as 4'0'' x 4'0'' to allow enough room for laying the brick. The depth of these piers isn't given, but we'll assume them to be the same depth as the foundation wall, 4'0'' deep.

The Cost Estimate

After the quantity estimate is done, we can easily make up a cost estimate. It can be on a separate sheet or on the same sheet as the quantity estimate. For this example we'll use a separate sheet, Figure 3-27. We'll transfer the quantities from the quantity estimate, and make these assumptions: Dirt is clay to be hauled four miles by truck. Trenches and piers will be excavated by hand with the dirt shoveled directly onto banks or into trucks. Leave enough dirt on the premises for backfill.

First we'll consider the general excavation. There are 244 cubic yards of general excavation to be moved by a 2 cubic yard crawler-mounted shovel. For light clay or heavy soil, it takes about 2.5 hours for 100 cubic yards, or 0.025 of an hour per cubic yard. That number comes from Figure 3-28, an example of a CEF you may want for your file. For 244 yards, that's 6.1 hours. Shovel rental at $30.00 rental plus $20.00 for the operator gives us our rate of $50.00 per hour. Total cost is $305.00.

The second item is loosening and loading clay from trenches. Use the rate of two hours per cubic yard from Figure 3-15. Two hours times three cubic yards gives you six hours of hand labor. At $10.00 per hour, that's $60.00.

The third item on the cost estimate is loosening and loading clay from piers. Use the rate of 2.25 hours per cubic yard. There are five cubic yards, so it would take 11.25 hours of hand shoveling at $10.00, or $112.50.

Next we estimate the hauling. In all we have 252 cubic yards to be hauled from the general excavation, the trench and the piers. Since this is bank measure, we'll increase it by 20%, or an additional 50.4 cubic yards. That's a total of 302.4 cubic yards to be hauled, or 8,164.8 cubic feet.

The truck we're using has a capacity of 135 cubic feet, so we'll divide 8,164.8 by 135 to find the number of loads it will take. It comes out to 60.5,

Labor for Bulk or Bank Excavation

Work element	Unit	Machine-hours per unit
1 CY backhoe		
Sand or gravel	100 CY	6.4
Light clay	100 CY	7.3
Heavy clay	100 CY	8.7
Hardpan	100 CY	9.0
1½ CY backhoe		
Sand or gravel	100 CY	5.5
Light clay	100 CY	6.0
Heavy clay	100 CY	7.9
Hardpan	100 CY	8.8
1½ CY crawler mounted shovel		
Sand or gravel	100 CY	3.7
Light clay	100 CY	4.1
Heavy clay	100 CY	4.5
Hardpan	100 CY	5.0
2 CY crawler mounted shovel		
Sand or gravel	100 CY	2.1
Light clay	100 CY	2.5
Heavy clay	100 CY	2.8
Hardpan	100 CY	3.5
1½ CY rubber tired front end loader		
Sand or gravel	100 CY	2.7
Light clay	100 CY	3.2
Heavy clay	100 CY	3.7
Hardpan	100 CY	4.4
2½ CY rubber tired front end loader		
Sand or gravel	100 CY	1.8
Light clay	100 CY	2.2
Heavy clay	100 CY	2.4
Hardpan	100 CY	3.5
1 CY tracked front end loader		
Sand or gravel	100 CY	2.9
Light clay	100 CY	3.7
Heavy clay	100 CY	3.9
Hardpan	100 CY	4.1
2 CY tracked front end loader		
Sand or gravel	100 CY	2.4
Light clay	100 CY	2.7
Heavy clay	100 CY	2.9
Hardpan	100 CY	3.3
2½ CY tracked front end loader		
Sand or clay	100 CY	1.5
Light clay	100 CY	1.8
Heavy clay	100 CY	2.2
3 CY tracked front end loader		
Sand or gravel	100 CY	1.1
Light clay	100 CY	1.3
Heavy clay	100 CY	1.5

These figures assume a selection of equipment appropriate for the work being done. The work includes digging and piling on site or loading on trucks. Use this table for bulk or bank excavation only.

CEF — Excavating, bulk or bank, by machine
Figure 3-28

rounded up to 61 loads. At the average rate of 15 miles per hour (allowing a low rate of speed to account for the time spent starting and stopping), it takes 32 minutes to travel the eight mile round trip. Then we add seven minutes per load for loading and dumping, for a total of 39 minutes, or 0.65 hours per load. So 61 loads at 0.65 hours per load will take 39.65 hours. At $26.00 per hour, that's $1,030.90.

We've written all of these costs in Figure 3-27. They total $1,508.40, of which $849.60 is labor. Now add 18% of the labor cost for payroll taxes and insurance (yours might be higher) and 10% for overhead. The subtotal is $1,812.17. Add 11% of this subtotal for your profit, and your contract price is $2,011.51.

This kind of detailed estimate is one of the most common forms of estimating. The cost estimate, however, is more of a system of bookkeeping than a system of estimating. It's tedious and difficult to make a long detailed estimate of a large building. You can get the same results more quickly and with less effort by using the *unit cost* method.

The Unit Cost Estimate
With unit costs, the estimator simply calculates the number of cubic yards involved and then multiplies the number of cubic yards by the unit cost to find the total cost.

The purpose of the unit cost method is to reduce the work of estimating. Its most valuable feature is the time and labor it saves compared to the usual cost estimate. But many estimators don't understand the unit cost method, so they don't use it. And I'll admit that establishing the unit costs can be a devilish task. But once you have them, they'll save you time on every estimate you do.

There are two distinct methods of compiling unit costs. In many ways they're similar. They are: first, the *estimated* or *computed* unit cost; and second, the *actual* unit cost obtained after the job is completed.

Finding the Estimated Unit Cost
To find the estimated unit cost, first make up a detailed cost estimate such as we just described. After you find the total cost of the work being estimated, divide the cost by the number of units represented. For example, we can get a unit cost for the *general excavating* for the estimate in Figure 3-27 by simply dividing the total cost of $2,011.51 by the cubic yards removed (bank measure). Figure 3-29 shows the computations for both the individual portions of the job and the total, $8.24 per cubic yard for general excavation.

Finding the Actual Unit Cost
The second method of finding the unit cost is to

		Per cubic yard
Cost of excavating and loading 244 CY for general excavation	$ 305.00	$1.25
Hand labor	$172.50	$0.71
Cost of hauling 302.4 CY or 61 loads in 39.65 hours @ $26.00 per hour	$1,030.90	$4.22
	$1,508.40	$6.18
18% "labor burden" ($849.60 payroll)	$152.93	$0.63
10% overhead	$150.84	$0.62
Subtotal	$1,812.17	$7.43
10% profit	$199.34	$0.81
Total	$2,011.51	$8.24

Finding the estimated unit cost
Figure 3-29

take certain factors from tables, estimating data or *previous records* to make up the unit cost of the labor per cubic yard.

The *actual unit cost* is, in most cases, a carefully prepared cost based on work *previously* done. It works if you've kept good cost records and can compute *unit costs* on similar jobs completed.

Sample Unit Cost Estimate
We'll work through an example to show how it's done. Let's do a take-off on the C. L. Foster house and work up an estimate on the excavation using the unit cost method. Study the excavation plan, Figure 3-30. Here are the specifications relevant to excavation:

1) *Excavation and Backfill:* Excavate the full depth of the soil where the building stands, plus a space 1'0" around the entire outside of the basement to allow adequate space for waterproofing the outside walls.

Separate the topsoil from the clay (subsoil) as far as practical and place in separate piles where directed on the lot. Excavate for all piers, trenches, and so on, as required by the drawings.

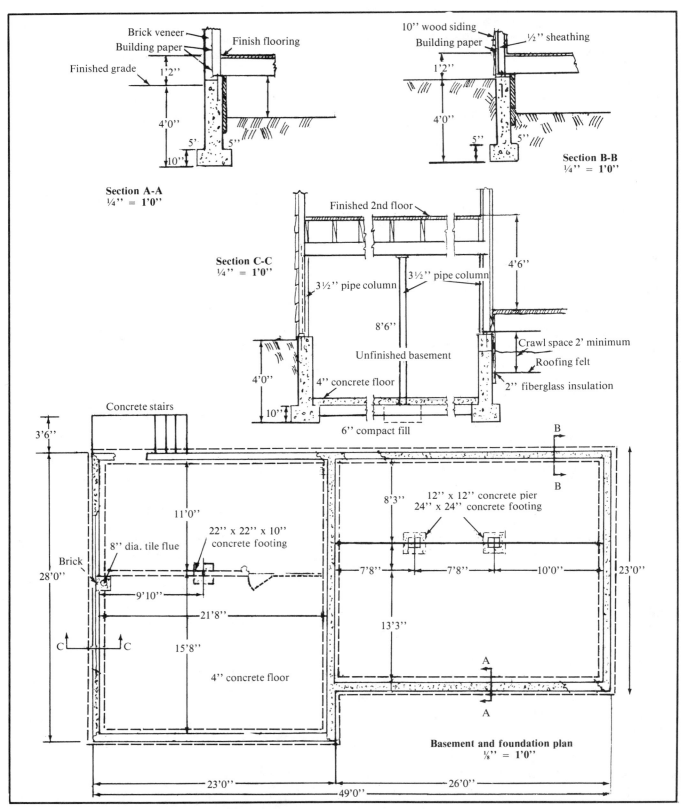

Excavation plan for Foster residence
Figure 3-30

Backfill around the wall up to finished grade line with a pitch away from the building. No other grading will be required of this contractor.

2) *Basement Floor:* The basement floor shall consist of 6 inches of well-tamped cinders or sand fill and then 4 inches of concrete floor consisting of one part portland cement, three parts clean dry sand, and five parts of crushed stone or gravel. Floors are to be as near level as possible with a slight pitch to drains.

We'll follow the specifications as nearly as possible. Assume that the top 8 inches is topsoil and the rest is clay. The estimate will have three parts: the quantity estimate (Figure 3-31), the unit cost sheet (Figure 3-32), and the summary or total cost sheet (Figure 3-33).

Topsoil excavation— The specifications state that the excavation is to extend 1'0'' around the outside of the basement. Dimensions for the main part of the basement are: length, 30'0'' (28'0'' plus 2'0'') and width, 25'0'' (23'0'' plus 2'0'').

The crawl space area is: length, 27'0'' (26'0'' plus 1'0'' because only one side is on the outside) and width, 25'0'' (23'0'' plus 2'0'').

The topsoil is 8 inches deep, which is 0.67 feet.

The projection on the north side for the stairs to the basement is by scale: length, 11'0'' (9'0'' plus 2'0'') and width, 4'6'' (3'6'' plus 1'0''). List it on the quantity sheet at 11.0' x 4.5'. The total volume of topsoil is 987.9 cubic feet, or 36.5 cubic yards. We'll round it out to 37 cubic yards.

General clay excavation— The masonry specifications state that 6 inches of cinders or sand is placed below the 4-inch concrete floor. The depth of general excavation to the bottom of the cinders from grade line is 4'0'' minus 4'', or 3'8'', but 8 inches of this is topsoil. This leaves 3'0'' of clay.

The volume of clay in the main part is 30'0'' x 25'0'' x 3'0'', or 2,250 cubic feet. That's 83.3 cubic yards.

The crawl space excavation is 27'0'' x 25'0'' x 0'11'' (0.92'). That's 621 cubic feet, or 23 cubic yards.

The total general excavation is 112 cubic yards, as shown in Figure 3-31.

Trenches— The trenches are to be dug 10 inches below the bottom of the concrete floor, or 4 inches below the cinders or general excavation. The outside of the trench extends 4 inches outside of the foundation wall. The length of the trench down one main section of wall is 4'' plus 23'0'' plus 4'' plus 26'0'', or 49'8''. Since there are two of these main wall sections (the other on the opposite side of the house), we'll double this figure and round it out to 100'0''.

The footing around the crawl space area consists of footings for different size walls. We'll have to figure these separately. Taking the 8-inch wall first, we find 25'8'' plus 4'' plus 20'10'', or 46'10''. For the 10-inch wall there are 26'0'' of footings.

The volume of the main section trenches is 100'0'' x 1'4'' x 4''. On the quantity sheet, it's 100'0'' x 1.3' x 0.3', or 39 cubic feet. Add the volume for the chimney (1' x 2' x 0.3'), which is 0.6 cubic feet. That's a total of 39.6 cubic feet, or 1.46 cubic yards.

Look at Figure 3-31 for our calculations for trench excavation for the 8-inch and 10-inch walls around the crawl space. The trench excavation totals 643.9 cubic feet, or 24 cubic yards.

Piers— The three column footings are small, so we'll leave them out.

Backfill— To find the volume of backfill, we total the volumes for the outside of the wall, the stairs and the crawl space. It comes to 862.4 cubic feet, or 32 cubic yards.

Compute the excavation costs on Figure 3-32 using current labor rates and the quantities from the quantity estimate sheet.

The specifications state that the dirt is to be piled on the site. We'll use a power shovel. The rate of excavating with a power shovel with a 3/8 CY dipper is about 320 to 375 cubic yards in eight hours (see Figure 3-19). We'll use the average of 350 cubic yards. The quantity per hour is 350 divided by 8, or 43.75, rounded out to 44 cubic yards per hour, or 0.023 hours per cubic yard.

Using our previous cost of $30.00 per hour shovel rental and $23.60 per hour for the operator (which includes 18% labor burden), we have a total of $53.60 per hour. Assume that the time required is 2.30 hours for 100 cubic yards at 0.023 hours per cubic yard.

The cost per hour for the shovel and operator is $53.60. The total unit cost is $532.60 times 0.023, or $1.23 per cubic yard.

Quantity Estimate Sheet

Building: C.L. Foster residence

Material: Excavation quantities   **Date:** _____

Notes or details: 8'' on top is black earth, other part clay piled on premises.

Contractor: _____

Description	Length	Width	Depth	Cubic feet	Cubic yards
Top Loam 8''					
Main part	30.0'	25.0'	67'	502.5	
Stair projection	11.0'	4.5'	.67'	33.1	
Crawl space area	27.0'	25.0'	.67'	452.3	
Total CF of loam				987.9	
Total CY of loam					37
General Clay Excavation					
Main part	30.0'	25.0'	3.0'	2250.0	
Crawl space area	27.0'	25.0'	.92'	621.0	
Stair projection	12.5'	4.5'	2.90'	146.5	
Total CF of clay				3017.5	
Total CY of clay					112
Trench Excavation					
Main wall footing	100.0'	1.3'	.3'	39.0	
Crawl space (8'' wall)	46.83'	3.3'	2.42'	374.0	
Crawl space (10'' wall)	26.0'	3.66'	2.42'	230.3	
Chimney	2.0'	1.0'	.3'	.6	
Total CF of trenches				643.9	
Total CY of trenches					24
Backfilling					
Outside	157.0'	1.0'	3.67'	576.2	
Stairs	16.0'	1.0'	3.67'	58.7	
Crawl space (outside)	94.0'	1.0'	2.42'	227.5	
Total CF of backfill				862.4	
Total CY of backfill					32

Excavation quantities for Foster residence
Figure 3-31

Unit Cost Sheet

Building: C.L. Foster residence

Material : Excavation unit cost

Date: _____

Notes or details: _____

_____

_____

Contractor: _____

Description	Hours	Rate	Unit Cost
Unit cost of excavating loam:			
Excavating 1 CY with tractor shovel	.023	$53.60	
Cost per cubic yard			$ 1.23
Unit cost of excavating clay:			
Excavating 1 CY with tractor shovel	.029	$53.60	
Cost per cubic yard			$ 1.55
Unit cost of trench excavation:			
Excavating and shoveling on bank	.08	$32.52	
Cost per cubic yard			$ 2.60
Excavating trenches by hand	1.4	$11.80	
Cost per cubic yard			$16.52
Unit cost of backfill:			
Backfill outside wall 1 CY	.015	$50.00	
Cost per cubic yard			$ .75
Backfill inside wall by hand	.67	$11.80	
Cost per cubic yard			$ 7.91

Excavation unit costs for Foster residence
Figure 3-32

We use the same method to find the unit cost of excavating clay. The clay, however, is deeper down than the topsoil. Therefore we find that it takes 2.90 hours for 100 cubic yards, or 0.029 hours per cubic yard. So the cost of excavating a cubic yard of clay is $53.60 times 0.029, or $1.55 per cubic yard of bank measure.

Look at Figure 3-32 again for the unit cost of trench excavation around the crawl space. We assume that the dirt is just thrown on the bank and later used for backfilling. A backhoe can excavate about 100 cubic yards a day as an average, or 0.08 hours per cubic yard. Assume the backhoe costs $16.00 per hour and the operator $14.00 per hour. Adding 18% for the labor burden makes a labor cost of $16.52 per hour. The labor and equipment cost together are $32.52. At 0.08 hours per cubic yard, that's $2.60 per cubic yard.

Trench excavation in the main section of the basement will be done by hand since it's too small an amount to do by machine. Look at Figure 3-15 under trenches up to 5 feet deep to find the rate. It

Cost Summary Sheet

Building: C.L. Foster residence

Material: Excavation cost summary

Date: _____

Notes or details: _____

_____

_____

Contractor: _____

Description	Cubic yards	Unit cost	Total cost
Top loam	37	$ 1.23	$ 45.51
Clay for general excavation	112	1.55	173.60
Trenches	22.4	2.60	58.24
Trenches by hand	1.46	16.52	24.11
Backfill	23.5	.75	17.63
Backfill by hand	8.4	7.91	66.44
Total cost			$385.53
Overhead 10%			38.55
Profit 10% (11% of total cost)			42.40
Total contract price			$466.48

Unit cost summary sheet
Figure 3-33

shows that to excavate one cubic yard of hard pan soil for trenches takes 1.4 hours. The rate is $10.00 per hour plus 18%, or $11.80. The cost is 1.4 times $11.80, or $16.52 per cubic yard.

Now figure the *unit cost of backfill* with a bulldozer. It takes 1.45 hours to backfill 100 cubic yards of loose dirt. This is about 0.015 hours per cubic yard. Assuming you can hire a 55 HP dozer and driver at $50.00 an hour, the cost per cubic yard is $50.00 times 0.015, or $0.75 per cubic yard.

In this particular case, part of the backfilling will have to be done with hand labor to avoid running heavy equipment over the concrete wall surround-ing the crawl space. Backfilling by hand requires 0.67 hours per cubic yard for heavy soil. Labor cost is $11.80 per hour. The unit cost is $11.80 times 0.67, or $7.91 per cubic yard.

Now look at Figure 3-33, the unit cost summary sheet. This is where you bring together all the quantities and unit costs to find the actual contract price. For excavation for the Foster residence, the total contract price is $466.48.

Now that you've estimated the cost of that hole in the ground, you can start figuring the cost of fill-ing it up. That's the subject of the next chapter, concrete estimating.

Chapter

Estimating Concrete

Most general construction contractors do some or all of their own concrete work, although there are subcontractors who specialize in it. This chapter explains how to estimate most residential jobs. We'll also talk a little about estimating larger reinforced concrete projects. But unless you're experienced in this area, you'll probably be better off leaving concrete on larger commercial and industrial projects to an experienced concrete sub.

Taking Off Concrete Quantities

Concrete is generally estimated by the cubic foot, or by the cubic yard, which contains 27 cubic feet. In floors or thin walls, however, it's often estimated by the square foot, by the square yard, or by the "square" (100 square feet). Concrete slabs are easy to estimate because the shape will nearly always be a square, rectangle, circle or a combination of the three. Slab dimensions are usually shown very clearly on the floor plan.

If there are different thicknesses, if different preparation is required, or if the top finishes vary, take off each part of the job separately. On your estimate sheet, note the length, width and depth of each portion, in decimals of a foot. Then multiply to find the number of cubic feet in that portion. Write the results in the "Total" column. Add the totals to find the total cubic feet for the job. Then divide by 27 to find the cubic yards of concrete needed for the slab.

Walls and footings are only slightly more difficult to estimate. First, look at the wall section to find the width and thickness of footing required for the main outside walls. If this footing has off-sets or is in two or more layers, find the size of each. Put these thicknesses and widths in the appropriate columns on your estimate sheet. Next, find the total length of this type of footing. If it continues entirely around the building, start at some convenient corner and scale off the perimeter dimensions. Enter this total on the estimate sheet, too.

Think about it for a minute and you'll realize that taking the building perimeter as the total footing length isn't really accurate — it counts each corner twice. This means you're estimating slightly more footing and wall than needed. Most concrete estimators ignore this excess or use it as an allowance for waste or excess footing depth.

Next, take all the cross walls and the longitudinal walls that have the same size. Put these down on the estimate sheet. Check for wall footings of other sizes, putting each size down with its total length.

After listing all the footings for walls, list the piers. Starting with the outside piers, list the number and sizes. Then do the same with the inside piers. Foundation walls, areas walls, and so on, come next. Copy down their size and length and tabulate them. Look for miscellaneous footings and walls, such as chimneys and machinery found-

	WIDTH	DEPTH	LENGTH		CUBIC FT	CY
MAIN FOOTING	2.0	.75	152.0		228.00	8.44
GARAGE FOOTING	1.5	.50	80.0		60.0	2.22
PORCH FOOTING	1.0	.50	30.0		15.0	.55
STEPS FOOTING	1.0	.50	14.0		7.0	.26
GARAGE SLAB	20.00	.33	20.0		132.0	4.89
PORCH SLAB	6.0	.33	18.0		35.64	1.32
DRIVEWAY	9.5	.33	40.0		125.4	4.64
					603.04	22.32
PIERS (6)	2.0	.75	2.0		18.0	.67
			TOTAL		621.04	23.0

Typical concrete estimate sheet
Figure 4-1

Labor and Materials for Slabs on Grade

	Material per Square Foot		Labor	
Thickness	C F of Concrete	S F per C Y of Concrete	Man-Hours Per 100 L F Forms and Screeds	Man-Hours Placing Concrete per 100 S F
2″	0.167	162		
3″	0.25	108	Averages	Aver-
4″	0.333	81	22 linear	ages
5″	0.417	65	feet	2
6″	0.50	54	per hour	hours

Placement includes finishing with topping. If topping is omitted, deduct 1.2 hours.
Placement labor is based on ready-mix concrete direct from chute. Add ½ hour per cubic yard if concrete is pumped into place and two hours if concrete is wheeled up to 40 feet into place.

CEF — Labor and material for slabs on grade
Figure 4-2

ations, and tabulate their size on the estimate sheet. After you've listed all the items carefully, go over the entire plans again and double-check each item. Your estimate sheet should look something like Figure 4-1.

Measure reinforced concrete work like you do plain concrete. It's more complicated, however, because columns and girders are often irregular shapes. In general, figure it in cubic feet, then con-vert the total to cubic yards. You can estimate floor areas and walls by the square foot or by the hundred square feet. Figure 4-2 can go in your CEF. It contains labor and material averages for slabs on grade.

Deduct openings unless they're small. Don't deduct for the space taken up by the reinforcement unless the amount is considerable. If heavy I-beams are encased in concrete, you might want to figure the volume of the beam and deduct that amount. Don't deduct for pipes or openings in concrete having a sectional area of less than one square foot.

Remember to take off concrete of different specifications (such as 5,000 PSI or high early strength) separately.

Designing Concrete Forms

Finding the amount of concrete required is a comparatively small part of the concrete estimate. A major part of the cost, both labor and material, will be formwork. And unlike most of the job, formwork isn't shown on the plans. You can design and build your own forms or use prefabricated wall forms built of wood, steel and wood, or aluminum. These forms come in many sizes and are assembled on the job site. Figures 4-3 through 4-11 show how one popular prefabricated wall forming system works. These forms can be rented in most communities.

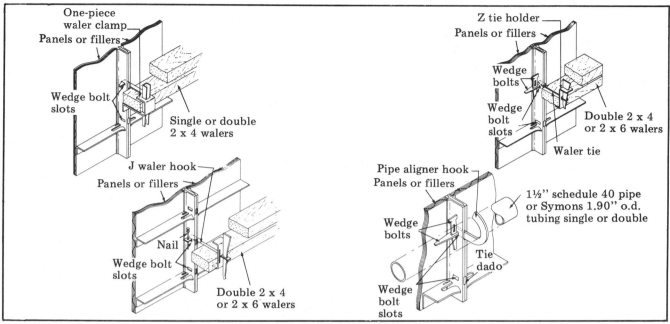

Courtesy of Symons Corporation

Installing walers
Figure 4-3

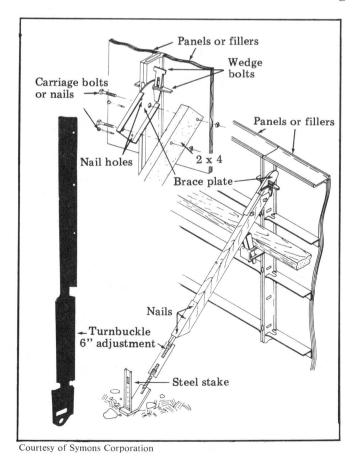

Courtesy of Symons Corporation

Plumbing and bracing to vertical alignment
Figure 4-4

Courtesy of Symons Corporation

Materials for quick and efficient bracing
Figure 4-5

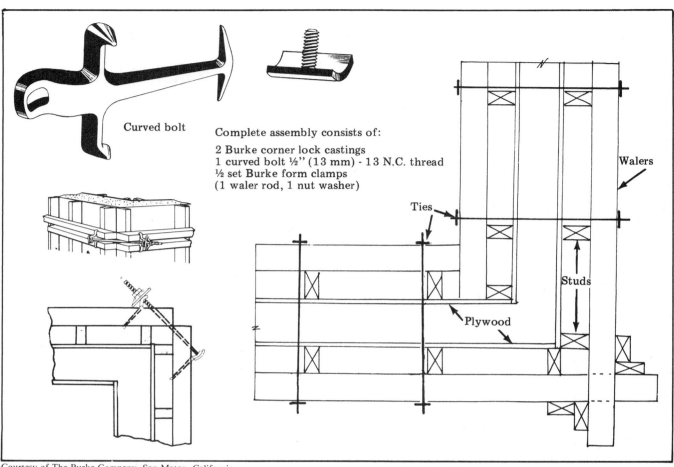

Curved bolt

Complete assembly consists of:

2 Burke corner lock castings
1 curved bolt ½" (13 mm) - 13 N.C. thread
½ set Burke form clamps
(1 waler rod, 1 nut washer)

Walers

Ties

Studs

Plywood

Courtesy of The Burke Company, San Mateo, California

Assembling wall form corners
Figure 4-6

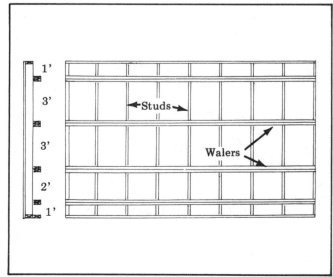

←Studs→

Walers

Typical wall form
Figure 4-7

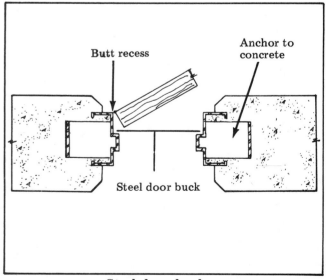

Butt recess

Anchor to concrete

Steel door buck

Steel door buck
Figure 4-8

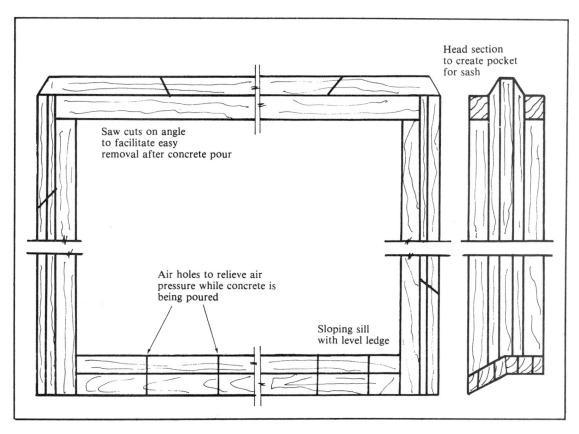

Formwork for window frame opening and recess
Figure 4-9

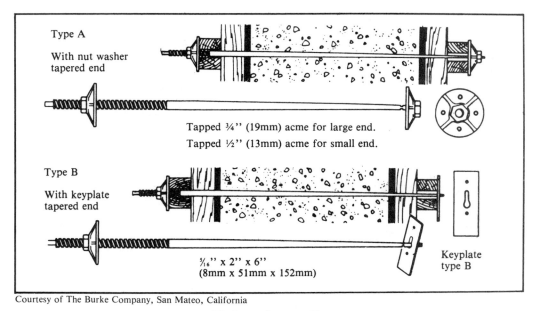

Courtesy of The Burke Company, San Mateo, California

Nut washer casting
Figure 4-10

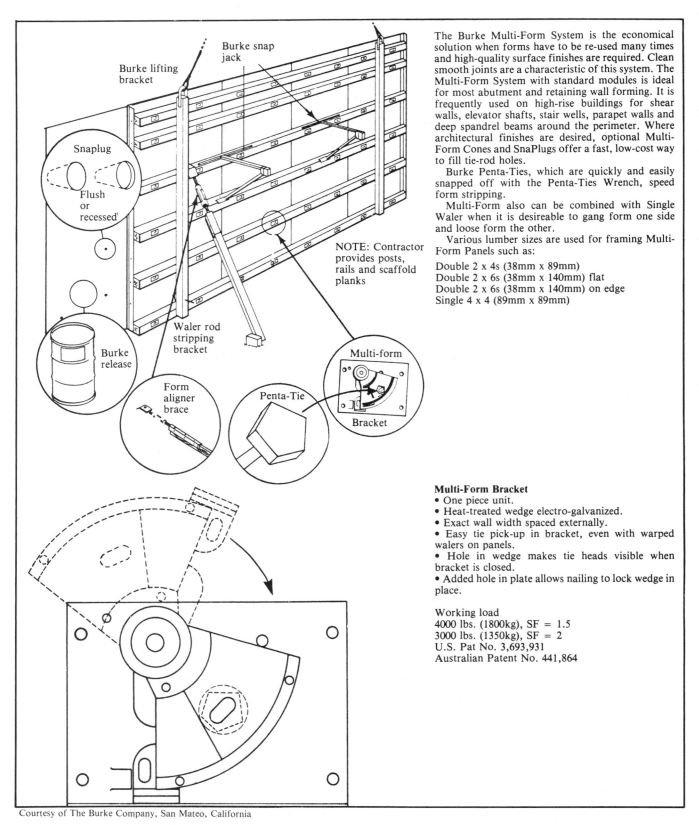

The Burke Multi-Form System is the economical solution when forms have to be re-used many times and high-quality surface finishes are required. Clean smooth joints are a characteristic of this system. The Multi-Form System with standard modules is ideal for most abutment and retaining wall forming. It is frequently used on high-rise buildings for shear walls, elevator shafts, stair wells, parapet walls and deep spandrel beams around the perimeter. Where architectural finishes are desired, optional Multi-Form Cones and SnaPlugs offer a fast, low-cost way to fill tie-rod holes.

Burke Penta-Ties, which are quickly and easily snapped off with the Penta-Ties Wrench, speed form stripping.

Multi-Form also can be combined with Single Waler when it is desireable to gang form one side and loose form the other.

Various lumber sizes are used for framing Multi-Form Panels such as:

Double 2 x 4s (38mm x 89mm)
Double 2 x 6s (38mm x 140mm) flat
Double 2 x 6s (38mm x 140mm) on edge
Single 4 x 4 (89mm x 89mm)

Multi-Form Bracket
• One piece unit.
• Heat-treated wedge electro-galvanized.
• Exact wall width spaced externally.
• Easy tie pick-up in bracket, even with warped walers on panels.
• Hole in wedge makes tie heads visible when bracket is closed.
• Added hole in plate allows nailing to lock wedge in place.

Working load
4000 lbs. (1800kg), SF = 1.5
3000 lbs. (1350kg), SF = 2
U.S. Pat No. 3,693,931
Australian Patent No. 441,864

Courtesy of The Burke Company, San Mateo, California

Multi-form ganged system
Figure 4-11

The erection cost of prefabricated wall forms will vary with the type of form used, the type of job, and the wall height. Walls less than 6 feet high require less labor because less bracing is needed. Straight walls without pilasters and box-outs will also reduce the amount of time required.

Forms for concrete construction must be stiff enough to support plastic concrete until it has hardened. They must be designed for all the weight they will be subjected to, including the dead load of the forms, the concrete in the forms, the weight of workers, weight of equipment and materials whose weight may be transferred to the forms, and the impact from vibration.

These factors vary with each project, but you can't afford to overlook any of them. Some forming systems go up quickly and are easy to strip. But the cost of erecting and stripping will be an important part of your concrete estimate.

When concrete is placed in the form, it's still liquid and exerts hydrostatic pressure on the forms. Once the concrete has started to set, the stress on the forms begins to decline. But the forms must be designed to withstand the maximum pressure developed as the concrete is placed. The actual pressure developed depends on the rate of placing and the temperature. The rate of placing affects the pressure because it determines how much hydrostatic head will be built up in the form. The hydrostatic head continues to increase until the concrete takes its initial set, usually in about 90 minutes. But at low temperatures, the initial set takes place much more slowly.

Designing Wall Forms

Being systematic when designing formwork will usually eliminate any major error. I suggest that you follow a consistent step-by-step procedure. For example, these would be the steps when designing wood forms for a concrete wall:

1) Check the material available for sheathing, studs, wales, braces, shoe plates and tie wires.

2) Determine the rate (vertical feet per hour) of placing the concrete in the form.

3) Make a reasonable estimate of the placing temperature of the concrete.

4) Find the maximum concrete pressure. Enter the rate of placing the concrete at the bottom of the graph in Figure 4-12. Draw a vertical line from that point until it intersects the correct concrete temperature curve. Read horizontally across from the point of intersection to the left side of the graph and read the maximum concrete pressure.

5) Calculate the maximum stud (vertical support) spacing. Enter the maximum concrete pressure you just found at the bottom of Figure 4-13. Draw a vertical line until it intersects the correct sheathing curve. Read horizontally across from the point of intersection to the left side of the graph. If the stud spacing isn't an even number of inches, round it down to the next lower even number of inches. For example, a stud spacing of 17.5" would be rounded down to 16".

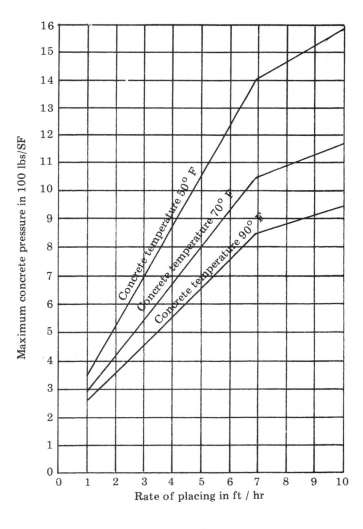

Maximum concrete pressure
Figure 4-12

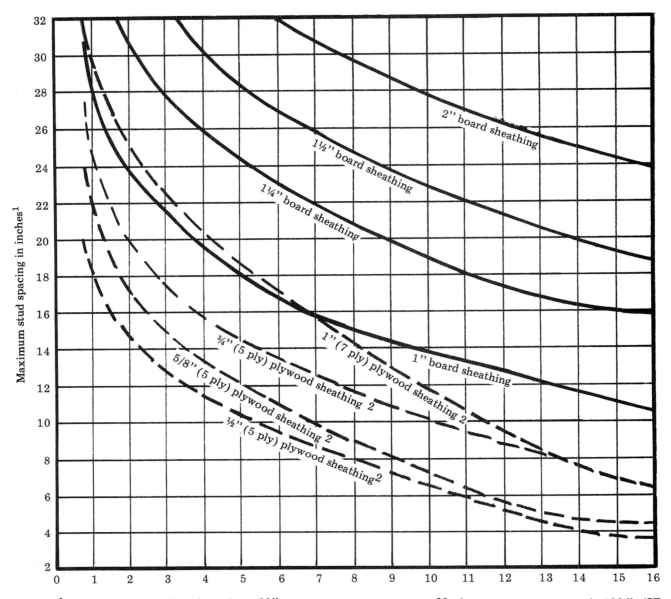

Maximum stud spacing
Figure 4-13

[1] Maximum allowable stud spacing = 32"
[2] Sanded face grain parallel to span

6) Find the uniform load on a stud by multiplying the maximum concrete pressure by the stud spacing.

Maximum concrete pressure (lb/sq ft) x stud spacing (ft) = uniform load on stud (lb/linear ft)

7) Find the maximum wale (horizontal support) spacing by entering the uniform load on a stud at the bottom of Figure 4-14. Draw a vertical line until it intersects the correct stud size curve. Read horizontally across from the point of intersection to the left side of the graph. If the wale spacing isn't an even number of inches, round it down to the next lower even number of inches.

Figure 4-15 shows the components of a wooden concrete wall form. It shows double wales (two similar members). Always use double wales for forms fabricated on the job.

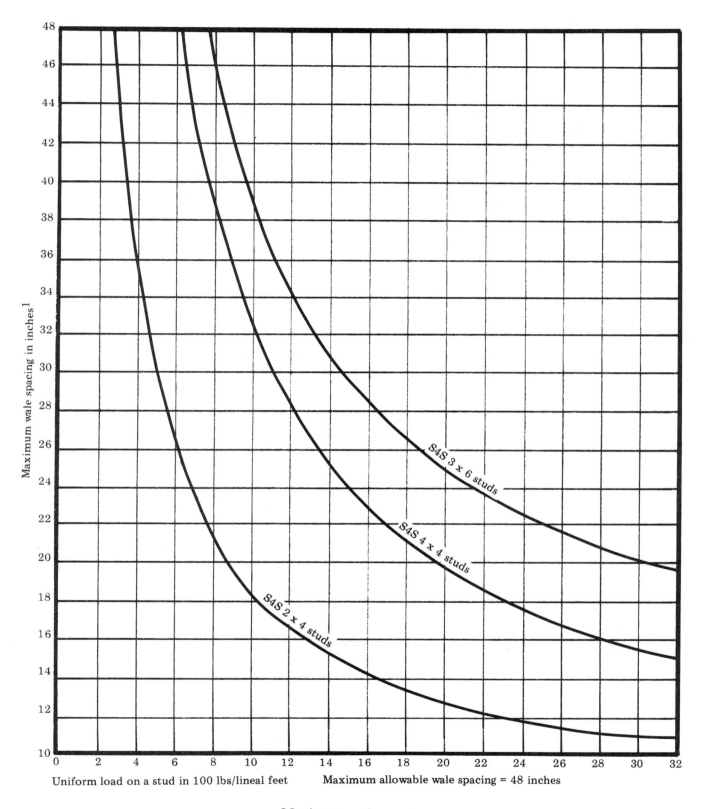

Maximum wale spacing
Figure 4-14

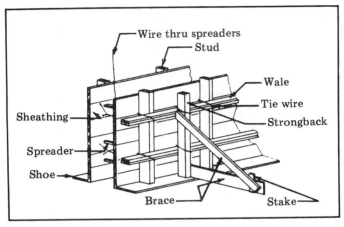

Forms for a concrete wall
Figure 4-15

8) Calculate the uniform load on a wale by multiplying the maximum concrete pressure by the wale spacing:

Maximum concrete pressure (lb/sq ft) x wale spacing (ft) = uniform load on wale (lb/linear ft)

9) Figure the tie wire spacing based on the wale size. Enter the uniform load on a wale at the bottom of Figure 4-16. Draw a vertical line until it intersects the correct double wale size curve. Read horizontally across from the point of intersection to the left side of the graph. If the tie spacing isn't an even number of inches, round it down to the next lower even number of inches.

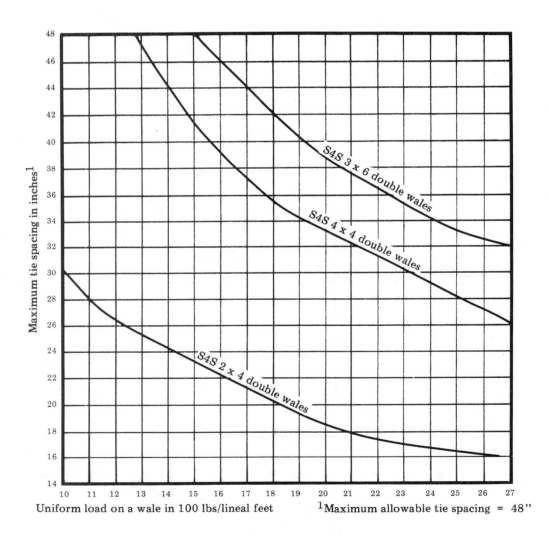

Uniform load on a wale in 100 lbs/lineal feet   [1]Maximum allowable tie spacing = 48"

Maximum tie wire spacing
Figure 4-16

Steel wire	
Size of wire (gauge number)	Minimum breaking load double strand (pounds)
8	1700
9	1420
10	1170
11	930

Breaking load of wire
Figure 4-17

10) Determine the tie wire spacing based on the tie wire strength: divide the tie wire strength by the uniform load on a wale. If the tie wire spacing isn't an even number of inches, round it down to the next lower even number of inches. If possible, use a tie wire size that will provide a tie spacing equal to or greater than the stud spacing. Always use a double strand of wire. If you don't know the strength of the available tie wire, Figure 4-17 gives the minimum breaking load for a double strand of wire.

Here's the formula to compute the spacing. The spacing in inches is equal to:

$$\frac{\text{Tie wire strength (lbs) x (12 in/ft)}}{\text{Uniform load on wale (lb/ft)}}$$

11) Determine the maximum tie spacing by selecting the smaller of the tie spacings based on the wale size and on the tie wire strength.

12) Compare the maximum tie spacing with the maximum stud spacing. If the maximum tie spacing is less than the maximum stud spacing, reduce the maximum stud spacing to equal the maximum tie spacing. If the maximum tie spacing is greater than the maximum stud spacing, they're both all right. In either case, tie at the intersections of the studs and wales.

13) Determine the number of studs for one side of a form by dividing the form length by the stud spacing. Add one to this number and round up to the next whole number.

$$\text{Number of studs} = \frac{\text{Length of form (ft) x 12 (in/ft)}}{\text{Stud spacing (inches)}} + 1$$

During form construction, place studs at the spacing calculated previously. The spacing between the last two studs may be less than the maximum allowable spacing.

14) Calculate the number of wales for one side of the form by dividing the form height by the wale spacing. Round up to the next whole number. Place the first wale one-half space up from the bottom and the remainder at the maximum wale spacing.

15) Figure the time required to place the concrete by dividing the height of the form by the rate of placing.

Wall Form Design Problem: Assume that you need forms for a concrete wall 40'0'' long, 2'0'' thick and 10'0'' high. Concrete will be delivered and placed at the rate of 192 cubic feet per hour. The forms will be made of 2 x 4's and 1'' board sheathing.

Assume that you decide to build the forms instead of using prefabricated forms. How do you estimate the materials? Here are the steps you'll use:

1) Plan area of forms: 40' x 2' = 80 sq ft

2) Rate of place: $\frac{192 \text{ cu ft/hr}}{80 \text{ sq ft}}$ = 2.4 ft/hr

3) Temperature of concrete: 70° F

4) Maximum concrete pressure (Figure 4-12): 460 lb/sq ft

5) Maximum stud spacing (Figure 4-13): 18+, so use 18''

6) Uniform load on studs: 460 lb/sq ft x $\frac{18 \text{ in}}{12 \text{ in/ft}}$ = 690 lb/ft

7) Maximum wale spacing (Figure 4-14): 23+, so use 22''

8) Uniform load on wales: 460 lb/sq ft x $\frac{22 \text{ in}}{12 \text{ in/ft}}$ = 843 lb/ft

9) Tie wire spacing based on wale size (Figure 4-16): > 30''

10) Tie wire spacing based on wire strength:

$$\frac{1420 \text{ lb} \times 12 \text{ in/ft}}{843 \text{ lb/ft}} = 20+, \text{ so use } 20''$$

11) Maximum tie spacing: 20"

12) Maximum tie spacing is greater than maximum stud spacing, therefore, reduce the tie spacing to 18" and tie at the intersection of each stud and double wale

13) Number of studs per side: $(40 \text{ ft} \times \frac{12 \text{ in/ft}}{18 \text{ in}}) = 26.7 + 1$, so use 28 studs

14) Number of double wales per side:

$10 \text{ ft} \times \frac{12 \text{ in/ft}}{22 \text{ in}} = 5+$, so use 6 double wales

15) Time required to place concrete:

$$\frac{10 \text{ ft}}{2.4 \text{ ft/hr}} = 4.17 \text{ hrs}$$

Now calculate the amount of lumber required:

• 1" board sheathing, 880 board feet (40' x 2 x 10' x 1.10). Add about 10% to allow for form shoes and waste.

• Studs, 374 board feet (28 x 10' x 2 x 0.667). Each linear foot of 2 x 4 has 0.667 board feet.

• Wales, 640 board feet (6 x 40' x 2 x 2 x 0.667).

• Stakes, braces and strongbacks, allow 300 board feet.

You'll need 2,194 board feet, or about 2.75 board feet per square foot of form. Most wall forms require from 2.5 to 2.75 board feet per square foot of form.

Column Forms

Use these steps to design wooden forms for a concrete column:

1) Check the materials available for sheathing, yokes, and battens. Standard materials for column forms are 2 x 4's and 1" sheathing.

2) Find the height of the column.

3) Find the largest cross-sectional dimension of the column.

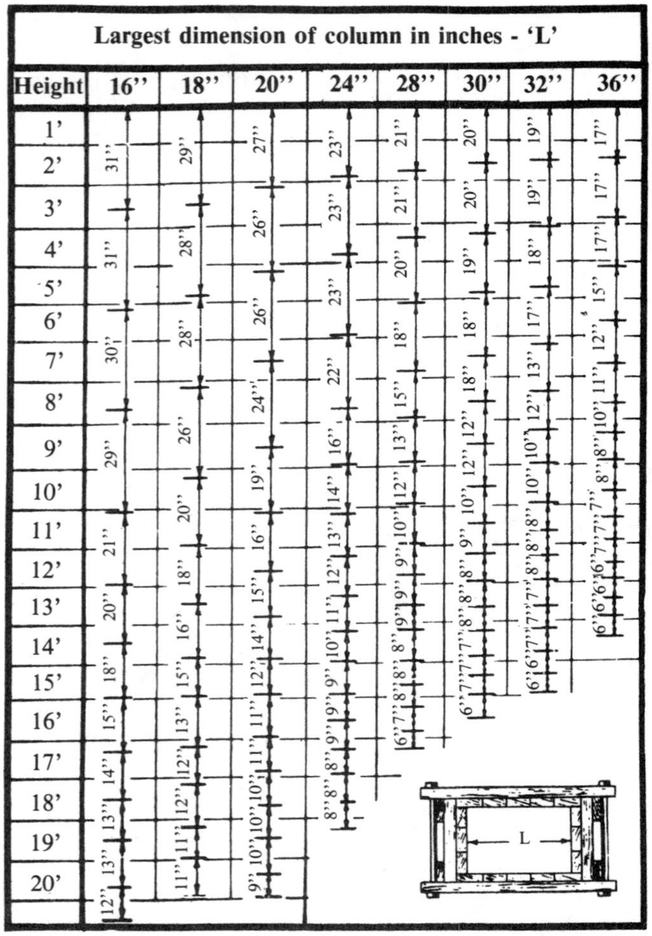

Column yoke spacing
Figure 4-18

4) Determine the yoke spacings by using Figure 4-18. Read down the first column until you reach the correct height of column. Then read horizontally across the page to the column headed by the largest cross-sectional dimension. The center-to-center spacing of the second yoke above the base yoke will be equal to the value in the lowest interval that is partly contained in the column height line. Find all the subsequent yoke spacings by reading up this column to the top. This gives you the maximum yoke spacings. Of course you can place the yokes closer together if you wish. Figure 4-18 is based on use of 2 x 4's and 1" sheathing.

Column Form Design Problem: Determine the yoke spacing for a 9' column whose largest cross-sectional dimension is 36". Assume that 2 x 4's and 1" sheathing are available.

Concrete slab thickness	Joist span						
	4'	5'	6'	7'	8'	9'	10'
4''	4'0''	4'0''	4'0''	4'0''	4'0''	3'0''	2'6''
5''	4'0''	4'0''	4'0''	3'6''	3'0''	3'0''	2'6''
6''	4'0''	3'0''	2'6''	2'0''	2'0''	2'0''	2'0''

Based only on joist strength, does not consider deflection of sheathing

**2 x 6 joist spacing
Figure 4-19**

Look at Figure 4-18 for the answers. You'll find that the maximum yoke spacings for the column starting from the bottom of the form are 8'', 8'', 10'', 11'', 12'', 15'', 17'', 17'', and 10''. The space between the top yokes has been reduced because of the limits of the column height.

Foundation Forms
Foundation forms include forms for large footings, wall footings, and column and pier footings. These foundations or footings are relatively low in height and have a primary function of supporting a structure. Since the concrete isn't very deep, the pressure on the form is relatively low. So you usually don't have to design a form based on strength considerations. Whenever possible, use the earth as a mold for the concrete footings.

Floor Forms
Design wooden forms for flat concrete slabs using these steps:

1) Determine material available for sheathing, cleats, joists, stringers, and shores. Typical materials are: 1'' tongue and groove, or 3/4'' plywood for sheathing; 1 x 4 cleats; 2 x 6 joists; 2 x 8, 4 x 4, or 4 x 6 stringers; and 4 x 4 shores.

2) Estimate total unit load on the floor form. The weight of ordinary concrete is about 150 pounds per cubic foot. Using this figure, the weight of concrete is 50 pounds per square foot for a 4'' slab, 63 pounds per square foot for a 5'' slab, and 75 pounds per square foot for a 6'' slab. You must also add a live load for men and construction materials. This is generally 50 pounds per square foot, or 75 pounds per square foot if powered concrete buggies are used.

3) Determine the spacing of floor joists. Figure 4-19 gives the joist spacing as a function of slab thickness and span of the joists for 2 x 6 joists. The spans are based solely on joist strength. The table doesn't take into consideration the deflection of the sheathing. If this is a concern, check it by separate calculation. You can shorten the joist span by adding stringers.

4) Determine the location of the stringers which support the joists. For short spans, it may not be necessary to use stringers.

5) Figure the spacing of the shores, or posts, which support the stringers. Maximum spans for stringers are given in Figure 4-20.

Concrete slab thickness	Spacing of stringers*					
	2'' x 8'' stringer			4'' x 6'' stringer		
	5'	6'	7'	5'	6'	7'
4''	5'6''	5'0''	4'6''	6'6''	6'0''	5'6''
5''	5'6''	5'0''	4'6''	6'0''	5'6''	5'0''
7''	--	--	--	5'6''	5'0''	4'6''
8''	--	--	--	5'0''	4'6''	4'6''

*Spacing based on live load of 50 lb. per SF. For a live load of 75 lb. per SF, increase slab thickness by 2'' and use the corresponding spacing.

**Maximum spans for stringers
Figure 4-20**

Floor Form Design Problem: A 5'' concrete floor slab has a span of 12 feet. Material available includes 1'' tongue and groove sheathing, 1 x 4's, 2 x 6's, 4 x 4's, and 2 x 8's. Determine the spacing of the joists, stringers, and shores.

• First find the total unit load (live load plus concrete): 50 plus 63 equals 113 pounds per square foot.

• For the spacing of joists, use 2 x 6's from Figure 4-19: Locate a stringer in the middle of the span, giving a joist span of 6 feet and a joist spacing of 4 feet.

• Stringer spacing, using 2 x 8's, is 6 feet (from above).

• Shore spacing, using 4 x 6's from Figure 4-20, is 5 feet.

Stair Forms

You can use various types of stair forms, including prefabricated forms. For moderate width stairs joining typical floors, design based on strength considerations is not generally necessary.

Estimating Concrete Forms

Forms are made from lumber, plywood, aluminum, steel and occasionally combinations of other materials. Use only plywood made with waterproof glue to withstand the moisture from the wet cement. Use 3/4'' ''plyform'' plywood where you want a smooth concrete surface. An assortment of panel forms of different lengths gives you considerable flexibility in fitting walls of various lengths. See Figure 4-21.

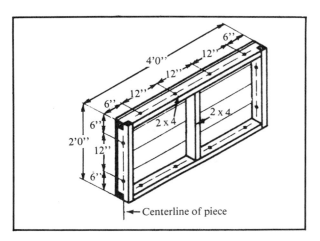

Concrete panel form
Figure 4-21

Space the studs used for bracing 12'' or 16'' on centers to work out with plywood sizes. When you're using plywood on fairly high walls, use 2 x 6's for studs. These walls must be exactly straight and 2 x 6 studs will usually make a straighter wall than 2 x 4 studs.

For the first use of the forms, you'll need from 10 to 20 pounds of nails per thousand board feet of lumber. If you can reuse the forms without rebuilding them entirely, you'll use from 5 to 10 pounds per MBF for each additional use.

Give the forms a coating of oil to prevent concrete from bonding to the form face. Brushing or spraying the faces with paraffin oil works well for the average job such as foundation walls, unexposed columns, or beams and slabs.

You can save by reusing form lumber later for roofing, bracing, or similar purposes. Choose lumber that's straight, structurally sound, strong, and only partially seasoned. Kiln-dried timber has a tendency to swell when soaked with water from the concrete. If the boards are tight-jointed, the swelling causes bulging and distortion. If you use green lumber, make allowance for shrinkage or keep the forms wet until the concrete is in place.

Softwoods such as pine, fir, and spruce make the best and most economical form lumber. They're light, easy to work with, and available in almost every region. Lumber that comes in contact with the concrete should be surfaced on at least one side and on both edges. Turn the surfaced side toward the concrete. The edges of the lumber may be square, shiplap, or tongue and groove. The latter makes a more watertight joint and tends to prevent warping.

Estimate forms for concrete footings, foundations and retaining walls by the square foot of contact area (surface in contact with the concrete). If the soil is hard and compact and the footings are below the basement floor, no forms are required. If the soil isn't compact, you'll need forms. Forms for most footings may be made of 1''-thick lumber planks which are cleated together, side by side, in the correct width. The sheathing is held in place by 2 x 4 studs and braces, spaced about 18'' apart around the outside of the forms. It's tied across the tops at 4'0'' to 6'0'' intervals with 1'' to 2'' spreader ties.

Here's an example: Calculate the number of square feet of forms required for a wall 40'0'' long and 4'0'' high. Forms will be required for both sides of the wall. So you multiply the length by the height (40' x 4' equals 160 square feet). Because you need forms on both sides, double it. The total is 320 square feet of forms.

Figure 4-22 illustrates a wall form for foundation walls and basement walls built in ground firm enough so that the earth side of the excavation will stand upright.

Footing Forms

Footings for columns and walls are made in many different ways. The size, design and construction depend on the load to be carried and the bearing strength of the soil. The fundamental purpose of a footing is to distribute the load imposed on the pier over a larger area of ground than would be covered by the pier or column itself.

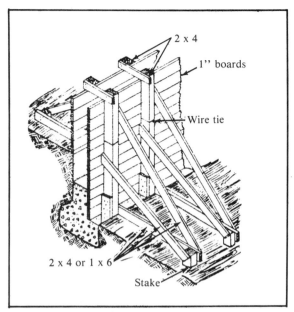

Wall form in firm ground
Figure 4-22

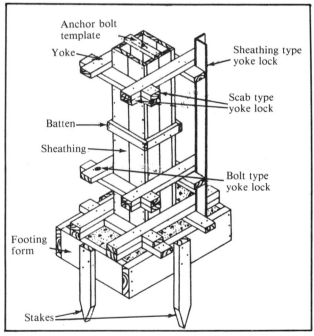

Form for a concrete column and footing
Figure 4-23

A footing should be made strong enough to support the dead weight of the wall or pier plus any live and dead weight imposed on the wall. In general, a footing for a foundation wall or pier for light frame construction should extend no more than 4'' beyond each side of the wall or pier, unless the footing is reinforced or tapered. The footing should be 6'' to 12'' thick.

Column and Pier Footings
Figure 4-23 shows a form for a column and its footing. The form illustrated has been made from 2 x 6 material, but the lumber dimension will vary with the height of the form. Figure 4-24 illustrates how the footing might be braced to prevent the forms from deflecting while the concrete is curing.

For each square foot of contact area, you'll need at least two board feet of lumber for form material. You'll also need stakes and perhaps diagonal braces. Plan on about 1/2 board foot of brace and stake material for each square foot of contact area. Except on large jobs, column or pier footing forms can be used only once because all the footings of the same size will be poured at the same time. Where the forms can be reused on the job, you can use them up to five times, or perhaps more.

The time required to form each footing will depend on the size of the footings. Larger footings (6'0'' or more on a side) will take less time per square foot of contact area. One carpenter and one laborer can install between 25 and 30 square feet of column footing forms in one hour, including bracing and oiling, depending on the size of the footing.

Wall Footings
Wall footing forms, like column footing forms, are usually made of 2'' lumber set on edge. Use stakes

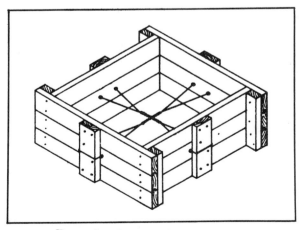

Braced column footing form
Figure 4-24

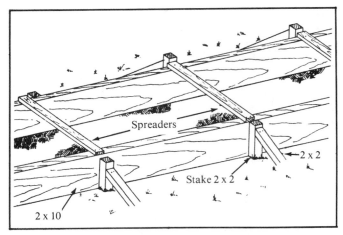

Wall footing form
Figure 4-25

Stepped footing
Figure 4-26

Wall forms	Height of wall (ft.)	BF per SF of forms	SF of forms in 8 hours	Carpenter hours per 100 SF	Labor hours per 100 SF	Removing forms	
						SF in 8 hours	Labor hours per 100 SF
Making only (sectional)	4 to 8	2½	280-320	2¾	¼	--	--
Erecting and removing (sectional)	3 to 4	2½	250-275	3	1/10	800	1
Erecting and removing (sectional)	5 to 8	2½	185-210	4	1/10	800	1
Foundation wall forms	4 to 6	2½	190-210	4	2	640	1-1/4
Foundation wall forms	7 to 8	2½	165-190	4½	2¼	550	1-1/2
Foundation wall forms	9 to 10	2½	150-160	5¼	2½	450	1-3/4
Foundation wall forms	11 to 12	3	135-150	5½	2¾	425	1-7/8
Retaining walls	16 to 20	3½	105-115	7¼	3½	325	2-1/2

Labor and material for 100 SF of wood wall forms
Figure 4-27

and braces to keep the forms in line while the concrete hardens. See Figure 4-25.

Where the building site slopes, the footing will usually be "stepped" as shown in Figure 4-26. The stepped footing corrects the tendency of the foundation to side down the slope when under load.

Each "step" should overlap the footing below by at least 8". Estimate 2½ board feet of lumber for each square foot of contact area. For labor, estimate about one manhour for each 40 square feet of form 10" high or less, or one manhour for each 30 square feet of form over 10" high. That includes erection, oiling, stripping and cleaning. If a 2 x 4 keyway is laid in the footing, another manhour will be required for each 75 feet of keyway. Be sure to include enough material for one linear foot of 2 x 4 keyway for each linear foot of footing since the key material is usually destroyed in removal.

For foundation and retaining walls, find the number of square feet for each side of each wall by multiplying the height by the length. The cost of labor and materials per square foot of forms varies with the overall height of the wall and the number of times the lumber can be used.

Most work requires 2½ board feet of lumber per square foot of form, based on use of 1" material for the forms. This includes all necessary studs, plates, wales and bracing. To compute the number of board feet of lumber required, multiply the total area (for both sides of a wall) by 2½. This will give you the total board feet of lumber required for both sides of the wall.

Figure 4-27 gives the quantities of lumber and labor hours required for 100 square feet of wall for different heights. Use these figures until you have more accurate records of what your crews can do

Labor and Materials for 100 S.F. of Foundation Wall Forms

Work Element	B.F. per S.F. of Forms	Make and Place Forms			Removing Forms	
		S.F. in 8 Hours	Carpenter Hours per 100 S.F.	Labor Hours per 100 S.F.	S.F. in 8 Hours	Labor Hours per 100 S.F.
Foundation wall forms						
4' to 6'	2.5	190-210	4.0	2.0	640	1.3
7' to 8'	2.5	165-190	4.5	2.3	550	1.5
9' to 10'	2.5	150-160	5.3	2.5	450	1.8
11' to 12'	3	135-150	5.5	2.8	425	1.9
Retaining walls						
16' to 20'	3.5	105-115	7.3	3.5	325	2.5

Labor and material for foundation wall forms
Figure 4-28

on the types of work you handle. Your CEF might include the information shown in Figure 4-28.

A carpenter will frame and erect about 500 board feet of lumber, or 200 square feet of forms, in eight hours. For each hour of carpenter time, allow about 1/2 hour of laborer time for carrying lumber. If there are no laborers and the carpenters handle and carry all the lumber, reduce the productivity figures. Estimate that each carpenter should handle, frame and erect about 265 to 300 board feet of lumber per eight-hour day.

A laborer can remove 1,600 board feet of lumber, or about 640 square feet of forms, in eight hours. High walls requiring work from a scaffold need 10 to 50 percent more time per square foot of forms.

Panel Forms
The labor cost of making panel forms will vary with the method used and the number made up at one time. Formwork labor and material can average from 30 to 50 percent of the total concrete wall costs, with labor averaging two to three times the material cost.

If the lumber is delivered cut to length, a carpenter should complete one panel of 28 to 32 square feet in 45 to 60 minutes. That's up to 10 panels in eight hours. Figures 4-22 and 4-29 show construction and use of wood panel forms.

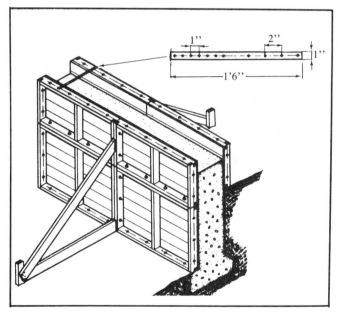

Sectional form for foundation wall
Figure 4-29

Figure 4-30 illustrates several sets of panel forms placed on top of each other for a foundation wall. The bottoms of the forms are held together by steel tie bands which pass under the concrete wall and are nailed to the bottom members of the panel

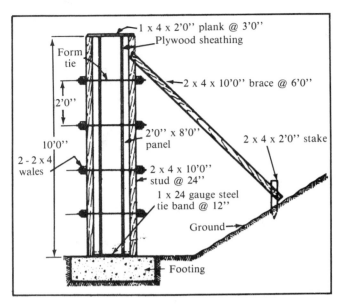

Steel section forms for foundation wall
Figure 4-30

forms. The tops of the forms may be held together with planks, as shown in Figure 4-30, or with steel tie bands.

Many contractors use foundation wall forms made up into panels 4' x 6', 4' x 7' or 4' x 8' in size. The frame for these sections is either 2 x 4 or 2 x 6 lumber, with intermediate braces spaced 1'6" to 2'2" on centers. The corners are reinforced with galvanized straps or strap iron, and the entire frame is sheathed with 1 x 6 sheathing or 3/4" plyform plywood.

Estimating Slabs

Concrete flatwork is relatively easy to estimate. Sidewalks, floors, driveways and patios involve four distinct operations: grading, setting the forms, handling the concrete into place, and finishing the surface. If there's anything extra, like a separate topping or a bed of cinders or gravel, add them to your estimate.

For nearly every slab, you'll need to do some grading to prepare the surface. The more soil you have to move, the more time it will take. If you place more than a few inches of fill on the surface, it will have to be compacted. This can be done by hand or with a rammer. For most soil types, compaction should be done while the soil is moist.

Where grading is done by hand with a shovel and wheelbarrow, a laborer should loosen, shovel and load 5 to 6 cubic yards in an eight-hour day. For a

larger area, a small bulldozer or tractor with a landscape bucket will do the job efficiently. Naturally, the time required will vary with the type of equipment, soil type and how far the dirt must be moved.

Forms for Slabs

You'll usually use 2 x 4 or 2 x 6 lumber for forms at each side of concrete walks, patios and driveways. The depth of the forms depends on the thickness of the concrete. On ordinary sidewalk work, double the linear feet of walk to find the linear feet of forms required. A sidewalk 4" thick will require 2 x 4 lumber; a 6" thickness needs 2 x 6 lumber. Figure 4-31 gives the number of board feet of lumber required to complete one square foot of sidewalk 4'0" to 10'0" wide and 4" to 8" thick.

	4" thick						
Width in feet	**4**	**5**	**6**	**7**	**8**	**9**	**10**
BF of lumber	$\frac{1}{2}$	$\frac{3}{8}$	$\frac{3}{10}$	$\frac{1}{4}$	$\frac{1}{4}$	$\frac{2}{10}$	
	6" thick						
BF of lumber	$\frac{5}{8}$	$\frac{1}{2}$	$\frac{4}{10}$	$\frac{3}{8}$	$\frac{1}{3}$	$\frac{3}{10}$	$\frac{1}{4}$
	8" thick						
BF of lumber	$\frac{7}{8}$	$\frac{2}{3}$	$\frac{6}{10}$	$\frac{1}{2}$	$\frac{4}{10}$	$\frac{3}{8}$	$\frac{1}{3}$

Board feet of lumber for 1 SF of walk
Figure 4-31

The quantities listed include the stakes needed to hold the forms in place. Figure the cost of the forms by calculating the cost of lumber required and dividing by the number of times it may be used on the job.

Where concrete floors are placed on the ground, you may need to set wood screeds 6'0" to 8'0" on centers so the floor will be level or have a uniform pitch. These are usually made of 2 x 4 lumber, with stakes placed 3 to 4 feet apart to hold the forms in place. The lumber required will vary with the spacing of the forms. Use the figures for 4" slabs from Figure 4-31. Figure 4-32 lists carpenter and laborer manhours required to form ground slabs and place screeds. Your CEF for labor to form slabs may show the breakdown by total manhours, as shown in Figure 4-33.

Item	Carpenter hours	Laborer hours
Ground slab forms		
4" or 6"	3.0	1.5
8"	3.8	1.9
12"	4.2	2.6
Slab depression forms	7.0	3.5
Screed for ground slab		
4"	.8	--
6"	1.0	--
8"	1.3	--
12"	2.2	--
Add for pitched screed	1.1	--

Includes layout, fabrication, erection, stripping and cleaning forms.

Labor hours for 100 LF of slab forms
Figure 4-32

Concrete for Slabs

First, check the plan to find the area of the surface to be covered. If there are different thicknesses, foundations requiring different preparation, or varying top finishes, set down each kind separately. On your estimate sheet, multiply the length, width and depth together (changing all inches to decimals of a foot). This will give you the number of cubic feet in that portion. Write the results in the "Total" column. Adding these totals will give you the total cubic feet in the job. Since concrete is usually measured by the cubic yard, divide the total by 27 to convert to cubic yards.

Grading and Tamping Fill

When you place cinders, sand, gravel or slag fill under concrete floors or walks, the cost will vary with the thickness of the fill and the job conditions. Will your crews have to use wheelbarrows to move the fill from the street, or can it be spread from piles? It costs as much to grade and tamp a bed of slag or gravel 3" thick as one 9" thick because only the surface is graded and tamped. The only additional labor on the thicker fill is for handling and spreading a larger quantity of material.

Where the fill varies from 3" to 4" thick, a man will spread, grade and tamp about 6 cubic yards in eight hours. If the fill varies from 5" to 6" thick, a worker will spread, grade and tamp about 9 cubic yards in a day. For fill from 7" to 9" thick, a worker will finish about 10 cubic yards in eight

hours. For a 10" to 12" fill, a worker will spread, grade and tamp about 12 cubic yards in eight hours.

When ordering material, remember that most types of fill shrink when compacted, Between 10 and 15% of the volume of sand, crushed rock or gravel is lost in compaction and up to 40% of the volume of cinders or slag is lost.

Reinforced Concrete

Estimate reinforced concrete in the same order and in the same units as you do the work on the job. This makes it easier to compare estimated and actual costs.

The first operation of any reinforced concrete job is erecting the temporary wood or metal forms to support the weight of the wet concrete until it has set enough to be self-supporting. Columns, girders, beams, spandrel beams, lintels, floor slabs, stairs and platforms usually require forms. Estimate each class of formwork separately.

After the forms are erected, place the reinforcing steel or wire mesh ready to receive the concrete. This may include unloading steel from cars or trucks at the job, cutting steel to length and bending to special shapes for columns, beams, girders and floor slabs, and placing it in position ready to receive the concrete.

After the reinforcing steel is in place, the job is ready for concrete placing. This may involve

Labor Forming Slabs on Grade

Work Element	Unit	Man-Hours Per Unit
Edge forms for slabs		
Up to 6 inches high	100 L.F.	4.5
Over 6 to 12 inches	100 L.F.	5.2
Over 12 to 24 inches	100 L.F.	6.8
Over 24 to 36 inches	100 L.F.	8.5
Keyed joint forms		
Up to 6 inches high	100 L.F.	6.8
Over 6 to 12 inches high	100 L.F.	7.5
Pad and base edge forms		
1 use	100 S.F.	6.8
3 uses	100 S.F.	6.2
5 uses	100 S.F.	5.7
Keyway forms, 4 uses		
2 x 4	100 L.F.	1.5
2 x 6	100 L.F.	1.6

Unit is per 100 linear feet or 100 square feet of contact area. Does not include excavation, backfill or trench clean-up.
Suggested Crew: 1 carpenter and 1 helper

CEF - Labor forming slabs on grade
Figure 4-33

preparing runways, hoisting, wheeling and placing it in position; then removing runways, removing old concrete drippings and spillage, patching, cleaning and washing equipment.

Work carefully! Don't omit any material or labor that's part of the job. Remember, the estimated cost of anything you forget is always zero. That's a guaranteed 100% miss.

Chapter 5

Estimating Masonry

Masonry includes both concrete block and clay brick. But of the two, concrete block is the more common. About two-thirds of the masonry walls built today are built with concrete block. It's used in schools, churches, apartment buildings, warehouses, industrial buildings, commercial, and office buildings. Chances are that most of your jobs will include some concrete masonry, either in structural walls, foundations, facing for wood-frame walls, or in chimneys.

Estimating Concrete Block

To estimate concrete block and partition units, first find the square footage of wall, then multiply by the number of blocks per 100 square feet. Use Figures 5-1 and 5-2 to find the number of blocks per 100 square feet. Of course you'll have to estimate walls of different thicknesses separately.

When estimating quantities, always take exact measurements. Don't count corners twice. Deduct in full for all openings, regardless of size. If you work carefully, you'll determine the actual number of blocks required for the job. Of course, you'll need to add a small percentage for waste and breakage, but knowing the exact number needed is good estimating practice.

A hollow load-bearing concrete block of 8" x 8" x 16" nominal size will weigh from 40 to 50 pounds if made from heavyweight aggregate such as sand, gravel, crushed stone or air-cooled slag. The same

block made with lightweight aggregate (expanded shale, clay, slag, or natural lightweight materials such as volcanic cinders and pumice) will weigh from 25 to 35 pounds.

Both heavyweight and lightweight units are used for all types of masonry construction. The choice of units depends on what's available at an attractive price, and the requirements of the structure being built. Lightweight units usually cost more to buy but less to install.

Block and Tile per 100 S.F. of wall

Size of Units	Wall Thickness	Unit Per 100 S.F.	C.F. of Mortar
8" x 12" x 16" block	12"	110	3.25
8" x 10" x 16" block	10"	110	3.25
8" x 8" x 16" block	8"	110	3.25
5" x 8" x 12" tile	8"	220	5.00
3½" x 8" x 12" tile	8"	300	6.00
5" x 6" x 12" tile	6"	220	4.00
3½" x 6" x 12" tile	6"	300	5.50
5" x 4" x 12" tile	4"	220	4.50
3½" x 4" x 12" tile	4"	300	5.50

These quantities are for load-bearing units. The table does not include corner or jamb blocks or lintels. Mortar quantities are based on a ⅜" joint plus 25% for waste.

Concrete units for 100 SF of wall area
Figure 5-1

Concrete blocks are manufactured in Grades N and S. Grade N is used in exterior walls below and above grade that may or may not be exposed to moisture penetration or weather, and for interior and backup walls. Grade S block is limited to use above grade in exterior walls with weather-protective coatings and in walls not exposed to the weather.

Concrete building units are made in sizes and shapes to fit different construction needs. Figure 5-3 shows the typical shapes and sizes. Concrete unit sizes are usually referred to by their nominal dimensions. A unit measuring 7⅝'' wide, 7⅝'' high and 15⅝'' long is referred to as an 8 x 8 x 16 unit. When it's laid in a wall with 3/8'' mortar joints, the unit will fill a space exactly 16'' long and 8'' high.

Lightweight Partition Block for 100 S.F. of Wall

Size of Unit	Wall Thickness	Units Per 100 S.F.	C.F. of Mortar
8'' x 6'' x 16''	6''	110	3.25
9'' x 4'' x 18''	4''	87	3.25
12'' x 4'' x 12''	4''	100	3.25
8'' x 4'' x 16''	4''	110	3.25
8'' x 4'' x 12''	4''	146	4.00
9'' x 3'' x 18''	3''	87	2.50
12'' x 3'' x 12''	3''	100	2.50
8'' x 3'' x 16''	3''	110	2.75
8'' x 3'' x 12''	3''	146	3.50

Lightweight partition units per 100 SF
Figure 5-2

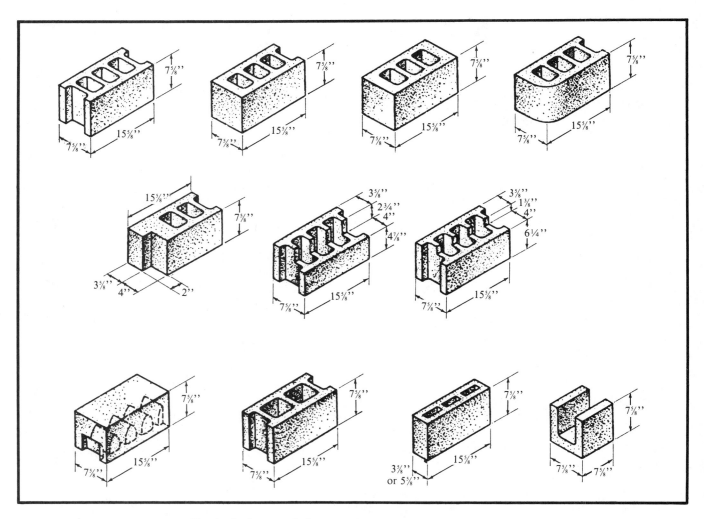

Typical sizes and shapes of concrete masonry units
Figure 5-3

LENGTH OF WALL	NO. OF UNITS	LENGTH OF WALL	NO. OF UNITS	LENGTH OF WALL	NO. OF UNITS	LENGTH OF WALL	NO. OF UNITS	LENGTH OF WALL	NO. OF UNITS	LENGTH OF WALL	NO. OF UNITS
0'-8''	½	20'-8''	15½	40'-8''	30½	60'-8''	45½	80'-8''	60½	100'-8''	75½
1'-4''	1	21'-4''	16	41'-4''	31	61'-4''	46	81'-4''	61	101'-4''	76
2'-0''	1½	22'-0''	16½	42'-0''	31½	62'-0''	46½	82'-0''	61½	102'-0''	76½
2'-8''	2	22'-8''	17	42'-8''	32	62'-8''	47	82'-8''	62	102'-8''	77
3'-4''	2½	23'-4''	17½	43'-4''	32½	63'-4''	47½	83'-4''	62½	103'-4''	77½
4'-0''	3	24'-0''	18	44'-0''	33	64'-0''	48	84'-0''	63	104'-0''	78
4'-8''	3½	24'-8''	18½	44'-8''	33½	64'-8''	48½	84'-8''	63½	104'-8''	78½
5'-4''	4	25'-4''	19	45'-4''	34	65'-4''	49	85'-4''	64	105'-4''	79
6'-0''	4½	26'-0''	19½	46'-0''	34½	66'-0''	49½	86'-0''	64½	106'-0''	79½
6'-8''	5	26'-8''	20	46'-8''	35	66'-8''	50	86'-8''	65	106'-8''	80
7'-4''	5½	27'-4''	20½	47'-4''	35½	67'-4''	50½	87'-4''	65½	107'-4''	80½
8'-0''	6	28'-0''	21	48'-0''	36	68'-0''	51	88'-0''	66	108'-0''	81
8'-8''	6½	28'-8''	21½	48'-8''	36½	68'-8''	51½	88'-8''	66½	108'-8''	81½
9'-4''	7	29'-4''	22	49'-4''	37	69'-4''	52	89'-4''	67	109'-4''	82
10'-0''	7½	30'-0''	22½	50'-0''	37½	70'-0''	52½	90'-0''	67½	110'-0''	82½
10'-8''	8	30'-8''	23	50'-8''	38	70'-8''	53	90'-8''	68	110'-8''	83
11'-4''	8½	31'-4''	23½	51'-4''	38½	71'-4''	53½	91'-4''	68½	111'-4''	83½
12'-0''	9	32'-0''	24	52'-0''	39	72'-0''	54	92'-0''	69	112'-0''	84
12'-8''	9½	32'-8''	24½	52'-8''	39½	72'-8''	54½	92'-8''	69½	112'-8''	84½
13'-4''	10	33'-4''	25	53'-4''	40	73'-4''	55	93'-4''	70	113'-4''	85
14'-0''	10½	34'-0''	25½	54'-0''	40½	74'-0''	55½	94'-0''	70½	114'-0''	85½
14'-8''	11	34'-8''	26	54'-8''	41	74'-8''	56	94'-8''	71	114'-8''	86
15'-4''	11½	35'-4''	26½	55'-4''	41½	75'-4''	56½	95'-4''	71½	115'-4''	86½
16'-0''	12	36'-0''	27	56'-0''	42	76'-0''	57	96'-0''	72	116'-0''	87
16'-8''	12½	36'-8''	27½	56'-8''	42½	76'-8''	57½	96'-8''	72½	116'-8''	87½
17'-4''	13	37'-4''	28	57'-4''	43	77'-4''	58	97'-4''	73	117'-4''	88
18'-0''	13½	38'-0''	28½	58'-0''	43½	78'-0''	58½	98'-0''	73½	118'-0''	88½
18'-8''	14	38'-8''	29	58'-8''	44	78'-8''	59	98'-8''	74	118'-8''	89
19'-4''	14½	39'-4''	29½	59'-4''	44½	79'-4''	59½	99'-4''	74½	119'-4''	89½
20'-0''	15	40'-0''	30	60'-0''	45	80'-0''	60	100'-0''	75	120'-0''	90

Note: Based on units 15-5/8'' long and half units 7-5/8'' long with 3/8'' thick head joints

Number of concrete blocks by length of wall
Figure 5-4

Use Figures 5-4 and 5-5 to estimate standard concrete block units. Figure 5-4 gives the number of units in walls of varying length. Figure 5-5 gives the number of courses of block in walls of varying height. Here's an example that shows how to use these tables.

We'll estimate the number of units required for a wall 76 feet long and 12 feet high. According to Figure 5-4, a 76-foot wall requires 57 units. Now look at Figure 5-5. A 12-foot wall requires 18 courses. So multiply 57 by 18 to find the total number of blocks required (1,026).

Let's try another. Estimate the concrete blocks required for a building 24 by 30 feet, and 11 courses high. First find the distance around the foundation — the total wall length. There are two 24-foot walls and two 30-foot walls, giving a total distance of 108 feet around the foundation. From Figure 5-4 we find a 108-foot wall requires 81 units. Multiply by 11 because the wall is 11 courses high. You need 891 blocks.

You can also use Figure 5-4 to lay out buildings on a modular basis to avoid having to cut blocks. If possible, lay out walls in lengths that take a whole

HEIGHT OF WALL	NO. OF UNITS	HEIGHT OF WALL	NO. OF UNITS	HEIGHT OF WALL	NO. OF UNITS	HEIGHT OF WALL	NO. OF UNITS
0'-8"	1	8'-8"	13	16'-8"	25	24'-8"	37
1'-4"	2	9'-4"	14	17'-4"	26	25'-4"	38
2'-0"	3	10'-0"	15	18'-0"	27	26'-0"	39
2'-8"	4	10'-8"	16	18'-8"	28	26'-8"	40
3'-4"	5	11'-4"	17	19'-4"	29	27'-4"	41
4'-0"	6	12'-0"	18	20'-0"	30	28'-0"	42
4'-8"	7	12'-8"	19	20'-8"	31	28'-8"	43
5'-4"	8	13'-4"	20	21'-4"	32	29'-4"	44
6'-0"	9	14'-0"	21	22'-0"	33	30'-0"	45
6'-8"	10	14'-8"	22	22'-8"	34	30'-8"	46
7'-4"	11	15'-4"	23	23'-4"	35	31'-4"	47
8'-0"	12	16'-0"	24	24'-0"	36	32'-0"	48

Note: Based on units 7-5/8" high and 3/8" mortar joints

Number of courses of concrete blocks by height of wall
Figure 5-5

number of blocks. For example, making a wall 41'4" instead of 42'0" will avoid cutting units and consequently eliminate waste.

As another example, assume that the distance between two openings has been designed at 2'9". Looking at Figure 5-4, we find that if the dimension is 2'8", we won't have to cut any units.

Figure 5-6 illustrates the use of 8 x 8 x 16 blocks as backup. Six stretcher courses of brick are used for each two block courses. Every seventh course is a header bond.

Mortar for Masonry
Masonry units are set on a mortar bed which creates a bond with adjoining units and permits adjustment of the level of each unit. Each block must be placed correctly so that the mortar will create a strong, weathertight wall. Mortar for reinforced masonry structures should be either Type M or Type S with minimum compressive strengths of 2,500 psi and 1,800 psi respectively.

Cement used in mortars may be portland cement or masonry cement. Portland cement called for in the specifications will generally be Type I or Type II. Type III is used when high early strength is required.

Labor Costs
The labor cost of laying concrete block will vary with the size and weight of the block, the weather, the class of work, and whether the walls are long and straight or cut up with several openings. Concrete masonry can be laid more economically if the walls are designed to use full and half-length units. That minimizes cutting and fitting of block.

Examine the length and height of the wall, width and height of openings and wall areas between doors, windows, and corners. Be sure you know which openings need only full-size and half-size units and which openings require cut block. All horizontal dimensions should be in multiples of nominal full-length masonry units. Vertical dimensions should be in multiples of 8 inches if you're using 8" high block.

Figure 5-7 is a labor estimating table for concrete masonry. It's based on average conditions with blocks laid in 1:1:6 cement-lime mortar. If portland cement mortar is used above grade, reduce the daily output by about 5%. All quantities are based on the output of one mason and one laborer or one hod carrier per hour. If

Concrete block as a backing
Figure 5-6

	Units laid per hour
Foundation - few openings 8 x 8 x 16 (112.5 blocks per 100 sq. ft.)	30
Superstructure - few openings 8 x 8 x 16 (112.5 blocks per 100 sq. ft.)	25
Superstructure - several openings 8 x 8 x 16 (112.5 blocks per 100 sq. ft.)	20 to 25
Partition - few openings 6 x 8 x 12 (150 blocks per 100 sq. ft.)	30 to 35
Partition - several openings 6 x 8 x 12 (150 blocks per 100 sq. ft.)	24 to 30
Add 10% to productivity for lightweight units.	

Note: This table assumes a crew of one mason and one helper.

Labor estimating table for concrete blocks
Figure 5-7

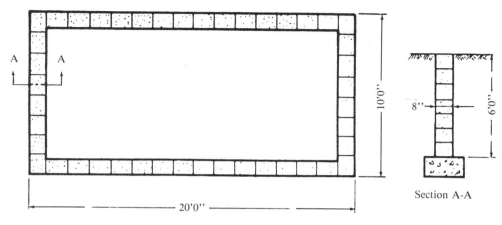

Foundation wall plan
Figure 5-8

hoisting time is required, add about 1/4 hour per 100 square feet.

Sample Concrete Block Estimate: Figure 5-8 is a plan and section view of a foundation wall to be built from 8 x 8 x 16 concrete block. Here are the steps to estimate the cost:

Area of wall: First, find the perimeter of the wall. That's 20' plus 20' plus 10' plus 10', or 60'. Now, to avoid counting the corners twice, deduct from your answer four times the thickness of the wall (4 x 8'' equals 32'', or 2.67'). That leaves 57.33'.

Multiply that by the height of the wall (6.0') to find the block wall surface area. Six times 57.33' equals 344 square feet.

Divide that by 100 to find the factor for 100 square feet. Your factor is 3.44. Figure 5-1 shows that 110 blocks are required for 100 square feet of 8 x 8 x 16 blocks. Multiply 110 by 3.44. You need 379 blocks.

Corners blocks: The height of the wall is 6'0'', or 72''. Divide 72'' by 8'' (the height of a block) to find the number of courses. There are nine courses of block. There are four corners in each course, or a total of 36 corner blocks. Now we have:

Total blocks	=	379
Corner blocks	=	+36
Stretcher blocks	=	415

Look at Figure 5-7. A mason with one helper, using 8 x 8 x 16 blocks in a foundation, can lay 30

an hour. Divide 415 by 30. It will take the crew 13.83 hours to lay 415 blocks. The total labor cost is:

13.83 mason hours at $16.00	=	$221.28
13.83 laborer hours at $10.00	=	138.30
		$359.58
18% labor burden		64.72
		$424.30
10% overhead		42.43
Subtotal		$466.73
10% profit		46.67
Contract price (labor)		$513.40

To find the labor cost per block, divide $513.40 by 415. That's $1.237 per block.

I should emphasize the point again: These are *my* labor figures. Your own labor figures will always be more accurate than anything printed in a book. There should be two profits in every job. One goes into your pocket. The other is what you learn to make the next job more profitable. Sharpen your estimating skills on every job. Note how long it takes to do every significant task and use those figures the next time similar work comes up.

Estimating Brick Masonry

Except for the non-modular "standard brick" (3¾'' x 2¼'' x 8'') and some oversize brick (3¾'' x 2¾'' x 8''), most of the brick produced and used in the United States is sized to fit the modular system. Even the "standard" brick is available in a modular size (nominal dimensions 4'' x 2⅔'' x 8''). Since non-modular brick is still available at many masonry supply houses, we'll include estimating information for non-modular as well as modular units.

Estimating Procedure

Because of its simplicity and accuracy, the most widely used estimating procedure is the *wall-area* method. It consists simply of multiplying known quantities of material required per square foot by the net wall area (gross areas less all openings).

Estimating material quantities is greatly simplified under the modular system. For a given nominal size, the number of modular masonry units per square foot of wall will be the same regardless of mortar joint thickness — assuming, of course, that the units are laid with the correct joint thickness. There are only three standard modular joint thicknesses: 1/4'', 3/8'' and 1/2''.

In contrast, the number of non-modular standard brick required per square foot of wall will vary with the thickness of the mortar joints.

Here's the estimating procedure. First, determine the net quantity of *all* material before adding any allowances for waste. Allowances for waste and breakage vary, but, as a general rule, add at least 5 percent to the net brick quantity and 10 to 25 percent to the net mortar quantity. Particular job conditions, or experience, may dictate different factors.

Nominal Size of Brick in. (t h l)	Number of Brick per 100 sq ft	Cubic Feet of Mortar			
		Per 100 Sq Ft		Per 1000 Brick	
		⅜-in. Joints	½-in. Joints	⅜-in. Joints	½-in. Joints
4 x 2⅔ x 8	675	5.5	7.0	8.1	10.3
4 x 3⅕ x 8	563	4.8	6.1	8.6	10.9
4 x 4 x 8	450	4.2	5.3	9.2	11.7
4 x 5⅓ x 8	338	3.5	4.4	10.2	12.9
4 x 2 x 12	600	6.5	8.2	10.8	13.7
4 x 2⅔ x 12	450	5.1	6.5	11.3	14.4
4 x 3⅕ x 12	375	4.4	5.6	11.7	14.9
4 x 4 x 12	300	3.7	4.8	12.3	15.7
4 x 5⅓ x 12	225	3.0	3.9	13.4	17.1
6 x 2⅔ x 12	450	7.9	10.2	17.5	22.6
6 x 3⅕ x 12	375	6.8	8.8	18.1	23.4
6 x 4 x 12	300	5.6	7.4	19.1	24.7

Note: No allowances for breakage or waste

Modular brick and mortar, single wythe walls, running bond
Figure 5-9

Estimating Tables

Figure 5-9 gives the net quantity of brick and mortar needed to build walls one wythe in thickness with various modular brick sizes and the two most common joint thicknesses (3/8'' and 1/2''). Mortar quantities are for full bed and head joints.

Figure 5-10 provides similar information for walls constructed only with non-modular brick.

The brick and mortar quantities in Figures 5-9 and 5-10 are for running (or stack) bonds which have no header courses. For bonds requiring full headers, apply the correction factors given in Figure 5-11.

When estimating quantities for multi-wythe walls, add the mortar quantities for interior vertical and longitudinal collar joints given in Figure 5-12.

Figure 5-13 shows the quantity of portland cement, hydrated lime and sand required for 1 cubic foot of four types of mortar. Although ASTM

Size of Brick in. (t h l)	With ⅜-in. Joints			With ½-in. Joints		
	Number of Brick per 100 Sq Ft	Cubic Feet of Mortar per 100 Sq Ft	Cubic Feet of Mortar per 1000 Brick	Number of Brick per 100 Sq Ft	Cubic Feet of Mortar per 100 Sq Ft	Cubic Feet of Mortar per 1000 Brick
2¾ x 2¾ x 9¾	455	3.2	7.1	432	4.5	10.4
2⅝ x 2¾ x 8¾	504	3.4	6.8	470	4.1	8.7
3¾ x 2¼ x 8	655	5.8	8.8	616	7.2	11.7
3¾ x 2¾ x 8	551	5.0	9.1	522	6.4	12.2

Note: No allowances for breakage or waste

Non-modular brick and mortar, single wythe walls, running bond
Figure 5-10

Bond	Correction Factor [1]
Full headers every 5th course only	1/5
Full headers every 6th course only	1/6
Full headers every 7th course only	1/7
English bond (full headers every 2nd course)	1/2
Flemish bond (alternate full headers and stretchers every course)	1/3
Flemish headers every 6th course	1/18
Flemish cross bond (Flemish headers every 2nd course)	1/6
Double-stretcher, garden wall bond	1/5
Triple-stretcher, garden wall bond	1/7

[1] Note: Correction factors are applicable only to those brick which have lengths of twice their bed depths. Add to facing and deduct from backing.

Correction factors for Figures 5-9 and 5-10
Figure 5-11

Cubic Feet of Mortar Per 100 Sq Ft of Wall		
¼-in. Joint	⅜-in. Joint	½-in. Joint
2.08	3.13	4.17

Note: Cubic feet per 1000 units = $\dfrac{10 \times \text{cubic feet per 100 sq ft of wall}}{\text{number of units per square foot of wall}}$

Cubic feet of mortar for collar joints
Figure 5-12

Material	Quantities by Volume				Quantities by Weight			
	Mortar Type and Proportions by Volume				Mortar Type and Proportions by Volume			
	M	S	N	O	M	S	N	O
	1:¼:3	1:½:4½	1:1:6	1:2:9	1:¼:3	1:½:4½	1:1:6	1:2:9
Cement	0.333	0.222	0.167	0.111	31.33	20.89	15.67	10.44
Lime	0.083	0.111	0.167	0.222	3.33	4.44	6.67	8.89
Sand	1.000	1.000	1.000	1.000	80.00	80.00	80.00	80.00

Material quantities per cubic foot of mortar
Figure 5-13

Average Specific Gravities and Unit Weights [1]

Material	Specific Gravity	Unit Weight in Pounds Per Cubic Foot
Portland Cement	3.15	94
Lime	2.25	40
Water	1.00	62.4

[1] Values for sand are not listed because they vary considerably. Obtain precise values from laboratory tests (or from supplier).

Average specific gravities and unit weights
Figure 5-14

Standard Specifications for Mortar for Unit Masonry (ASTM Designation C 270) permit a range of proportions for each mortar type, the quantities in Figure 5-13 have been based on a single set of proportions for each of these types. For your convenience, quantities in the table are based on both weight and volume. On the job, mortar is generally proportioned by volume, although proportioning by weight is more accurate.

Mortar Yield

For given volumes of materials, mortar yield depends on proportions, water content and air content. Water content will vary with sand gradation, lime and cement content, and, quite often, the judgment of the brick mason.

Mortar yield calculations are based on absolute volume. To determine yield, first find:

1) Unit weights and specific quantities of all materials. These are given in Figure 5-14.

2) Total volume of water used in mortar mix, including mixing water and the water present in the sand.

Sand: When a relatively small amount of water (4 to 10 percent) is added to dry sand, it *bulks;* that is, it increases in volume far in excess of the volume of water added. This increase can be as much as 50 percent, depending largely on the grade of the sand.

Because sand volume varies with the moisture content, measuring sand by volume isn't very accurate. Weighing sand is more accurate. But it isn't nearly as convenient as measuring volume. For proportion specifications, ASTM C 270 assumes that 1 cubic foot of damp, loose sand (bulked sand) is equal to 80 pounds of dry sand (and that it has bulked approximately 38 percent).

Specific Gravity of Sand: The yard that supplies

your sand will probably have an estimate of the specific gravity of the sand they sell. The procedure for determining specific gravity of sand is given in ASTM C 128, Standard Method of Test for Specific Gravity and Absorption of Fine Aggregate. An average for silica sands is 2.65.

Moisture Content of Sand: In any given sand, the moisture content may vary from day to day or even from hour to hour. This variation isn't as critical in mortar as it is in portland cement concrete. Most damp, loose sand has between 1/2 and 1 gallon of water per cubic foot. Usually it's safe to assume this amount of water. If you need more accuracy, ASTM C 70, Standard Method of Test for Surface Moisture in Fine Aggregate is the accepted procedure.

Weight of Sand: Mortar yield depends on moisture content, specific gravity and unit weight (bulked and dry). To determine bulked unit weight, weigh 1 cubic foot of bulked sand. The standard method for determining unit weight of aggregate is given in ASTM C 29, Standard Method of Test for Unit Weight of Aggregate. When you know the moisture content and bulked weight, you can easily compute the weights of water and dry sand.

Water: Use the total weight of water in yield calculations; that is, the sum of the weights of mixing water and water present in the sand. If you know the moisture content of the sand, it's a simple matter to calculate the weight of water in the sand. Add the weight of water present in the sand to the weight of mixing water added to the batch at the mixer. To convert gallons of mixing water to pounds, multiply by 8.33.

Absolute Volume: The absolute solid volume of a material is the volume of its solid portion only; voids are not included. Thus:

$$V_m = \frac{W_m}{62.4G_m} \quad \cdots \cdots \cdots \quad (1)$$

In Eq (1):

V_m = absolute solid volume of any given material in cubic feet
W_m = batch weight of the material in pounds
G_m = specific gravity of the material

The asbolute volume of mortar is the sum of the absolute volumes of all ingredients (cement, lime, sand and water):

$$V_M = V_c + V_L + V_S + V_W \quad (2)$$

Mortar Yield: Mortar yield is equal to its absolute volume plus the volume of entrapped air. When the air content of mortar is known, the yield may be found from the relationship:

$$Y = \frac{100 \ V_M}{100\text{-a}} \quad \cdots \cdots \cdots \cdots \quad (3)$$

In Eq (3):

Y = = volume of mortar yield in cubic feet
V_M = absolute solid volume of mortar in cubic feet from Eq (2)
a = air content of freshly mixed mortar in per cent

Air contents of freshly mixed portland cement-lime mortars are in the order of 10 percent.

Mortar Yield by "Rule-of-Thumb": If the specs don't say anything about mortar proportions or yield, use the following rule-of-thumb:

For each 1 cubic foot of damp loose sand, the mortar yield will be 1 cubic foot.

Here's a sample mortar yield calculation:
Batch proportions by volume: 1 cubic foot portland cement, 1 cubic foot lime, 6 cubic feet damp loose sand, and 8⅓ gallons of mixing water. For all materials except sand, specific gravities and unit weights are as shown in Figure 5-14. The specific gravity of the sand is 2.65. Moisture content of sand is assumed to be 8 percent at a damp loose weight of 87 pounds per cubic foot. Batch weight of sand is 6 times 87 pounds, or 522 pounds. Assume the mixed mortar contains 10 percent air.

Step 1. Determine the total weight of each material in the batch, using Figure 5-14:

Cement	W_C	= (1) (94)	= 94 lb.
Lime	W_L	= (1) (40)	= 40 lb.
Sand	W_S	= $\frac{(100\text{-}8) (522)}{100}$	= 480.2 lb.

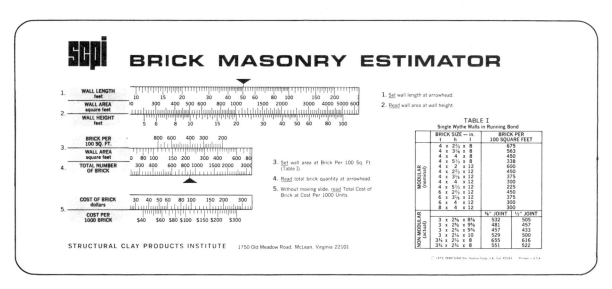

Brick Masonry Estimator
Figure 5-15

Water in sand $= \dfrac{(8)(522)}{100} = 41.8$ lb.

Mixing water $= (8\text{-}1/3)(8.33) = \underline{69.4}$ lb.

Total water W_W $= 111.2$ lb.

Step 2. Using the batch weights determined in Step 1, calculate the absolute solid volume for each material from Equation (1) and the total from Equation (2):

Cement $V_C = \dfrac{94}{(3.15)(62.4)} = 0.478$ CF

Lime $V_L = \dfrac{40}{(2.25)(62.4)} = 0.285$

Sand $V_S = \dfrac{480.2}{(2.65)(62.4)} = 2.904$

Water $V_W = \dfrac{111.2}{(1)(62.4)} = \underline{1.782}$

Total $V_M = \hspace{2cm} 5.449$ CF

Step 3. Determine mortar yield from Equation (3):

$Y = \dfrac{5.449(100)}{(100-10)} = 6.05$ CF

Note: By this rule-of-thumb method, the yield is 6

cubic feet. But if this mix had required more or less mixing water, or had it varied in air content, the yield would have varied also. Conceivably, this difference will be significant only if you're handling large volumes.

Ways to Cut Labor Costs
The higher the labor cost, the more important it is to find ways to save manhours. One good way to cut manhours for bricklaying is to use larger size brick.

Figure 5-16 lists brick equivalent factors for popular modular brick units. The standard modular unit is used as the basis for comparison. You can see that the 8" jumbo, the largest unit listed (8" x 4" x 12") is equivalent to 4½ standard modular units. Every time your mason sets one of these jumbo units, it's like placing 4½ standard units. That can cut labor costs considerably.

Brick Estimator Slide Chart
The only completely accurate way to estimate masonry is to calculate the units required and figure the manhours. But to check your figures or to make a ballpark estimate, consider using The Brick Institute of America (BIA) Brick Masonry Estimator. It costs about a dollar and is small enough to carry in your pocket. If you know the length and height of the walls, the size and cost of the brick, it will give you the brick cost.

The BIA Estimator, shown in Figure 5-15, includes data for 19 modular and non-modular brick

Unit Designation	Nominal Dimensions, in.	Brick Equivalents[1]
Standard Modular	4 x 2 2/3 x 8	1.0
Engineer	4 x 3 1/5 x 8	1.2
Economy or Jumbo Closure	4 x 4 x 8	1.5
Double	4 x 5 1/3 x 8	2.0
Roman	4 x 2 x 12	1.13
Norman	4 x 2 2/3 x 12	1.5
Norwegian	4 x 3 1/5 x 12	1.8
Economy 12 or Jumbo Utility	4 x 4 x 12	2.25
Triple	4 x 5 1/3 x 12	3.0
SCR	6 x 2 2/3 x 12	2.25
6-in. Norwegian	6 x 3 1/5 x 12	2.7
6-in. Jumbo	6 x 4 x 12	3.38
8-in. Jumbo	8 x 4 x 12	4.5

[1]Based on nominal face dimensions only.

Brick equivalent factors for modular brick
Figure 5-16

sizes. It also helps make mortar estimates. After you've determined the brick quantity, use the Estimator to find the quantity of portland cement, lime and sand for four types of mortars, types M, S, N and O. It's simple to use. Each of the 12 steps is clearly printed beside the corresponding scale and window. The slide chart has a wide range of estimating applications: up to one million brick units and a $300,000 order.

You can order one from:

Brick Institute of America
1750 Old Meadow Road
McLean, Virginia 22101

Sample Estimating Problem 1

First, let's estimate a brick veneer, three-bedroom, single-family house with a carport and an outside storage room. Here are the specs:

- 1,441 SF (net) brick veneer

- Brick units, 4 x 2⅔ x 8, laid in 1/2 running bond pattern

- Head and bed joints are 3/8'' thick

- No collar joints

- Mortar type: ASTM C 270, type N (1:1:6)

portland cement, hydrated lime and masonry sand

Solution: First, use Figure 5-9 to find the number of brick needed for 1,441 square feet of wall:

$$\frac{1441}{100} \times 675 = 9{,}727 \text{ units}$$

+ 5% waste	= 486
Total brick required	10,213 units

Next, use Figure 5-9 again to find the quantity of mortar needed for 3/8'' joints:

$$\frac{9727}{1000} \times 8.1 = 79 \text{ CF}$$

+ 25% waste	= 20
Total mortar required	99 CF

Finally, use Figure 5-13 to find the material quantities of portland cement, hydrated lime and sand for type N mortar, 1:1:6. Multiply the cubic feet of mortar required by the pounds of material per cubic foot. Divide each answer by the weight of that material per purchase unit, to convert to the number of sacks of cement (94 lbs) or lime (50 lbs) needed, or tons of sand.

Mortar joint	Wall thickness	Material Brick 100 SF wall	Brick SF per 1000 brick	Wall ties Per 100 SF	Mortar CF per 100 SF	Labor 100 square feet wall Mason	Labor 100 square feet wall Laborer
1/4	4″	698	143	100	4.48		
3/8	4″	655	153	93	6.56	6½	5
1/2	4″	616	162	88	8.34	hours	hours
5/8	4″	581	172	83	10.52	average	average
3/4	4″	549	182	78	12.60		

Note: Mortar includes 20% waste for all head and bed joints.

CEF — Brickwork material and labor
Figure 5-17

Portland cement $= \dfrac{99 \text{ CF} \times 15.67 \text{ lbs}}{94 \text{ lbs per sack}} = 16.5 \ (17 \text{ sacks})$

Hydrated lime $= \dfrac{99 \text{ CF} \times 6.67 \text{ lbs}}{50 \text{ lbs per sack}} = 13.2 \ (14 \text{ sacks})$

Sand $= \dfrac{99 \text{ CF} \times 80 \text{ lbs}}{2,000 \text{ lbs per ton}} = 3.96 \ (4 \text{ tons})$

Figure 5-17 shows material and labor figures for brickwork. This is one of the tables you'll want in your CEF. It shows that a 1,441 square foot brick veneer requires 93.66 mason hours and 72.05 laborer hours.

Sample Estimating Problem 2
Here's a more complicated estimate. It's 30-unit complex of two-story townhouses. Here are the specs:

- 13,095 (net) SF of 10″ brick cavity

- 24,780 (net) SF of 8″ solid brick (double wythe)

- Brick units, 4 x 2⅔ x 12, laid in 1/3 running bond pattern

- Head and bed joints are 3/8″ thick

- Collar joints are 1/2″

- Mortar type: ASTM C 270, type N (1:1:6) portland cement, hydrated lime and masonry sand

Solution:
1) Net wall area of 4″ single wythe: For the 10″ cavity, 13,095 SF. For the 8″ solid, 24,780 SF. 13,095 plus 24,780 equals 37,875. Doubled, that's 75,750 SF.

2) Brick quantity for 4 x 2⅔ x 12 units (Figure 5-9):

$\dfrac{75,750}{100} \times 450 = 340,875 \text{ units}$
+ 5% waste $= \underline{17,044}$
Total brick required $\quad 357,919 \text{ units}$

3) Mortar quantity, 3/8″ head and bed joints (Figure 5-9):

$\dfrac{340,875}{1,000} \times 11.3 = 3,852 \text{ CF}$

4) Mortar quantity, 1/2″ collar joints (Figure 5-12):

$\dfrac{24,780}{100} \times 4.17 = 1,033 \text{ CF}$

5) Add steps 3 and 4:

Head and bed joint mortar $= 3,852$ CF
Collar joint mortar $= \underline{1,033}$
$\qquad\qquad\qquad\qquad\quad 4,885$ CF
+ 25% waste $= \underline{1,221}$
Total mortar required $= 6,106$ CF

6) Mortar material quantities of portland cement, hydrated lime and sand for type N mortar, 1:1:6, including waste (Figure 5-13):

$$\text{Portland cement} = \frac{6{,}106 \times 15.67}{94 \text{ lbs.}} = 1{,}018 \text{ sacks}$$

$$\text{Hydrated lime} = \frac{6{,}106 \times 6.67}{50 \text{ lbs.}} = 815 \text{ sacks}$$

$$\text{Sand} = \frac{6{,}106 \times 80}{2{,}000 \text{ lbs.}} = 244 \text{ tons}$$

Use Figure 5-17 to compute the labor cost.

Estimating Glass Building Blocks

Glass blocks are available in three sizes: 5¾" x 5¾", 7¾" x 7¾", and 11¾" x 11¾", all 3⅞" high. Four blocks are required per square foot of wall for the 5¾" size, 2.25 blocks for the 7¾" size, and one block for the 11¾" size. Allow 2.5 percent for breakage.

Measure the mortar materials for laying glass blocks by volume. You can assume that 25 pounds of quick lime or 40 pounds of hydrated lime yield one cubic foot. The mortar is made of one part waterproof portland cement, one part lime paste and four parts of well-graded sand. Mix it to a stiff consistency and apply.

The average joint is 1/4" thick. Use expansion joint strips 1/2" thick, 4⅛" wide and 25" long at the heads, jambs and mullions of all openings. The strips are held in place by an adhesive such as asphalt emulsion. Butt them together end to end to form a continuous cushion around the edges of panels up to 10' x 10' in size.

To find the required opening size for glass block panels, use the dimensions in Figure 5-18, adding 1/2" for width and 3/8" for height. Use 1/4" for mortar joints.

After the panels have been laid and the mortar has set, ram nonstaining oakum tightly between the sides of the block and the sides of the "chase." Force the oakum back 3/8" from the finished surface. Caulk the 3/8" recess with a nonhardening waterproof caulking material to a depth not less than 3/8".

Reinforcing wall ties are run continuously with ends lapped 6 inches. Place them in the center of horizontal mortar joints in every fourth course for

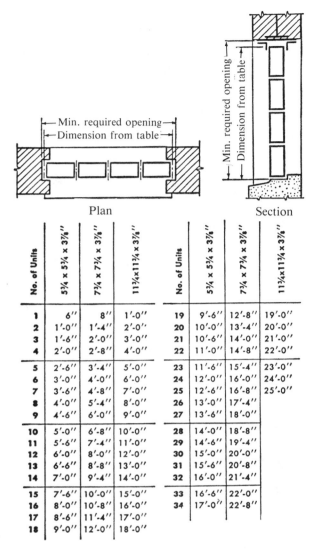

No. of Units	5¾ x 5¾ x 3⅞	7¾ x 7¾ x 3⅞	11¾ x 11¾ x 3⅞	No. of Units	5¾ x 5¾ x 3⅞	7¾ x 7¾ x 3⅞	11¾ x 11¾ x 3⅞
1	6"	8"	1'-0"	19	9'-6"	12'-8"	19'-0"
2	1'-0"	1'-4"	2'-0"	20	10'-0"	13'-4"	20'-0"
3	1'-6"	2'-0"	3'-0"	21	10'-6"	14'-0"	21'-0"
4	2'-0"	2'-8"	4'-0"	22	11'-0"	14'-8"	22'-0"
5	2'-6"	3'-4"	5'-0"	23	11'-6"	15'-4"	23'-0"
6	3'-0"	4'-0"	6'-0"	24	12'-0"	16'-0"	24'-0"
7	3'-6"	4'-8"	7'-0"	25	12'-6"	16'-8"	25'-0"
8	4'-0"	5'-4"	8'-0"	26	13'-0"	17'-4"	
9	4'-6"	6'-0"	9'-0"	27	13'-6"	18'-0"	
10	5'-0"	6'-8"	10'-0"	28	14'-0"	18'-8"	
11	5'-6"	7'-4"	11'-0"	29	14'-6"	19'-4"	
12	6'-0"	8'-0"	12'-0"	30	15'-0"	20'-0"	
13	6'-6"	8'-8"	13'-0"	31	15'-6"	20'-8"	
14	7'-0"	9'-4"	14'-0"	32	16'-0"	21'-4"	
15	7'-6"	10'-0"	15'-0"	33	16'-6"	22'-0"	
16	8'-0"	10'-8"	16'-0"	34	17'-0"	22'-8"	
17	8'-6"	11'-4"	17'-0"				
18	9'-0"	12'-0"	18'-0"				

Glass block quantities
Figure 5-18

the smallest blocks, in every third course for the middle-size blocks, and in every course for the largest blocks. Ties should be two number 9 galvanized wires spaced 2 inches apart with welded number 14 gauge cross wire. These come in 8-foot long strips and should be not more than 0.20 inches thick at the weld.

Compile your CEF for labor and materials for a glass block wall, as shown in Figure 5-19.

Estimating Stonework

Wherever stone is required on a job, your first questions should be about size and placement of the stone. Then you can figure the cost of the dif-

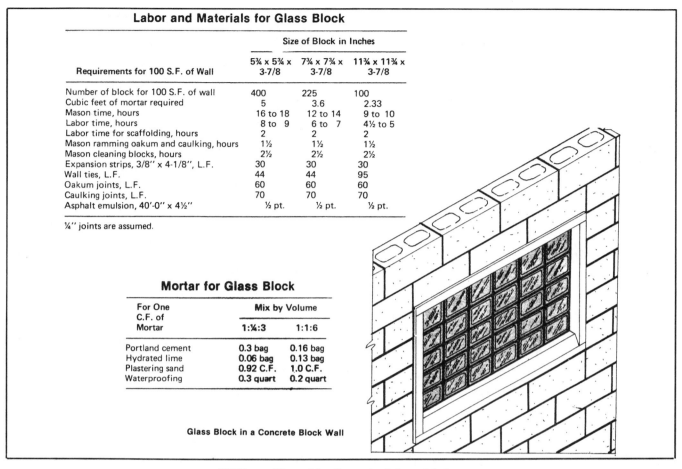

Labor and Materials for Glass Block

Requirements for 100 S.F. of Wall	Size of Block in Inches		
	5¾ x 5¾ x 3-7/8	7¾ x 7¾ x 3-7/8	11¾ x 11¾ x 3-7/8
Number of block for 100 S.F. of wall	400	225	100
Cubic feet of mortar required	5	3.6	2.33
Mason time, hours	16 to 18	12 to 14	9 to 10
Labor time, hours	8 to 9	6 to 7	4½ to 5
Labor time for scaffolding, hours	2	2	2
Mason ramming oakum and caulking, hours	1½	1½	1½
Mason cleaning blocks, hours	2½	2½	2½
Expansion strips, 3/8" x 4-1/8", L.F.	30	30	30
Wall ties, L.F.	44	44	95
Oakum joints, L.F.	60	60	60
Caulking joints, L.F.	70	70	70
Asphalt emulsion, 40'-0" x 4½"	½ pt.	½ pt.	½ pt.

¼" joints are assumed.

Mortar for Glass Block

For One C.F. of Mortar	Mix by Volume	
	1:¼:3	1:1:6
Portland cement	0.3 bag	0.16 bag
Hydrated lime	0.06 bag	0.13 bag
Plastering sand	0.92 C.F.	1.0 C.F.
Waterproofing	0.3 quart	0.2 quart

Glass Block in a Concrete Block Wall

**CEF — Glass block material and labor
Figure 5-19**

ferent kinds of stone used, the amount and cost of tooling or cutting necessary to complete the job, and the higher cost of doing any special type of work.

The major variables that affect the costs are:

1) The *kind* and *quality* of the stone used and the selection of the proper grade for the work.

2) The cost of *machining* the stone. This should include an analysis of the various sizes and the best procedures to follow. Look for a reasonable amount of uniformity in size and texture with the surrounding work.

3) The *design* of the stone units. Consider carefully the sizes to be used, the way the stone are to be stacked, the pattern required and the bonding

system. All of these influence cost. An experienced stonemason who knows good placement and jointing technique can reduce the amount of cutting required.

General contractors usually don't estimate stonework unless it's a very small amount. Instead, they get bids from regular stone contractors who specialize in the working and setting of stone. For custom stonework, get bids from experienced stone contractors.

Stonework may be classified into three different groups as far as general construction work is concerned. These groups are:

- Rough or rubble stone

- Dimensioned or ashlar stone

- Cobble or field stone

Rubble Stone

Rubble stone may be quarried from local quarries or broken out of large field boulders. This class of stone comes in random sizes and shapes and usually requires very little trimming except to knock off the rough angles. It's used for foundation walls, exterior walls and chimneys where a rustic effect is wanted.

The term "rubble" can also apply to stone quarried and cut into irregular shapes to form a random pattern in the wall. Because of the odd shapes, it may appear that such walls are made up of miscellaneous pieces of stone. But this isn't always true. Stones for "rubble effect" are often sawed out of certain definite rough sizes or widths with the face left rough. These lengths are then broken or cut to the desired size by the stone mason. "Rough squared stone" may be a better term for this kind of stone.

The stonework can be broken down into coursed and uncoursed rubble masonry. Figure 5-20 shows a typical wall of uncoursed rubble stone.

Uncoursed rubble stone
Figure 5-20

Rubble stone may be measured by the cubic foot, cubic yard, or the ton — but the cubic yard is most common. A cubic yard of stone weighs about 3,500 pounds, but this can vary considerably, depending on the kind of stone used.

Quantity take-off is done the same way as with brickwork. First figure the quantities in outside wall footings and outside foundation walls, using the outside measurements. Tabulate the width, height and length. Then take all the inside walls and list their dimensions. Don't deduct for openings less than four square feet. After figuring all the stone you can see on the plans, go back one more time to see that you've included every item

and listed the correct size.

The volume of the stone required will be less than the volume of wall because of space taken by the mortar. Mortar may form from 15 to 35 percent of the volume in rubble masonry, 10 to 20 percent in squared stone and 5 to 10 percent in ashlar masonry. The actual quantity of mortar required for rubble stone depends on the size and shape of the stone and thickness of the wall. If stones are large and regular, about 5.5 cubic feet of mortar per 27 cubic feet of stone is enough.

Rubble masonry is usually set in cement mortar consisting of one part portland cement to three parts clean sand. If lime is used to increase the workability of the mortar, replace 10 to 15 percent of the volume of the cement with an equal amount of lime. A cement-lime mortar which won't stain light-colored stones consists of one part hydrated lime, one part of non-staining cement, and six parts of clean sand.

Figure 5-21 illustrates several special forms of anchors such as:

a) a metal key plate with anchor bolt.

b) an anchor bolt with a metal dowel for use between ends of stone.

c) an anchor bolt with metal dowel for use in beds of stone.

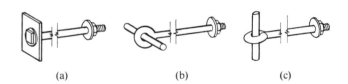

(a)   (b)   (c)

Anchors to fasten stone to steel framing members
Figure 5-21

Like brickwork, stonework should be washed with cold water and strong fiber brushes and then the joints pointed.

The thickness, height and size of stone determines the quantity a stone mason can lay in one day. For ordinary work, the figures in Figure 5-22 will apply. Figure 5-23 shows additional labor output for stone masonry. You'll probably want both of these tables in your CEF. Rustic faced rubble stonework is used in high-class buildings and on residences of all kinds. The stones vary in size from

Kind	Thickness	Mason time	Labor time
Light rubble	16" to 18"	3 hours	3 hours
Heavy rubble	2'0" to 3'6"	2 hours	2 hours
Cobble or rustic	6"	5¼ hours	5¼ hours
Cobble or rustic	12"	2½ hours	2½ hours

Labor for rubble stone in hours per cubic yard
Figure 5-22

4" to 18". The face of walls and piers are of stone backed with common brick or concrete block. The smaller the stone, the more it will cost to lay. An 18" stone can be laid almost as quickly as a 4" stone. Estimate this kind of work by the square foot of surface, by the cubic foot, or by the cubic yard.

Kind Of Work	Crew	Output
Preparing Stone		
Rough Squaring	1 Stonecutter	3 to 6 C.F. Per Hour
Smoothing Beds And Builds	1 Stonecutter	2 to 3 S.F. Per Hour
Hammering	1 Stonecutter	1.5 S.F. Per Hour
Laying Stone By Hand		
Rubble	1 Mason, 1-3 Helpers	6 to 7 C.F. Per Hour
Cut Stone Or Ashlar	1 Mason, 1-3 Helpers	4 to 6 C.F. Per Hour
Cut Stone Veneer	1 Mason, 1-3 Helpers	6 to 7 C.F. Per Hour
Laying Stone With Aid Of Hand Hoist, Or Derrick		
All Kinds Of Masonry	1 Mason, 1-3 Helpers	7 to 10 C.F. Per Hour
Cut Stone Veneer	1 Mason, 1-3 Helpers	8 to 10 S.F. Per Hour
Pointing		
Simple	1 Mason, 1 Helper	30 to 40 S.F. Per Hour
Cleaning	1 Mason	30 to 50 S.F. Per Hour

Labor output for stone masonry
Figure 5-23

Lannon or Split Faced Ashlar Stone

Ashlar work generally refers to a thin covering of stonework or a facing used on the outside of buildings. Ashlar work is also defined as stone cut to dimension, or *dimensioned stone*. Actually, ashlar stone means both. It's stonework that has been carefully dimensioned or cut to size (reducing the joints to a minimum thickness) and used as a facing over a building. Another distinction should be drawn, however, between ashlar and ordinary cut stone. Ashlar refers to stonework where the entire face of a building is covered with this facing stone.

In a sense, all cut stone would include any kind of stone that has been machined or cut to a special size or shape. It could also mean any stone that has been tooled or planed on the surface, or to smooth stones of any kind. The common usage of the term "cut stone," however, applies to "cut stone trim" or "stone trimming." This includes sills, band courses or cornices.

When ashlar is used for the elevation of buildings, it's laid up in blocks which vary in size according to the architect's specifications and drawings. The average is between 8" to 16" high and 18" to 36" long. In most cases, these stones are only 4" to 8" thick and are used as a veneer over an 8" or 12" masonry wall. Each stone, especially of the larger size, should be anchored into the masonry wall with two heavy iron anchors. A bond stone should extend through the wall every 6 or 8 square feet.

Many brick buildings have stonework of some kind for ornamental or trimming purposes. These may consist of base or sill courses, lintels, pilasters, or quoins. The surface may be "natural rough," planed, rubbed or tooled. Where molds are continuous, they may be formed by a planer. Where molds have offsets, they must be done by hand, except when the stone is cast to a definite size and according to specifications.

Where a cornice is made by a number of horizontal molds and contains medallions, the molds are made up of short pieces and the ornaments or medallions are inserted. Stone used for this purpose is usually sold by the linear foot.

Lannon and similar stones quarried in different parts of the country are used in better churches, colleges and residences. The stone comes 4" to 8" thick and 1" to 15" high. When backed up with

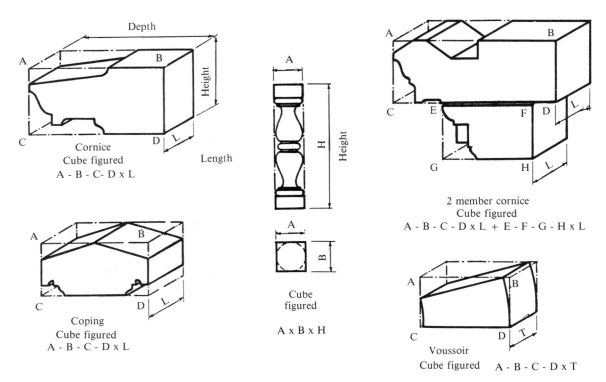

Figuring cubic feet of cut stone
Figure 5-24

brick or block, the stone is usually 4" to 8" thick so it will bond with the masonry. When used as a veneer on residences, it's generally 4" thick and is tied to the wall with metal ties.

The cost of setting this stone varies greatly, depending on the size of the stone and the amount of cutting required. Lannon or split ashlar stone is usually sold by the ton and comes with rough beds, backs and joints. It weighs about 160 pounds to the cubic foot, so there are 12 to 12½ cubic feet of stone to the ton.

Stone 4 inches thick will make about 36 square feet of wall to the ton. With a 1/2" mortar joint, estimate about 40 square feet of wall. Stone 5 inches thick will make about 30 square feet of wall to the ton, or 33 square feet with a 1/2" mortar joint. If backed with brick or block and based on an average thickness of 6 inches, one ton will lay about 25 square feet of wall. With a 1/2" mortar joint, one ton will lay up 26 to 28 square feet of wall.

Figure 5-23 includes labor output figures for setting ashlar stone.

Estimating Cut Stone

Cut stone is usually estimated by the cubic foot.

But you'll have to figure all ornamental moldings, carvings and panel work separately. The illustrations in Figure 5-24 show how to find the cubic feet in a one-piece cornice and coping, a baluster and a two-piece cornice.

When taking off cut stone quantities, use an estimate sheet like the one in Figure 5-25. Take off each quantity and draw a rough sketch of the stone at the side of the sheet. This gives the stone cutter a clear idea of the stone you need. List the length, width, height and number of pieces needed.

Because of the variation in size of cut stone, there's no firm figure for how much mortar is required per 100 cubic feet of stone. A good average is 4 to 5 cubic feet of mortar for 100 cubic feet of stone of average size, 18" x 30" x 4".

Back plastering of stones will require 4 cubic feet of mortar for each 100 cubic feet of stone, or 1 cubic foot for 40 to 50 square feet of surface. Use white, nonstaining portland cement for setting and back plastering limestone or other porous stones. This prevents stains on the face of the stone. Don't use ordinary portland cement for setting limestone.

The mortar for cut stone is generally the same as the mortar used in other kinds of masonry. But

Cut Stone Estimate

Job George H. Harvey   Date_____

Material Buff, Rubbed

Sketch	Pieces	Length	Depth	Height	Description	Number	Cu. Ft.
	1	1' 0''	6''	6''	Base	F 1	.25
	1	7' 4''	6''	6''	Base	F 2	1.83
	1	4'10''	6''	6''	Base	F 3	1.20
Base	1	4' 2''	6''	6''	Base	F 4	1.04
	1	3' 8''	6''	6''	Base	F 5	.91
	1	1' 0''	6''	6''	Base	F 6	.25
	1	1' 6''	5''	6''	Sill course	F 7	0.31
	1	5' 4''	5''	6''	Sill course	F 8	1.11
	1	1' 8''	5''	6''	Sill course	F 9	.34
	4	5' 1''	5''	6''	Window sills	F10	4.23
	1	4' 0''	5''	6''	Window sills	F11	.83
	1	3' 4''	5''	6''	Win. sills bsmt.	F12	.69
	1	3'10''	5''	6''	Win. sills bsmt.	F13	.79
	1	4' 4''	5''	6''	Win. sills bsmt.	F14	.90
	1	3' 4''	5''	6''	Win. sills bsmt.	F15	.69
	1	3' 4''	5''	6''	Win. sills 1st flr.	F16	.69
	1	2'10''	5''	6''	Win. sills 1st flr.	F17	.59
	1	3'10''	5''	6''	Win. sills 1st flr.	F18	.79
	1	4' 4''	5''	6''	Win. sills 1st flr.	F19	.90
	1	2' 4''	5''	6''	Win. sills 1st flr.	F20	.48
	1	4' 0''	5''	6''	Win. sills 1st flr.	F21	.83
	1	3' 4''	5''	6''	Win. sills 2nd flr.	F22	.69
	1	2'10''	5''	6''	Win. sills 2nd flr.	F23	.59
	1	3'10''	5''	6''	Win. sills 2nd flr.	F24	.79
	1	4' 4''	5''	6''	Win. sills 2nd flr.	F25	.90
	1	4' 0''	5''	6''	Win. sills 2nd flr.	F26	.83
	1	2' 4''	5''	6''	Win. sills 2nd flr.	F27	.48

Cut stone estimate
Figure 5-25

there are a few important points to consider when setting stone. Generally, as with all other masonry, the mortar will be either lime mortar, lime-cement mortar or cement mortar.

Lime mortar is made of pure lime and sand. Cement mortar is made of pure cement and sand. It can be confusing, though, because these terms are also used for mortar with a slight mixture of other materials.

A cement-lime mortar contains considerable cement but little lime. For example, about two sacks of cement to one-half sack of hydrated lime (one-half cubic foot of lime paste or 15 pounds of dry lump lime) would be a cement-lime mortar.

A lime-cement mortar is a mortar containing more lime than the regular cement-lime mortar. It's often a so-called 50-50 mix: a mixture of about 0.8 of a bag of lime (hydrated) to one sack of cement and 5 cubic feet of sand. Stainless or white portland cement is used in all of these mixtures. Ordinary portland cement isn't satisfactory.

When laying ashlar or cut stone masonry, keep

the mortar out of the joints to a depth of about 3/4", or rake it out to that depth. Fill this space with a special mortar to make a tighter, good looking joint. This process is called "pointing." The different kinds of joints formed by pointing are illustrated in Figure 5-26. Pointing mortar is generally made much thicker and richer than ordinary mortar. Nearly as much hydrated lime is used in pointing mortar as in a so-called cement-lime mortar. Less sand is used.

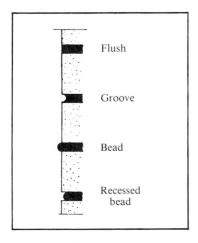

Kinds of stone joints
Figure 5-26

Estimating Chimneys and Fireplaces

The usual way to find the number of brick in a chimney is to find the number of brick per foot of chimney and then multiply by the total height. To find the number of brick required per foot of height, first find out how many brick are required for one course of the chimney and then how many courses there are in 1 foot of height.

Figure 5-27 shows various sizes of chimney and flues and how the brick are bonded in each case. For each different size of flue and chimney, it shows two layers of courses so you can see how the courses will bond. In some cases, one course will require more brick than the alternating course. Chimney (f) is an example. It takes 12 brick for one course and 13 brick for the next course. This makes an average of 12½ brick per course.

Figure 5-28 shows how many courses make up 1 foot in height. The standard brick is 2¼" thick and the average joint is 1/2" thick, making a 2¾" height for every course. In 12" of height, there are 12 divided by 2.75, or 4.36 courses.

The chimney shown in (a) in Figure 5-27 has six brick to each course and requires 6 x 4.36, or 26.16 (round out to 27) brick per foot of height. The chimney (f), which we found above to have an average of 12½ brick per course, requires 12.5 x 4.36, or 54.5 brick per foot in height. Round it to 55.

To find the total brick in a chimney, multiply the brick per foot of height by the total height. Thus, a chimney like type (a), requiring 27 brick per foot of height, built 30 feet high, would require 30 times 27, or 810 brick.

The type of chimneys shown by (m), (n) and (o) in Figure 5-27, are built as part of a wall, so they don't require as many extra brick. Type (m) requires four extra brick for each course. Type (n) requires only two extra brick.

Figure 5-29 gives the number of brick required per foot in height for different size chimneys. The letters refer to the illustrations of the different sizes in Figure 5-27. Standard size brick and 1/2" mortar joints are assumed. The quantity of mortar includes mortar for setting the flue lining.

Figure 5-30 lists the standard flue lining sizes and weights.

How to Estimate Fireplaces

Fireplaces or chimneys of irregular shapes that can't be figured by linear feet of height can be figured by the cubic foot method. Figure the total cubic feet of brickwork required and deduct the openings for flue and ashpit.

Figure 5-31 shows a fireplace requiring face brick on the outside. First, let's see how to figure the common brickwork in this fireplace.

The portion under the first floor is 6 feet by 2 feet, or 12 square feet, as shown in Section A-A. The openings as illustrated by Section A-A are about 1 foot wide. The combined length is about 4 feet. (This is the length of the flue openings combined.) The section area of the opening would be 4 feet by 1 foot, or 4 square feet. The net area of brickwork is 12 square feet minus 4 square feet. That equals 8 square feet.

It's 8 feet high. The total cubic feet of brickwork would be 8 x 8, or 64 cubic feet. Assuming 21 brick per cubic foot, we have 64 x 21, or 1,344 common brick.

The arch carrying the hearth is about 2' x 6', or 12 square feet in area. At seven brick per square foot, it would require 7 x 12, or 84 brick.

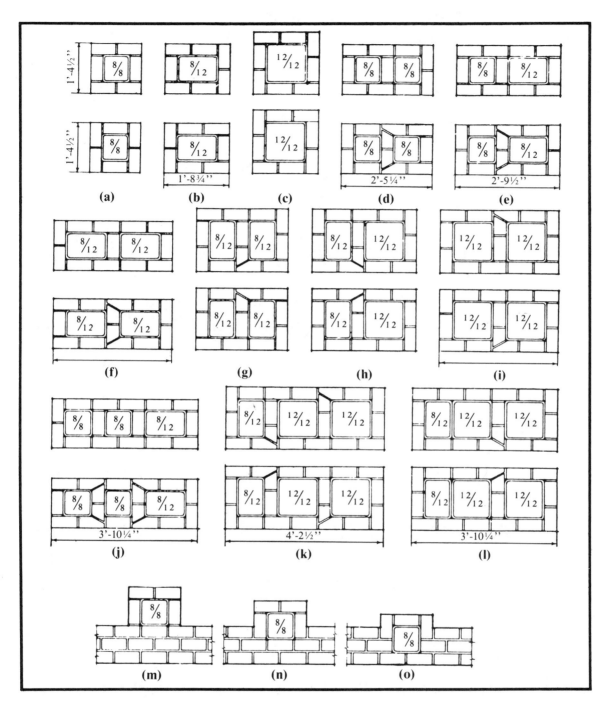

Sizes of chimneys and flues, with their brick bonds
Figure 5-27

Section B-B, C-C and D-D drawn above the first floor show how many brick would be required per course. In Section B-B we see that the outer portion requires face brick. Likewise, the face of the fireplace calls for face brick. So the only part that takes common brick is the filling-in brick and the brick on the ends that will probably be plastered over.

By actual count, we would have 10 to 12 brick per course in Section B-B. In Section C-C we would have about nine brick per course. In Section D-D we also have approximately nine brick per

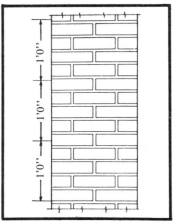

Number of courses in 1 foot of height
Figure 5-28

Size and number of flues	Number of brick	Cubic feet mortar
(a) 1 - 8" x 8" flue	27	0.5
(b) 1 - 8" x 12" flue	31	0.5
(c) 1 - 12" x 12" flue	35	0.6
(d) 2 - 8" x 8" flue	46	0.8
(e) 1 - 8" x 8" and 1 - 8" x 12" flue	51	0.9
(f) 2 - 8" x 12" flue	55	0.10
(g) 2 - 8" x 12" flue	53	0.9
(h) 1 - 8" x 12" and 12" x 12" flue	58	1.0
(i) 2 - 12" x 12" flue	62	1.1
(j) 2 - 8" x 8" and 1 - 8" x 12" flue	70	1.2
(k) 1 - 8" x 12" and 2 - 12" x 12" flue	83	1.4
(l) 1 - 8" x 12" and 2 - 12" x 12" flue	70	1.2
(m) 1 - 8" x 8" extending 12" from face of wall	18	0.4
(n) 1 - 8" x 8" extending 8" from face of wall	9	0.3
(o) 1 - 8" x 8" extending 4" from face of wall	0	0.0

Number of brick for chimneys per foot of height
Figure 5-29

Nominal size of chimney	Outside dimensions of flue linings	Weight per feet	Length of piece
4" x 8"	4½" x 8½"	14 lbs.	2 feet
4" x 13"	4½" x 13"	20 lbs.	2 feet
8" x 8"	8½" x 8½"	18 lbs.	2 feet
8" x 12"	8½" x 13"	30 lbs.	2 feet
8" x 16"	8½" x 18"	36 lbs.	2 feet
12" x 12"	13" x 13"	38 lbs.	2 feet
12" x 16"	13" x 18"	45 lbs.	2 feet
16" x 16"	18" x 18"	62 lbs.	2 feet

Size of flue linings
Figure 5-30

course. An approximate average would be 10 brick per course from the first floor to the top of the roof. This would make 10 x 4.36, or 43.6 brick per foot of height. Round it to 44. The distance from the first floor to the top of the roof is 25 feet. This gives a total of 25 x 44, or 1,100 common brick.

Above the roof, the entire chimney will be built of face brick, except for the portion between the two flues. If we include this in the estimate of common brick, we may assume two brick per course: 2 x 4.36 equals 8.72, or 9 brick per foot of height. The chimney extends 6'0" above the roof. This portion, then, would require 6 x 9, or 54 common brick.

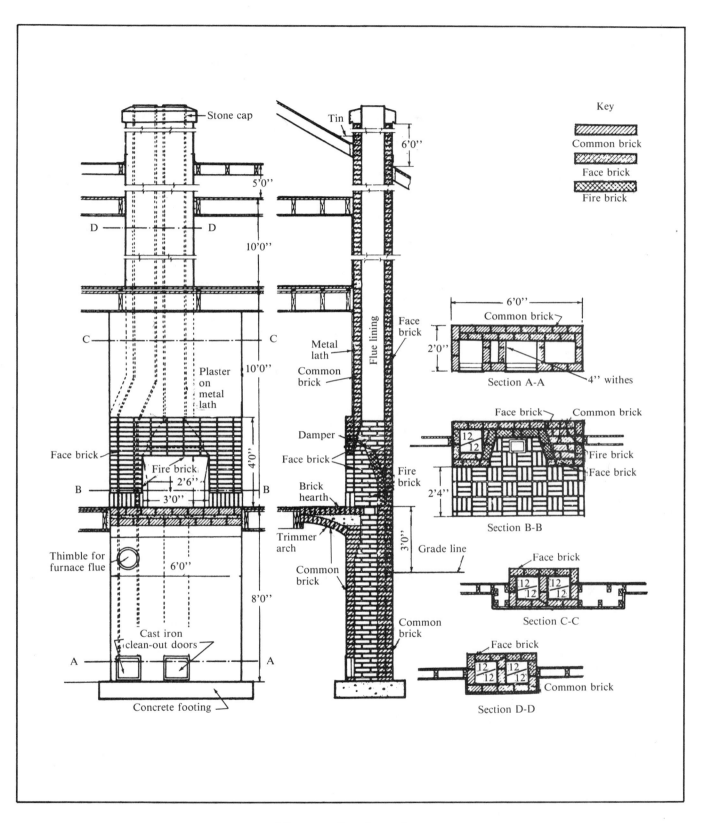

Masonry fireplace plan
Figure 5-31

Adding up the estimate of common brick:

Below the first floor:	1,344
The arch supporting the hearth:	84
First floor to top of the roof:	1,100
Above the roof:	54
Total common bricks	2,582

We'll add a margin for breakage and estimate a total of 2,600 common brick.

In estimating the fire brick required for the lining of the fireplace, we may either figure the square feet of area or calculate them by the courses. Section B-B shows that between six and seven fire brick are required per course. The fireplace opening requiring firebrick is approximately 2'6'' high. At seven brick per course, we have 7 x 4.36, or 30.52 brick per foot of height. Round it to 31. So 31 x 2.5 equals 77.5, or 78 fire brick for the fireplace lining.

Find the brick required for the hearth by the square foot method. The main portion of the hearth is 6'0'' x 2'4''. Since 2'4'' converts to 2⅓', multiply 6 x 2⅓. That equals 14 square feet. The part of the hearth that's in the fireplace is 2'6'' x 1'6'', or 3¾ square feet. So the total square feet of hearth is 14 plus 3¾, or 17¾ square feet. We'll round it up to 18. At seven brick per square foot, we need 7 x 18, or 126 brick for the hearth.

The brick for the front of the fireplace may also be found by the square foot method. The front measures 6'0'' x 4'0'', or 24 square feet. From this deduct the fireplace opening. It's 3'0'' x 2'6'', or 7½ square feet. The total area requiring face brick for the face of the fireplace is 24 less 7½, or 16½ square feet. Make it 17. This would require 17 x 7, or 119 face brick.

The exposed area of the chimney on the outside of the building also requires face brick. The number of face brick required per course is found in Section B-B, Section C-C, and Section D-D. In Section B-B, we find that it requires 10 brick per course. Ten times 4.36 is 43.6, or 44 brick per foot of height. About 10 feet of the chimney is this width up to 10 feet above grade. This portion requires 10 x 44, or 440 face brick. From this point to the roof line is about 18 feet. This part requires face brick as shown in Section C-C and Section D-D. Six brick are required for each course: 6 x 4.36 equals 26.16, or 26½ brick per foot of height.

For 18 feet, it takes 18 x 26½, or 477 face brick.

The part above the roof is 6'0'' and is built as shown in Section D-D. Face brick is required for all four sides of the chimney. So we need 12 brick per course: 12 x 4.36 equals 52.32 or 53 brick per foot of height. This will make 53 x 6, or 318 face brick. The total number of face brick required for the outside of the chimney is:

440
447
318
1,205 face brick

Here's a summary of the brick required for this fireplace:

Common brick:	2,600
Fire brick for lining:	78
Face brick for hearth:	126
Face brick for fireplace face:	119
Face brick for outside of chimney:	1,205

Labor for Fireplaces

A bricklayer and laborer will lay an average of 500 common brick a day on short chimneys. If they extend considerably above the roof and scaffolds are needed, reduce the estimate by about 25%.

Estimate flue lining for chimneys by the linear foot. It's furnished in 2-foot lengths. A mason will place about 30 pieces of 8'' x 8'' flue lining or 60 linear feet per eight-hour day; estimate about 25 pieces of 12'' x 12'' and 18 pieces of 18'' x 18''.

It's hard to estimate face brick for a fireplace by the thousand. If the fireplace is elaborate, with panels, or laid in Flemish bond, a mason will lay fewer brick per hour. A mason should complete an ordinary fireplace containing 200 to 225 face brick in about eight hours.

Lining the back of the fireplace with fire brick will take about four hours of mason work. If there's only one mason working on the job, he'll need one laborer to help him. Only one-half the time is required for laborers if there are several masons on the job.

If brick hearths and backhearths are laid in square or basketweave pattern, a mason should finish one backhearth in two to three hours. The front hearth will take about the same amount of time, or four to six hours for both. If laid in herringbone pattern with cut brick at all edges, it will take three to four hours mason time for each, or six to eight mason hours for both.

Labor on Arches
Allow extra time for brickwork if there are a number of large brick arches which have to be laid with wood centers. Arches with a diameter of 4 to 6 feet will take four to six hours to build. Arches 7 to 8 feet in diameter will require about eight hours of masonry labor.

The material and manhour estimates in this chapter may or may not apply on your jobs. If you have no other figures, use the estimates on the preceding pages. They're based on BIA Technical Notes No. 10 and BIA Brick Builder Notes No. 3 and will apply on many jobs. Your work may be different. Modify the figures I've given here once you have enough data to make comparisons.

In the next chapter, we'll move on to the next stage of construction, rough carpentry.

Chapter 6

Estimating Rough Carpentry

Rough carpentry includes all the framing and sheathing on any building. It's the part of the building that's usually not exposed to view — but don't think that because it's hidden it's all right to do sloppy work. The strength of the building depends on the quality and integrity of the "skeleton," the framing. So choose the lumber and the framing method carefully. Insist on quality work by your carpenters. And take pride in the houses you build.

Lumber Grading

Depending on the species, framing lumber and plywood are graded according to rules established by various associations, inspection authorities, and government agencies. Some of these groups both define the rules under which a species will be graded and provide certified inspectors to inspect the wood at the mill. These inspectors grade stamp each piece or certify grades by lot. Some grade stamps or certifications not only identify the grade,

	Thickness	Width
Board lumber	1"	2" or more
Light framing	2" to 4"	2" to 4"
Studs	2" to 4"	2" to 6", 10' and shorter
Structural light framing	2" to 4"	2" to 4"
Structural joists and planks	2" to 4"	5" and wider
Beams and stringers	5" and thicker	More than 2" greater than thickness
Posts and timbers	5" x 5" and larger	Not more than 2" greater than thickness
Decking	2" to 4"	4" to 12" wide
Siding		Thickness expressed by dimension of butt edge
Moldings		Size at thickest and widest points

Lengths of lumber generally are 6' and longer in multiples of 2'.
Source: Western Wood Products Association

Typical lumber product classifications
Figure 6-1

The thicknesses apply to all widths.

Thicknesses			Face Widths		
Nominal	Minimum Dressed		Nominal	Minimum Dressed	
	Dry* Inches	Green* Inches		Dry* Inches	Green* Inches
BOARDS**					
			2	1½	1 9/16
			3	2½	2 9/16
1	¾	25/32	4	3½	3 9/16
			5	4½	4⅝
			6	5½	5⅝
			7	6½	6⅝
1¼	1	1 1/32	8	7¼	7½
			9	8¼	8½
			10	9¼	9½
			11	10¼	10½
			12	11¼	11½
1½	1¼	1 9/32	14	13¼	13½
			16	15¼	15½
DIMENSION					
			2	1½	1 9/16
2	1½	1 9/16	3	2½	2 9/16
			4	3½	3 9/16
2½	2	2 1/16	5	4½	4⅝
			6	5½	5⅝
3	2½	2 9/16	8	7¼	7½
			10	9¼	9½
			12	11¼	11½
3½	3	3 1/16	14	13¼	13½
			16	15¼	15½
DIMENSION					
			2	1½	1 9/16
			3	2½	2 9/16
4	3½	3 9/16	4	3½	3 9/16
			5	4½	4⅝
			6	5½	5⅝
			8	7¼	7½
			10	9¼	9½
4½	4	4 1/16	12	11¼	11½
			14		13½
			16		15½
TIMBERS					
5 & thicker	–	½ off	5 & wider	–	½ off

*These are minimum dressed sizes. American Softwood Lumber Standard, PS 20-70.
**Boards less than the minimum thickness for 1 inch nominal but ¼ inch or greater thickness dry (11/16 inch green) may be regarded as American Standard Lumber, but such boards shall be marked to show the size and condition of seasoning at the time of dressing. They shall also be distinguished from 1-inch boards on invoices and certificates.

Nominal size chart for softwood lumber
Figure 6-2

but also the species, mill number, intended use, and whether the piece was dry to within a specified moisture content when surfaced. Since grading is based on visual inspection and the judgment of the grader, however, it isn't an exact science.

Framing or structural lumber is graded according to how well it will perform an intended use. Major consideration is given to possible defects that affect strength. When structural or framing members will be exposed, select them for both strength and appearance.

Finish lumber for siding, paneling and other finishing work is graded on the basis of appearance. Items that'll be clear-finished require a higher appearance grade than those you'll cover with opaque paint. Finish grade lumber is milled in 2 inches or less nominal thickness.

Laminated members are graded on both strength and appearance. They may also be graded for strength and encased in appearance-grade lumber.

Appearance-grade structural lumber is expensive. And any piece that warrants a high-appearance grade will generally be as strong as any piece in that species. But you have to be careful. Most appearance grades are judged from one side only. Serious defects on the reverse side might make it unsuitable as a structural member. Check *both* sides before installing appearance-grade lumber where strength is also important.

Figure 6-1 shows typical lumber product classifications. Figure 6-2 shows nominal and actual sizes for softwood lumber.

American Softwood Lumber Standard, PS 20-70

The American Softwood Lumber Standard, PS 20-70, establishes a uniform set of rules for grading framing lumber. Before PS 20-70, every regional species had its own set of regional rules and standards. These standards took into consideration the unique characteristics of the species and the expected use in that particular region. This didn't create any problems, as long as you bought your lumber from the local mill. But lumber purchased from different areas or different mills wasn't comparable. You needed a set of span tables for each grade on each job.

PS 20-70 changed that. It established a National Grading Rule Committee "to maintain and make fully and fairly available grade strength ratios, nomenclature, and descriptions of grades for dimension lumber." These grades and grade requirements, as developed, are now used by all regional grade-writing agencies. Nearly all framing lumber is graded by grade-writing agencies that are certified by the American Lumber Standards Committee.

The National Grading Rule separates dimension lumber into *width* categories. Pieces up to 4 inches wide are graded as *Structural Light Framing*, *Light Framing* and *Studs*. Pieces 6 inches and wider are graded as *Structural Joists* and *Planks*. Figure 6-3 shows these dimension lumber grades.

Where a fine appearance and high bending strength are required, the National Grading Rule also provides an *Appearance Framing* grade.

Structural Light Framing grades are for those uses where higher bending strength is required. The four grades included in this category are *Sel Str* (Select Structural), *No. 1, No. 2,* and *No. 3.*

Light Framing grades are used where good appearance at a lower design level is satisfactory. Grades in this category are called *Const* (Construction), *Std* (Standard), and *Util* (Utility).

A single *Stud* grade is also provided under the National Grading Rule. It's intended specifically for use as a vertical bearing member in walls and partitions to a maximum height of 10 feet.

Structural Joist and Plank grades are available in widths 6 inches and wider for use as joists, rafters, headers, built-up beams, and similar uses. Grades in this category are *Sel Str* (Select Structural), *No. 1, No. 2,* and *No. 3.*

Not all grades described in the National Grading Rule and listed in Figure 6-3 will be available in all species or all regions. The Sel Str and No. 1 grades are frequently used for truss construction when high strength is required. For general construction, you'll use No. 2, No. 3 and Btr, or Std and Btr. Use No. 3 and Util grades where less strength is needed.

Lumber that's embedded in the ground or often exposed to damp conditions must be cut from naturally durable species or pressure treated to stop

2"- 4" thick, 2"- 4" wide	
Structural light framing	Sel Str (Select Structural) No. 1 No. 2 No. 3
Studs	Stud
Light framing	Const (Construction) Std (Standard) Util (Utility)

2"- 4" thick, 6" and wider	
Structural joists and planks	Sel Str (Select Structural) No.1 No. 2 No. 3

2"- 4" thick, 2" and wider	
Appearance framing	A (Appearance)

**Dimension lumber grades
(National Grading Rule)
Figure 6-3**

deterioration. This also includes lumber used for sills resting on a concrete slab which is in direct contact with the earth, and joists which are closer than 18 inches to the ground.

The durable species most frequently used are California redwood, western red cedar and tidewater red cypress. But not all redwood and red cedar are adequate for these purposes. Be sure to specify foundation grades of redwood and red cedar. In cypress, select a heart structural grade.

Milling Lumber

The sawmill first cuts the logs into rough-sawn lumber. Then it's air or kiln dried before being milled into its final form or dimension. If it's surfaced when still green, it will shrink after it's milled, resulting in uneven dimensions.

Most framing lumber is surfaced on four sides (S4S). But it might be unsurfaced, or surfaced on one side and one edge (S1S1E) or other combinations such as S2S (surfaced two sides). Surfacing of lumber is more for uniformity in width and thickness than for appearance.

Resawn lumber refers to surfaced lumber that has been split, generally by a band saw, into roughly equal thickness, or lumber that has a saw texture on one surface of a milled piece. Resawn lumber will not generally have the thickness uniformity of S1S lumber, which is surfaced on one side after rough sawing. Rough and resawn lumber may be specified band sawn or circular sawn, depending on the texture desired.

Lumber Characteristics

As a construction estimator, you should be familiar with these lumber characteristics:

• Vertical grain refers to the run of the annual growth rings to the flat side or face of a wood member. Wood cut with the growth rings at 45 to 90 degrees to the long dimension face (side) is called *quarter sawn* or *vertical grain*. Wood sawed with the growth rings at an angle of 0 to 45 degrees to the side of the piece is called *plain sawed* or *flat grain*. Don't confuse vertical grain with straight grain, a term used to indicate that the wood fibers (or growth rings) are parallel to the long dimension of the piece.

• Trees grow rapidly in the spring. As the season progresses, the growth rate slows until it virtually stops in the fall. Spring growth is characterized by soft, light fibers, and summer growth by hard, dense fibers. Closely spaced annual growth rings indicate slow growth, typical of dense virgin forests where trees must compete for sunlight and nutrients. Lack of sunlight to the lower limbs causes them to die and fall off. This is called "natural pruning." Wood from these trees is strong, dense and relatively knot-free. Widely spaced annual growth rings indicate rapid growth, typical of second growth stands and managed tree farms. Wood from these trees is less dense, less strong, and tends to have more knots. The majority of lumber milled today is from rapid growth trees.

• Wood shrinks as it loses moisture, and swells as it absorbs moisture. This change in volume is most pronounced in the direction of the annual growth rings, less in the direction of the rays (across the rings), and very little in the direction parallel to the grain. Initial shrinkage takes place when the natural free water is removed from the cell cavities

and intercellular spaces of the wood by air or kiln drying.

• Air-dried lumber has many advantages over green lumber: reduced weight and shrinkage, less checking and warping, increased strength and nail holding properties, less susceptibility to the attack of fungi and insects, improved paint-holding capability, and better absorption of preservatives. (During peak building activity, the lumber shipped to your jobs may be greener than lumber shipped in fall or winter months. In fact, I've seen lumber arrive at the job that looked like it was still growing!)

• Kiln-dried lumber has these advantages over air-dried: less weight, more uniform drying that reduces warping, a shorter drying time, far less fungi and insects, resins that are fully set, and reduced loss of quality during seasoning.

Expressing Dimensions

There's a standard way of expressing the size of lumber on estimates: Give the dimensions of the cross section first, then the length in feet. For example, a joist written as 2 x 8 x 16'0" has a nominal cross section of 2 x 8, and is 16'0" long.

Board Feet

All rough lumber is sold by the number of board feet. A board foot is a piece of lumber 1 inch thick, 12 inches wide and 12 inches long. The term is abbreviated *B.F.* or *BF*. You can easily calculate the number of board feet in a piece of lumber by writing the standard dimensions over 12, in the form of a fraction. For example, the board feet in a joist 2 x 8 x 16'0" equals:

$$\frac{2 \times 8 \times 16}{12} = \frac{256}{12} = 21\text{-}1/3 \text{ BF}$$

When more than one piece is needed, simply multiply the amount found in one piece by the number of pieces. For example, if there are 40 joists 2 x 8 x 16'0", the number of board feet would be:

$$40 \times \frac{2 \times 8 \times 16}{12} = 853\text{-}1/3 \text{ BF}$$

For every linear foot of a 2 x 8 piece, there are 1⅓ board feet. For every linear foot of 2 x 6, there is 1 board foot. For every linear foot of 2 x 4, there is 2/3 board foot.

It's easy to see how we got these numbers:

$$\frac{2 \times 8}{12} = 1\text{-}1/3$$

$$\frac{2 \times 6}{12} = 1$$

$$\frac{2 \times 4}{12} = 2/3$$

Figure 6-4 is a chart with the board feet figured out for different sizes of lumber. It also shows how to do the board foot conversions for these and other sizes of boards. Use it as a short-cut method for figuring board feet. For example, find the number of board feet in a piece of 1 x 8 x 14'0":

$$14 \times 2/3 = 9\text{-}1/3 \text{ BF}$$

Figure 6-4 shows that 1 foot of 2 x 4 contains 2/3 BF. To find the number of board feet in a 14' piece, multiply 14 by 2/3. Or just look it up in the chart.

Thousand Board Feet
The cost of framing lumber is usually figured by the 1,000 board feet, which is abbreviated *M.B.F.* or *MBF*. To find the cost per board foot when you know the cost per MBF, just divide the cost by 1,000. Let's try it. Find the cost of 40 pieces of 2 x 4 x 16'0" at $325.00 per MBF.

$$\frac{40 \times 2 \times 4 \times 16}{12} = 426.66 \text{ BF}$$

$$\frac{\$325}{1,000} = \$0.325 \text{ per BF}$$

$$426.66 \times \$0.325 = \$138.66 \text{ for } 426\text{-}2/3 \text{ BF}$$

If a certain job requires a given number of board feet of lumber and you know the price, you'll first have to find the number of pieces of each size required. For example, find the number of pieces of 2 x 6 x 20'0" which should be delivered in an order of 2,500 BF:

$$\frac{2 \times 6 \times 20}{12} = 20 \text{ BF in each piece}$$

$$\frac{2,500}{20} = 125 \text{ pieces}$$

Linear Feet
Smaller size lumber is generally figured in linear feet. If an estimate calls for 135 board feet of 1 x 3 cross-bridging and you need to order it by linear feet, here's how to convert it:

$$\frac{1 \times 3}{12} = \frac{1}{4} \text{ BF in each linear foot of material}$$
$$135 \text{ divided by } \frac{1}{4} = 135 \times \frac{4}{1} = 540 \text{ linear feet}$$

Taking Off Lumber Estimates
When taking off quantities from a plan, follow the sequence of construction: foundation, floor, walls, roof. This reduces the chance of omitting something. In practice, you'll probably group together all lumber of the same description on the estimate sheet so that the cost and labor can be calculated quickly.

Always use a checklist when taking off material quantities. No one can be expected to remember the hundreds of items that go into most construction projects. Even if you have an excellent memory, the penalty for forgetting something is too great. Start with a checklist something like Figure 6-5. Modify and improve it until it's perfected for the types of work you handle.

You'll probably want separate checklists for different kinds of jobs. There can be a list for ordinary frame buildings, another for brick and masonry buildings, another for reinforced concrete buildings. But there's a limit on how much faith you can put in a simple reminder list. The list is just that: a reminder. The plans and specifications govern what items must be included. Somewhere on your checklist it should say: "Read the plans and specs carefully."

Board Feet Content

Size of Timber In Inches	10'	12'	14'	16'	18'	20'	22'	24'
1 x 2	1⅔	2	2⅓	2⅔	3	3⅓	3⅔	4
1 x 3	2½	3	3½	4	4½	5	5½	6
1 x 4	3⅓	4	4⅔	5⅓	6	6⅔	7⅓	8
1 x 5	4⅙	5	5⅚	6⅔	7½	8⅓	9⅙	10
1 x 6	5	6	7	8	9	10	11	12
1 x 8	6⅔	8	9⅓	10⅔	12	13⅓	14⅔	16
1 x 10	8⅓	10	11⅔	13⅓	15	16⅔	18⅓	20
1 x 12	10	12	14	16	18	20	22	24
1 x 14	11⅔	14	16⅓	18⅔	21	23⅓	25⅔	28
1 x 16	13⅓	16	18⅔	21⅓	24	26⅔	29⅓	32
1 x 20	16⅔	20	23⅓	26⅔	30	33⅓	36⅔	40
1¼ x 4	4⅙	5	5⅚	6⅔	7½	8⅓	9⅙	10
1¼ x 6	6¼	7½	8¾	10	11¼	12½	13¾	15
1¼ x 8	8⅓	10	11⅔	13⅓	15	16⅔	18⅓	20
1¼ x 10	10⅓	12½	14½	16⅔	18⅔	20⅚	22⅚	25
1¼ x 12	12½	15	17½	20	22½	25	27½	30
1½ x 4	5	6	7	8	9	10	11	12
1½ x 6	7½	9	10½	12	13½	15	16½	18
1½ x 8	10	12	14	16	18	20	22	24
1½ x 10	12½	15	17½	20	22½	25	27½	30
1½ x 12	15	18	21	24	27	30	33	36
2 x 4	6⅔	8	9⅓	10⅔	12	13⅓	14⅔	16
2 x 6	10	12	14	16	18	20	22	24
2 x 8	13⅓	16	18⅔	21⅓	24	26⅔	29⅓	32
2 x 10	16⅔	20	23⅓	26⅔	30	33⅓	36⅔	40
2 x 12	20	24	28	32	36	40	44	48
2 x 14	23⅓	28	32⅔	37⅓	42	46⅔	51⅓	56
2 x 16	26⅔	32	37½	42⅔	48	53⅓	58⅔	64
2½ x 12	25	30	35	40	45	50	55	60
2½ x 14	29⅙	35	40⅚	46⅔	52½	58⅓	64⅙	70
2½ x 16	33⅓	40	46⅔	53⅓	60	66⅔	73⅓	80
3 x 6	15	18	21	24	27	30	33	36
3 x 8	20	24	28	32	36	40	44	48
3 x 10	25	30	35	40	45	50	55	60
3 x 12	30	36	42	48	54	60	66	72
3 x 14	35	42	49	56	63	70	77	84
3 x 16	40	48	56	64	72	80	88	96
4 x 4	13⅓	16	18⅔	21⅓	24	26⅔	29⅓	32
4 x 6	20	24	28	32	36	40	44	48
4 x 8	26⅔	32	17⅓	42⅔	48	53⅓	58⅔	64
4 x 10	33⅓	40	46⅔	53⅓	60	66⅔	73⅓	80
4 x 12	40	48	56	64	72	80	88	96
4 x 14	46⅓	56	65⅓	74⅔	84	93⅓	102½	112

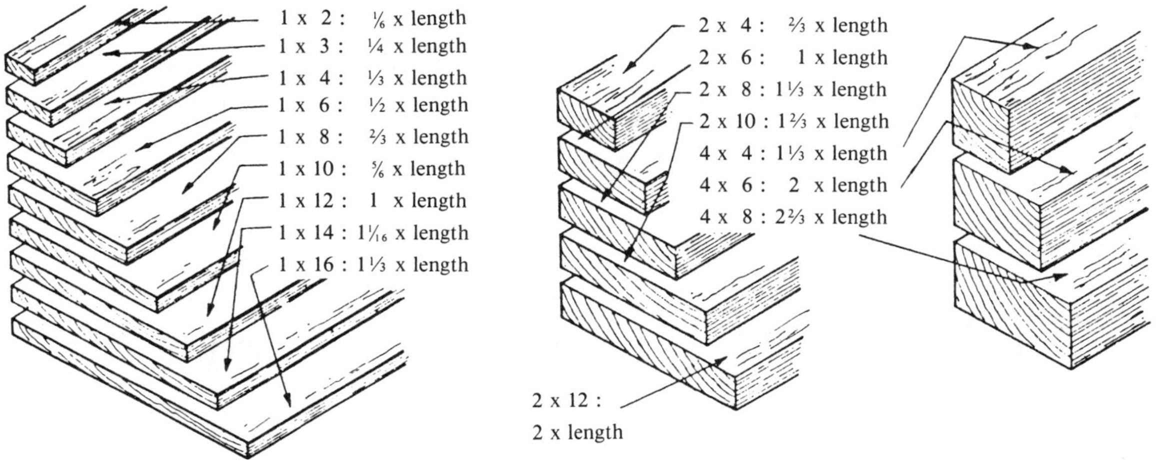
1 x 2 : ⅙ x length
1 x 3 : ¼ x length
1 x 4 : ⅓ x length
1 x 6 : ½ x length
1 x 8 : ⅔ x length
1 x 10 : ⅚ x length
1 x 12 : 1 x length
1 x 14 : 1 1/16 x length
1 x 16 : 1⅓ x length
2 x 4 : ⅔ x length
2 x 6 : 1 x length
2 x 8 : 1⅓ x length
2 x 10 : 1⅔ x length
4 x 4 : 1⅓ x length
4 x 6 : 2 x length
4 x 8 : 2⅔ x length
2 x 12 :
2 x length

Converting linear feet to board feet
Figure 6-4

Excavation
- [ ] Backfilling
- [ ] Clearing the site
- [ ] Compacting
- [ ] Dump fee
- [ ] Equipment rental
- [ ] Equipment transport
- [ ] Establishing new grades
- [ ] General excavation
- [ ] Hauling to dump
- [ ] Pit excavation
- [ ] Pumping
- [ ] Relocating utilities
- [ ] Removing obstructions
- [ ] Shoring
- [ ] Stripping topsoil
- [ ] Trenching

Demolition
- [ ] Cabinet removal
- [ ] Ceiling finish removal
- [ ] Concrete cutting
- [ ] Debris box
- [ ] Door removal
- [ ] Dump fee
- [ ] Dust partition
- [ ] Electrical removal
- [ ] Equipment rental
- [ ] Fixtures removal
- [ ] Flooring removal
- [ ] Framing removal
- [ ] Hauling to dump
- [ ] Masonry removal
- [ ] Plumbing removal
- [ ] Roofing removal
- [ ] Salvage value allowance
- [ ] Siding removal
- [ ] Slab breaking
- [ ] Temporary weather protection
- [ ] Wall finish removal
- [ ] Window removal

Concrete
- [ ] Admixtures
- [ ] Anchors
- [ ] Apron
- [ ] Caps
- [ ] Cement
- [ ] Columns
- [ ] Crushed stone
- [ ] Curbs
- [ ] Curing
- [ ] Drainage
- [ ] Equipment rental
- [ ] Expansion joints
- [ ] Fill
- [ ] Finishing
- [ ] Floating
- [ ] Footings
- [ ] Foundations
- [ ] Grading
- [ ] Gutters
- [ ] Handling

- [ ] Mixing
- [ ] Piers
- [ ] Ready mix
- [ ] Sand
- [ ] Screeds
- [ ] Slabs
- [ ] Stairs
- [ ] Standby time
- [ ] Tamping
- [ ] Topping
- [ ] Vapor barrier
- [ ] Waterproofing

Forms
- [ ] Braces
- [ ] Caps
- [ ] Cleaning for reuse
- [ ] Columns
- [ ] Equipment rental
- [ ] Footings
- [ ] Foundations
- [ ] Key joints
- [ ] Layout
- [ ] Nails
- [ ] Piers
- [ ] Salvage value
- [ ] Slab
- [ ] Stair
- [ ] Stakes
- [ ] Ties
- [ ] Walers
- [ ] Wall

Reinforcing
- [ ] Bars
- [ ] Handling
- [ ] Mesh
- [ ] Placing
- [ ] Tying

Masonry
- [ ] Arches
- [ ] Backing
- [ ] Barbecues
- [ ] Cement
- [ ] Chimney
- [ ] Chimney cap
- [ ] Cleaning
- [ ] Clean-out doors
- [ ] Dampers
- [ ] Equipment rental
- [ ] Fireplace
- [ ] Fireplace form
- [ ] Flashing
- [ ] Flue
- [ ] Foundation
- [ ] Glass block
- [ ] Handling
- [ ] Hearths
- [ ] Laying
- [ ] Lime
- [ ] Lintels
- [ ] Mantels
- [ ] Marble

- [ ] Mixing
- [ ] Mortar
- [ ] Paving
- [ ] Piers
- [ ] Repair
- [ ] Reinforcing
- [ ] Repointing
- [ ] Sand
- [ ] Sandblasting
- [ ] Sills
- [ ] Steps
- [ ] Stonework
- [ ] Tile
- [ ] Veneer
- [ ] Vents
- [ ] Wall ties
- [ ] Walls
- [ ] Waterproofing

Rough Carpentry
- [ ] Area walls
- [ ] Backing
- [ ] Beams
- [ ] Blocking
- [ ] Bracing
- [ ] Bridging
- [ ] Building paper
- [ ] Columns
- [ ] Cornice
- [ ] Cripples
- [ ] Door frames
- [ ] Dormers
- [ ] Entrance hoods
- [ ] Fascia
- [ ] Fences
- [ ] Flashing
- [ ] Framing clips
- [ ] Furring
- [ ] Girders
- [ ] Gravel stop
- [ ] Grounds
- [ ] Half timber work
- [ ] Hangers
- [ ] Headers
- [ ] Hip jacks
- [ ] Insulation
- [ ] Jack rafters
- [ ] Joists, ceiling
- [ ] Joists, floor
- [ ] Ledgers
- [ ] Nails
- [ ] Outriggers
- [ ] Pier pads
- [ ] Plates
- [ ] Porches
- [ ] Posts
- [ ] Rafters
- [ ] Ribbons
- [ ] Ridges
- [ ] Roof edging
- [ ] Roof trusses
- [ ] Rough frames
- [ ] Rough layout

Material estimate checklist
Figure 6-5

☐ Scaffolding
☐ Sheathing, roof
☐ Sheathing, wall
☐ Sills
☐ Sleepers
☐ Soffit
☐ Stairs
☐ Straps
☐ Strong backs
☐ Studs
☐ Subfloor
☐ Timber connectors
☐ Trimmers
☐ Valley flashing
☐ Valley jacks
☐ Vents
☐ Window frames

Finish Carpentry
☐ Baseboard
☐ Bath accessories
☐ Belt course
☐ Built-ins
☐ Cabinets
☐ Casings
☐ Caulking
☐ Ceiling tile
☐ Closet doors
☐ Closets
☐ Corner board
☐ Cornice
☐ Counter tops
☐ Cupolas
☐ Door chimes
☐ Door hardware
☐ Door jambs
☐ Door stop
☐ Door trim
☐ Doors
☐ Drywall
☐ Entrances
☐ Fans
☐ Flooring
☐ Frames
☐ Garage doors
☐ Hardware
☐ Jambs
☐ Linen closets
☐ Locksets
☐ Louver vents
☐ Mail slot
☐ Mantels
☐ Medicine cabinets
☐ Mirrors
☐ Molding
☐ Nails
☐ Paneling
☐ Rake
☐ Range hood
☐ Risers
☐ Roofing
☐ Room dividers

☐ Sash
☐ Screen doors
☐ Screens
☐ Shelving
☐ Shutters
☐ Siding
☐ Sills
☐ Sliding doors
☐ Stairs
☐ Stops
☐ Storm doors
☐ Threshold
☐ Treads
☐ Trellis
☐ Trim
☐ Vents
☐ Wallboard
☐ Watertable
☐ Window trim
☐ Wardrobe closets
☐ Weatherstripping
☐ Windows

Flooring
☐ Adhesive
☐ Asphalt tile
☐ Carpet
☐ Cork tile
☐ Flagstone
☐ Hardwood
☐ Linoleum
☐ Marble
☐ Nails
☐ Pad
☐ Rubber tile
☐ Seamless vinyl
☐ Slate
☐ Tack strip
☐ Terrazzo
☐ Tile
☐ Vinyl tile
☐ Wood flooring

Plumbing
☐ Bathtubs
☐ Bar sink
☐ Couplings
☐ Dishwasher
☐ Drain lines
☐ Dryers
☐ Faucets
☐ Fittings
☐ Furnace hookup
☐ Garbage disposers
☐ Gas service lines
☐ Hanging brackets
☐ Hardware
☐ Laundry trays
☐ Lavatories
☐ Medicine cabinets
☐ Pipe
☐ Pumps

☐ Septic tank
☐ Service sinks
☐ Sewer lines
☐ Sinks
☐ Showers
☐ Stack extension
☐ Supply lines
☐ Tanks
☐ Valves
☐ Vanity cabinets
☐ Vent stacks
☐ Washers
☐ Waste lines
☐ Water closets
☐ Water heaters
☐ Water meter
☐ Water softeners
☐ Water tank
☐ Water tap

Heating
☐ Air conditioning
☐ Air return
☐ Baseboard
☐ Bathroom
☐ Blowers
☐ Collars
☐ Dampers
☐ Ducts
☐ Electric service
☐ Furnaces
☐ Gas lines
☐ Grilles
☐ Hot water
☐ Infrared
☐ Radiant cable
☐ Radiators
☐ Registers
☐ Relocation of system
☐ Thermostat
☐ Vents
☐ Wall units

Roofing
☐ Adhesive
☐ Asbestos
☐ Asphalt shingles
☐ Built-up
☐ Canvas
☐ Caulking
☐ Concrete
☐ Copper
☐ Corrugated
☐ Downspouts
☐ Felt
☐ Fiberglass shingles
☐ Flashing
☐ Gravel
☐ Gutters
☐ Gypsum
☐ Hip units
☐ Insulation

Material estimate checklist
Figure 6-5 (continued)

☐ Nails
☐ Ridge units
☐ Roll roofing
☐ Scaffolding
☐ Shakes
☐ Sheet metal
☐ Slate
☐ Tile
☐ Tin
☐ Vents
☐ Wood shingles

Sheet Metal
☐ Access doors
☐ Caulking
☐ Downspouts
☐ Ducts
☐ Flashing
☐ Gutters
☐ Laundry chutes
☐ Roof flashing
☐ Valley flashing
☐ Vents

Electrical Work
☐ Air conditioning
☐ Appliance hook-up
☐ Bell wiring
☐ Cable
☐ Ceiling fixtures
☐ Circuit breakers
☐ Circuit load adequate
☐ Clock outlet
☐ Conduit
☐ Cover plates
☐ Dimmers
☐ Dishwashers
☐ Dryers
☐ Fans
☐ Fixtures
☐ Furnaces
☐ Garbage disposers
☐ High voltage line
☐ Hood hook-up
☐ Hook-up
☐ Lighting
☐ Meter boxes
☐ Ovens
☐ Panel boards
☐ Plug outlets
☐ Ranges
☐ Receptacles
☐ Relocation of existing lines
☐ Service entrance
☐ Switches
☐ Switching
☐ Telephone outlets
☐ Television wiring
☐ Thermostat wiring
☐ Transformers
☐ Vent fans
☐ Wall fixtures
☐ Washers

☐ Water heaters
☐ Wire

Plastering
☐ Bases
☐ Beads
☐ Cement
☐ Coloring
☐ Cornerite
☐ Coves
☐ Gypsum
☐ Keene's cement
☐ Lath
☐ Lime
☐ Partitions
☐ Sand
☐ Soffits

Painting and Decorating
☐ Aluminum paint
☐ Cabinets
☐ Caulking
☐ Ceramic tile
☐ Concrete
☐ Doors
☐ Draperies
☐ Filler
☐ Finishing
☐ Floors
☐ Masonry
☐ Paperhanging
☐ Paste
☐ Roof
☐ Sandblasting
☐ Shingle stain
☐ Stucco
☐ Wallpaper removal
☐ Windows
☐ Wood

Glass and Glazing
☐ Breakage allowance
☐ Crystal
☐ Hackout
☐ Insulating glass
☐ Mirrors
☐ Obscure
☐ Ornamental
☐ Plate
☐ Putty
☐ Reglaze
☐ Window glass

Indirect Costs
☐ Barricades
☐ Bid bond
☐ Builder's risk insurance
☐ Building permit fee
☐ Business license
☐ Cleaning floor
☐ Cleaning glass
☐ Clean-up
☐ Completion bond

☐ Debris removal
☐ Design fee
☐ Equipment floater insurance
☐ Equipment rental
☐ Estimating fee
☐ Expendable tools
☐ Field supplies
☐ Job shanty
☐ Job phone
☐ Job signs
☐ Liability insurance
☐ Maintenance bond
☐ Patching after subcontractors
☐ Payment bond
☐ Plan checking fee
☐ Plan cost
☐ Protecting adjoining property
☐ Protection during construction
☐ Removing utilities
☐ Repairing damage
☐ Sales commission
☐ Sales taxes
☐ Sewer connection fee
☐ State contractor's license
☐ Street closing fee
☐ Street repair bond
☐ Supervision
☐ Survey
☐ Temporary electrical
☐ Temporary fencing
☐ Temporary heating
☐ Temporary lighting
☐ Temporary toilets
☐ Temporary water
☐ Transportation equipment
☐ Travel expense
☐ Watchman
☐ Water meter fee
☐ Waxing floors

Administrative Overhead
☐ Accounting
☐ Advertising
☐ Automobiles
☐ Depreciation
☐ Donations
☐ Dues and subscriptions
☐ Entertaining
☐ Interest
☐ Legal fees
☐ Licenses and fees
☐ Office insurance
☐ Office phone
☐ Office rent
☐ Office salaries
☐ Office utilities
☐ Pensions
☐ Postage
☐ Profit sharing
☐ Repairs
☐ Small tools
☐ Taxes
☐ Uncollectable accounts

Material estimate checklist
Figure 6-5 (continued)

Here's a schedule of carpentry work that will apply to many frame buildings:

Basement: Posts, girders, plates.

Floors: Joists, bridging, subfloors, insulation, finish floors.

Exterior wall framing: Sill, bottom and top plates, studs, door and window frames, ribbons, bracing, firestops, nails, sheathing, insulation.

Partitions: Bottom and top plates, studs, firestops, nails, door frames.

Roof framing: Plates (if separate from plates over studs), trusses or ceiling joists, rafters, collar beams, bracing, sheathing, building paper, nails, insulation, gable end studs.

Exterior wall covering: Building paper, sheathing, siding, shingles, furring for stucco, lath for stucco, grounds for stucco, nails.

Roof covering: Shingles or other covering and flashing.

Exterior trim: Water table, corner boards, cornice, molding, brackets, doors, windows, blinds and shutters, nails.

Porch work: Framing, subfloors, finish floors, columns, cornice, ceiling, brackets, rails, balusters, steps.

Rough interior work: Lath, grounds, furring, wall or plasterboard, rough stairs.

One of the most important steps in estimating is arranging and listing the items in a comprehensive and systematic way. Mark and place all the items on the estimate sheet so you can find them at any time without going through the whole plan or estimate sheet.

Often, changes are made in plans which will require refiguring a portion of the work. If items are clearly marked, they can be revised without going through the entire estimate. The estimate should be so written that anyone can check each item with the actual amount of material and the cost for that particular part of the work. A well-arranged estimate sheet saves money and time. Figure 6-6 is a form you can use as an estimate sheet. Use your office copier or have a print shop print up a few hundred and bind them into pads.

You'll always want to include at least the following information on a material list for rough carpentry work:

1)   The number of pieces of each size

2)   The thickness of the lumber

3)   The width of each piece

4)   The length of each piece

5)   The grade of lumber

6)   The kind of wood

7)   Whether it's rough, partly smooth, or surfaced on all four sides.

The most common errors made by estimators are crowding their estimates, and not making each item clear. Allow enough space for totaling figures. Clearly mark each item with size and amount. This is a slow and tedious process, but it must be done right if you (or anyone else who looks at the estimate) want a clear picture of what's required for the job.

Estimating Posts, Girders and Sills

Find the number of posts in the basement on the foundation plan. The length of the posts may vary and the size depends on the load to be carried. Multiply the average length of each post by the number of posts. Order that many linear feet of lumber in lengths which will cut without waste. In some cases cast iron, steel or reinforced concrete posts are used. The specs should define what material is required.

Wood girders that support the first floor joists are usually shown by dotted lines on the basement plan. Wood girders may be built-up, solid or laminated. Size depends on the load to be carried and the spacing. Find the length of girders by scaling off the distance between posts with an architect's scale.

When putting these items on your estimate sheet, give the length and size of each piece. Common sizes of wooden posts are 6 x 6, 6 x 8, and 8 x 8.

Description	Quantity	Unit	Material		Labor		Total

Project ——————
Date ——————

Estimated By ——————
Checked By ——————

Sheet No. ——————

**Estimate form
Figure 6-6**

Built-up girders		
Size of girder	BF per LF	Nails per 1000 BF
4 x 6	2	53
4 x 8	2.66	40
4 x 10	3.33	32
4 x 12	4	26
6 x 6	3	43
6 x 8	4	32
6 x 10	5	26
6 x 12	6	22
8 x 8	5.33	30
8 x 10	6.66	24
8 x 12	8	20

Built-up girders
Figure 6-7

Length of span	Spacing of joists									
	12"	16"	20"	24"	30"	36"	42"	48"	54"	60"
6	7	6	5	4	3	3	3	3	2	2
7	8	6	5	5	4	3	3	3	3	2
8	9	7	6	5	4	4	3	3	3	3
9	10	8	6	6	5	4	4	3	3	3
10	11	9	7	6	5	4	4	4	3	3
11	12	9	8	7	5	5	4	4	3	3
12	13	10	8	7	6	5	4	4	4	3
13	14	11	9	8	6	5	5	4	4	4
14	15	12	9	8	7	6	5	5	4	4
15	16	12	10	9	7	6	5	5	4	4
16	17	13	11	9	7	6	6	5	5	4
17	18	14	11	10	8	7	6	5	5	4
18	19	15	12	10	8	7	6	6	5	4
19	20	15	12	11	9	7	6	6	5	5
20	21	16	13	11	9	8	7	6	5	5
21	22	17	14	12	9	8	7	6	6	5
22	23	18	14	12	10	8	7	7	6	5
23	24	18	15	13	10	9	8	7	6	6
24	25	19	15	13	11	9	8	7	6	6
25	26	20	16	14	11	9	8	7	7	6
26	27	21	17	14	11	10	8	8	7	6
27	28	21	17	15	12	10	9	8	7	6
28	29	22	18	15	12	10	9	8	7	7
29	30	23	18	16	13	11	9	8	7	7
30	31	24	19	16	13	11	10	9	8	7
31	32	24	20	17	13	11	10	9	8	7
32	33	25	20	17	14	12	10	9	8	7
33	34	26	21	18	14	12	10	9	8	8
34	35	27	21	18	15	12	11	10	9	8
35	36	27	22	19	15	13	11	10	9	8
36	37	28	23	19	15	13	11	10	9	8
37	38	29	23	20	16	13	12	10	9	8
38	39	30	24	20	16	14	12	11	9	9
39	40	30	24	21	17	14	12	11	10	9
40	41	31	25	21	17	14	12	11	10	9

One joist has been added to each of the above quantities to take care of extra joist required at end of span.
Add for doubling joists under all partitions.

Number and spacing of wood joists for any floor
Figure 6-8

Allow about 5 percent for waste. Use Figure 6-7 to figure board feet and nails required for built-up girders.

After listing the girders, tabulate the sills, beginning at one corner and working around the building. List the sizes, number and lengths on the estimate sheet. The result is the linear feet of sill required. Thickness and width measurements of the lumber will be given in the specifications or on the plan. Sill plates, which may be 2 x 6, 2 x 8, 2 x 10 or 4 x 6 in size, are bored for anchor bolts.

Estimating Floor Joists

After estimating the sills, examine the floor joists. Look over the plan and put down the largest joists first. These are usually under partitions, or are trimmers and headers around stairs or chimney openings. Next, take off the regular floor joists. Begin at one side or top of the plan according to the way the joists run, and complete each bay before starting on the next. Do this on all floors.

The size and direction are generally given on the plan, but you may have to look at the wall sections through the main wall. Usually the size, spacing and direction of the first floor joists are given on the basement plan.

The second-story floor joists appear on the first-floor plan. The attic joists appear on the second-floor plan. A note on a plan referring to the joists usually refers to the joists overhead. Joists are carried by girders and exterior walls on the first floor, and by bearing partitions and exterior walls on the second floor. Joists should span the short dimension of a room if possible.

Joists are available in lengths of an even number of feet. The usual lengths are from 8 feet to 16 feet. Longer lengths up to 24 feet may be available, but the cost is often prohibitive.

Consider also the type of framing when figuring the lengths of joists. Is the girder built-up? Can the joists be butted together at the girder? Or should they overlap at the girder and be spiked together? What does your code say about this?

Also, study the details at the exterior wall. In western framing, there's generally a header to which the joists frame. In balloon framing, the joists extend to the outside edge of the exterior wall

Material					Nails
Board feet required for 100 SF of surface area					Per 1000 BF
Size of joist	12" o.c.	16" o.c.	20" o.c.	24" o.c.	Pounds
2 x 6	128	102	88	78	10
2 x 8	171	136	117	103	8
2 x 10	214	171	148	130	6
2 x 12	256	205	177	156	5

Lumber needed for floor joists
Figure 6-9

studding. For brick supporting walls, the joists must be set at least 4 inches into the brickwork.

To find the number of joists required for any floor area, divide the spacing of the joists into the length of the wall that carries the floor joists. Then add one to allow for the extra joist required at the end of the span.

For conventional construction, add one extra piece for every partition that runs parallel to the joists, unless the plan or specs indicate otherwise.

A room 28 feet long, with joists spaced 16" on centers (one joist every 1⅓ feet), will require 3/4 as many joists as the length of the room in feet, plus one extra joist. Three-quarters of 28 is 21, plus one extra joist at the end, makes 22 joists. Figure 6-8 gives the number of joists required for any standard spacing. Figure 6-9 gives the board feet of lumber for common floor joist jobs.

Header joists, as shown in Figure 6-10, must also be estimated in addition to the number of regular joists. In the modern braced framing method, blocking of the same size stock as the joists is used between the joists, and must be included in the take-off.

Estimating Bridging

Bridging includes solid wood blocking, wood cross-bridging (Figure 6-11) or metal cross-bridging (Figure 6-12).

When required, bridging is usually placed in all spans of joists 8'0" or longer, and at intervals of 6'0" to 8'0" in longer spans.

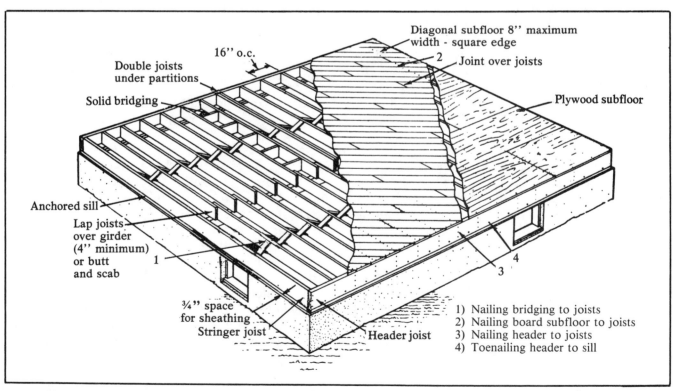

Floor framing
Figure 6-10

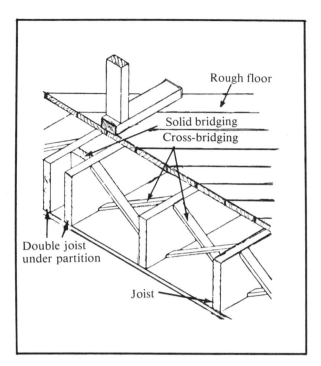

Solid and cross-bridging
Figure 6-11

Metal cross-bridging
Figure 6-12

Wood bridging material is usually cut from 1 x 3 or 1 x 4 lumber. Joists 10 to 14 feet long will require one row of bridging, or two pieces to each joist, as shown in Figure 6-11. Joists 16 to 20 feet long will require two double rows of bridging or four pieces to each joist. Drive two 7d or 8d nails at each end. To estimate the quantity of bridging required for a floor, measure the length of the space in which bridging is to be placed and multiply it by 3. For example: In a house 24 feet wide and 54 feet long, the girder is placed 12 feet from each sidewall. The span of the joists on each side is 12 feet, so each span would require one double row of bridging. Thus, each row of bridging would be 54 feet long.

54 x 2 = **108** total feet to be bridged
108 x 3 = **324** linear feet of bridging stock

Metal bridging is usually more expensive than wood bridging, but it's faster to install. The time saved in labor may make it cheaper overall.

Estimating Subflooring
Subflooring is installed over the floor joists to form a working platform and base for underlayment and finish flooring. It's usually made of plywood, but lumber is still used occasionally.

Plywood subflooring reduces waste and saves manhours. Figure 6-13 describes the panels used for subflooring.

Figure 6-14 explains what the American Plywood Association (APA) Registered Trademarks mean. The span rating "32/16" means the rafter span should not exceed 32 inches center to center, and the floor joist span should not exceed 16 inches center to center.

In estimating plywood, take the square feet of the floor area and divide by the square feet in one piece of plywood. Round it off to the next whole number.

Example: If there's 1,116 square feet of floor area and 4' x 8' (32 square feet) plywood is used, the number of pieces required will be:

$$\frac{1{,}116 \text{ SF}}{32} = 34.87 \text{ pieces. Round to 35.}$$

Deduct only large openings (4' x 8' or larger) when estimating plywood. If the floor dimensions for the width and length are on a 4-foot module (20', 24', 28', etc.) there won't be any waste. If the

Grade Designation	Description & Common Uses	Typical Trademarks	Most Common Thicknesses (inches)			
			3/8	1/2	5/8	3/4
APA rated sheathing EXP 1 or 2	Specially designed for subflooring and wall and roof sheathing, but can also be used for a broad range of other construction and applications. Can be manufactured as conventional veneered plywood, as a composite, or as a nonveneered panel for special engineered applications, including high load requirements and certain industrial uses, veneered panels conforming to PS1 may be required. Specify Exposure 1 when construction delays are anticipated.	APA RATED SHEATHING 32/16 1/2 INCH SIZED FOR SPACING EXPOSURE 1 000 PS 1-74 C-D INT/EXT GLUE NRB-108	•	•	•	•
APA rated Sturd-I-Floor EXP 1 or 2	For combination subfloor-underlayment. Provides smooth surface for application of resilient floor covering and possesses high concentrated and impact load resistance. Can be manufactured as conventional veneered plywood, as a composite. Available square edge or tongue-and-groove. Specify Exposure 1 when construction delays are anticipated.	APA RATED STURD-I-FLOOR 24 OC 23/32 INCH SIZED FOR SPACING T&G NET WIDTH 47-1/2 EXPOSURE 1 000 INT/EXT GLUE NRB-108 FHA-UM-66			• 19/32	• 23/32

Performance-rated panels for subflooring
Figure 6-13

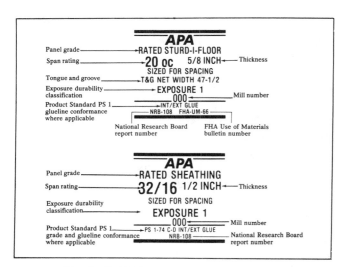

Typical APA registered trademarks
Figure 6-14

floor dimensions are on a 2-foot module (20', 22', 24', etc.) waste will be 3 to 5 percent.

Board flooring usually consists of square-edge or tongue-and-grooved boards no wider than 8 inches and not less than 3/4-inch thick (nominal). The flooring may be applied either diagonally (the most common method) or at right angles to the joists.

The right-angle method requires less cutting than the diagonal method, so it reduces labor and waste.

In estimating board flooring, allow extra material for coverage and waste. For 1 x 8 flooring, add 16 percent. If the flooring is laid diagonally, add 25 percent for 6-inch boards and 21 percent for 8-inch boards.

Estimating Building Paper
There are several grades and kinds of building paper. The standard width is 36 inches. A roll of 15-pound asphalt saturated felt usually covers 500 square feet. Special kinds of paper or felt vary in width, number of square feet and weight per roll. 15-pound felt is commonly laid over board or plywood subflooring. To find the number of rolls of 15-pound felt required, divide the area of the floor to be covered by 500. Count any part of a roll that's left as a whole roll.

Estimating Studs for Western Framing
To figure the linear feet of stud walls, start with the basement plan and measure the number of feet of inside stud partitions. Next, estimate the first-floor plan. Starting at one corner of the building, continue around the exterior walls. Measure the linear feet of inside stud partitions running in one direction, then measure the partitions running at right

Studs spaced 16'' on centers, with single top and bottom plates					
Length partition in feet	Number studs required	8'	Ceiling heights in feet		
			9'	10'	12'
2	3	1.25	1.167	1.13	1.13
3	3	0.833	.812	.80	.80
4	4	0.833	.812	.80	.80
5	5	0.833	.812	.80	.80
6	6	0.833	.812	.80	.80
7	6	0.833	.75	.75	.80
8	7	0.75	.75	.75	.70
9	8	0.75	.75	.75	.70
10	9	0.75	.75	.75	.70
11	9	0.75	.70	.70	.67
12	10	0.75	.70	.70	.67
13	11	0.75	.70	.70	.67
14	12	0.75	.70	.70	.67
15	12	0.70	.70	.70	.67
16	13	0.70	.70	.70	.67
17	14	0.70	.70	.70	.67
18	15	0.70	.70	.67	.67
19	15	0.70	.70	.67	.67
20	16	0.70	.70	.67	.67
For dbl. plate, add per S.F.		0.13	.11	.10	.083
For 2x8 studs, double above quantities.					
For 2x6 studs, increase above quantities 50%.					

Board feet of lumber required per square foot of wood stud partition using 2 x 4 studs
Figure 6-15

angles to those just figured. Finally, figure the exterior walls and partitions of the second story the same way.

Rule for walls without openings— Multiply the length of the wall or partition by 3/4, or 0.75, and add one piece. The answer is the number of studs 16 inches o.c. Use Figure 6-15 to estimate studs and plates for studs 16 inches o.c. Figure 6-16 gives coverage figures for all standard stud walls.

Additional pieces for corners and T's may be required, depending on your method of framing.

Rule for walls with openings— When studs are spaced 16 inches o.c., estimate one stud for each linear foot of wall or partition. This is usually sufficient to allow doubling for studs at corners, for framing window and door openings, and for trussing over the heads of large openings, as shown in Figure 6-17.

If studs are placed 24 inches o.c.— Many builders space studs 24 inches o.c. instead of 16 inches. That reduces the cost of construction and holds down the price of houses. As you can see from Figure 6-16, considerably less material is required for studs placed 24 inches o.c.

If you're placing studs 24 inches o.c., estimate the stud requirements for a partition wall without openings by dividing the length of the wall by two and adding one piece. For a wall 16 feet long, how many studs would you need?

$$16 \div 2 = 8 + 1 = 9 \text{ studs}$$

Figure a partition wall with openings the same way, except add two extra pieces for each opening.

For exterior walls, add one piece for each corner. So a 60-foot exterior wall with two windows and one door would be:

60' ÷ 2	=	30
	+	1
Add for two corners		2
Add for two window openings		4
Add for door opening		2
Total studs required (24'' o.c. spacing)		39

You may need additional pieces for corners and T's, depending on your method of framing.

Rule for top and bottom plates— Multiply the linear feet of all walls and partitions by 2 if a top and bottom plate are required. The result will be the linear feet of plates required. For a double plate

Exterior-Wall Studs			
Size of studs	Spacing on center	BF per SF of area	Lbs. nails per 1000 BF
2 x 3	12''	.83	30
	16''	.78	
	20''	.74	
	24''	.71	
2 x 4	12''	1.09	22
	16''	1.05	
	20''	.98	
	24''	.94	
2 x 6	12''	1.66	15
	16''	1.51	
	20''	1.44	
	24''	1.38	
Includes an allowance for corner bracing.			

Partition Studs			
Size of studs	Spacing on center	BF per SF of area	Lbs. nails per 1000 BF
2 x 3	12''	.91	25
	16''	.83	
	24''	.76	
2 x 4	12''	1.22	19
	16''	1.12	
	24''	1.02	
2 x 6	16''	1.48	16
	24''	1.22	
Includes an allowance for top and bottom plates, end studs, blocks, backing, framing around openings and normal waste.			

Board feet per square foot for wall studs
Figure 6-16

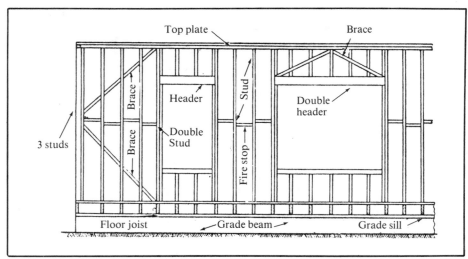

Wall framing
Figure 6-17

consisting of two top members and a single bottom plate, multiply the linear feet of walls and partitions by 3.

Let's work a problem involving studs spaced 16 inches o.c. Find the amount of lumber required to build a stud partition 20' long, 8'6" high, and having single top and bottom plates.

20' x ¾ = 15 studs. Add 1 extra for end of span = 16 studs, 8'6" long. Two plates (1 bottom, 1 top) 20'0" long equal 40 linear feet.

16 pieces of 2 x 4 at 8'6" =	90-2/3 BF
2 pieces of 2 x 4 at 20'0" =	26-2/3 BF
Total	117-1/3 BF

Firestops— A firestop is a piece of framing stock placed horizontally or at a slight angle about half way between the bottom and top plates. It's the same size as the stud material. Subtract from the total linear feet of walls and partitions the total or combined width of all openings. The result equals the linear feet of firestops required.

Estimating Balloon Framing
In balloon framing, the studding extends from the sill to the top plates of the second story. Estimate the number of studs required in the exterior wall the same way as for western framing, noting the length requirement. Shorter pieces may be used under and over openings. Add two full-length studs for every opening. Inside partition studs are figured like partitions in western framing.

A balloon frame requires a piece of 1-inch stock notched into the inside of exterior sidewall studs which are two stories high. This piece, which forms a support for the second floor joists, is called a *ribbon*. See Figure 6-18. The width will vary according to the plans and specs. The ribbon is usually 1 x 4 or 1 x 6. To figure the amount of ribbon, determine the linear feet of exterior supporting walls. The direction of the second floor or ceiling joists will determine which are support walls. Any wall parallel to the joists does not require a ribbon. You might wish to compare western framing (Figure 6-19) and modern braced framing (Figure 6-20) with balloon framing.

Estimating Corner Bracing
Corner bracing adds rigidity to the structure by protecting against lateral forces such as wind. You may use:

1) Wood sheathing boards installed at a 45-degree angle in opposite directions from each corner

2) 1 x 4 or wider boards let into the outer face of studs, sole plate and top plate located near each corner and set at a 45-degree angle

3) 4' x 8' plywood applied with the long dimension vertical

4) Structural fiberboard in 4' x 8' panels applied with the long dimension vertical

5) Steel bands installed at corners at a 45-degree angle

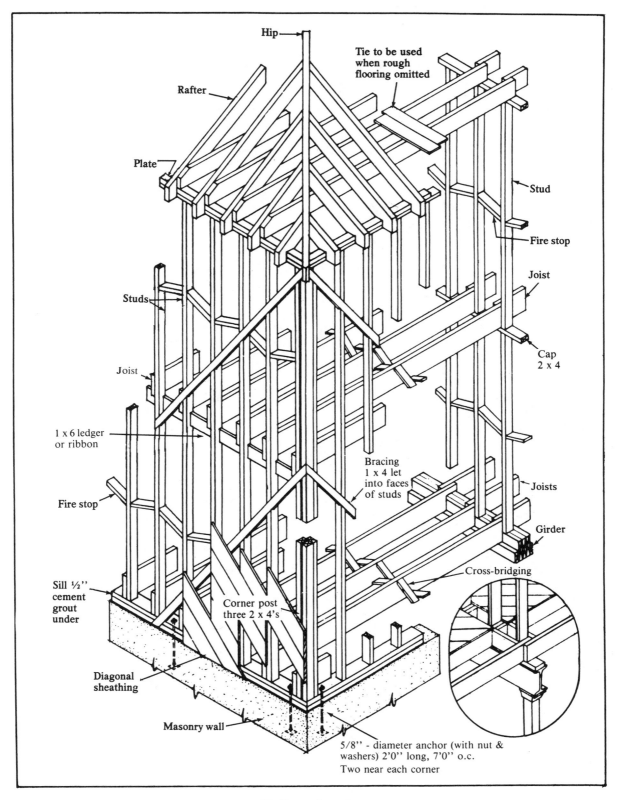

Balloon framing
Figure 6-18

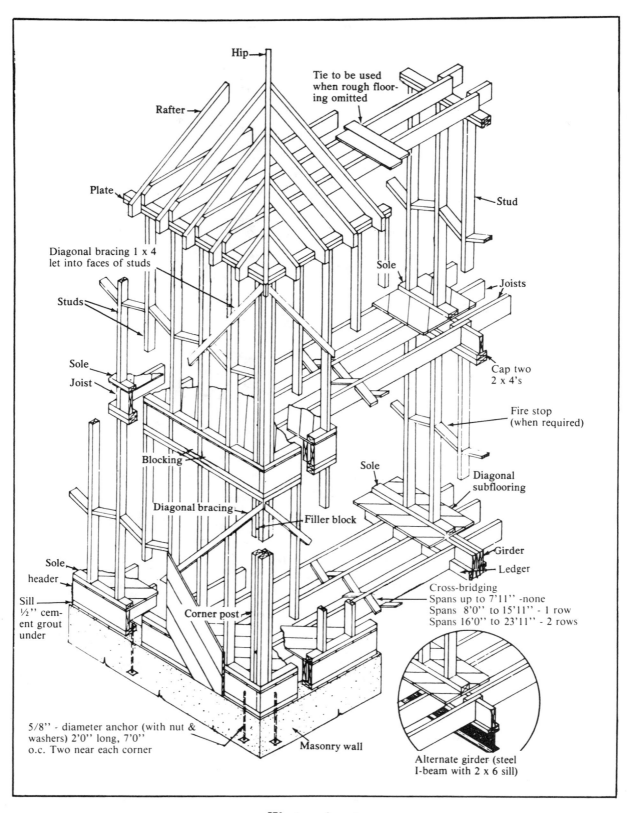

Western framing
Figure 6-19

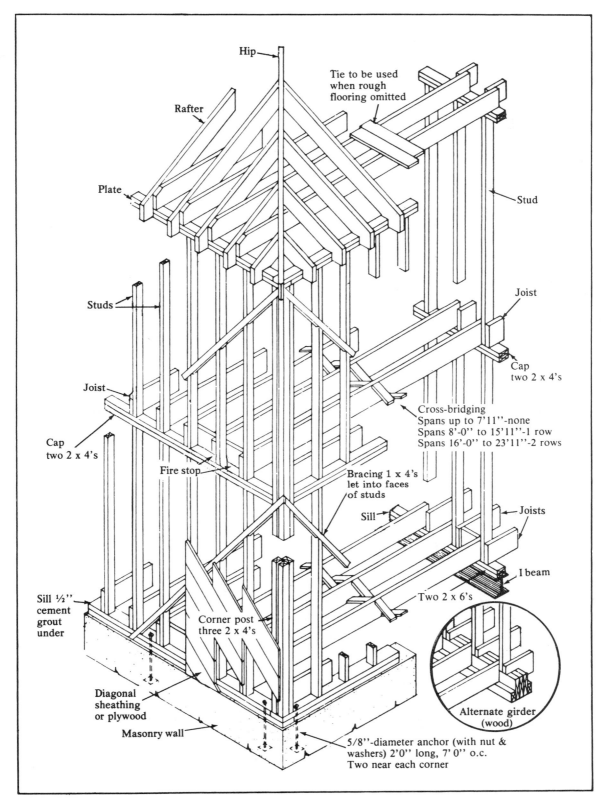

Hip

Tie to be used
when rough
flooring omitted

Rafter

Plate

Stud

Studs

Joist

Joist

Cap
two 2 x 4's

Cap
two 2 x 4's

Cross-bridging
Spans up to 7'11''-none
Spans 8'-0'' to 15'11''-1 row
Spans 16'-0'' to 23'11''-2 rows

Fire stop

Bracing 1 x 4's
let into faces
of studs

Sill

Joists

I beam

Two 2 x 6's

Sill ½''
cement
grout
under

Corner post
three 2 x 4's

Diagonal
sheathing
or plywood

Masonry wall

Alternate girder
(wood)

5/8''-diameter anchor (with nut &
washers) 2'0'' long, 7' 0'' o.c.
Two near each corner

Modern braced framing
Figure 6-20

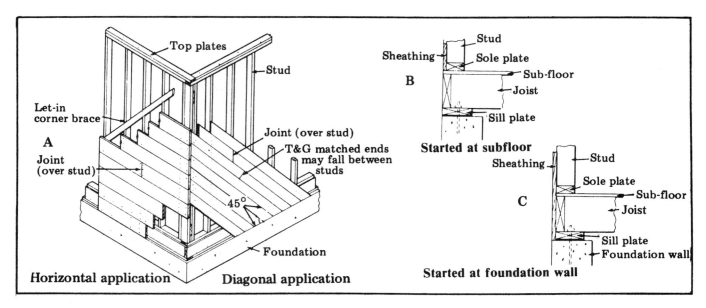

Applying board wall sheathing
Figure 6-21

Panel span rating	Maximum stud spacing (inches)	Nail size***	Nail spacing***	
			Panel edges	**Intermediate**
12/0, 16/0, 20/0 or wall-16 o.c.	16*	6d for panels ½'' thick or less, 8d for thicker panels		
24/0, 24/16, 32/16	24**		6''	12''

APA-rated sheathing panels continuous over two or more spans

*Apply plywood panels less than ⅜'' thick with face grain across studs when exterior covering is nailed to sheathing
**Apply 3-ply plywood panels with face grain across studs 24'' o.c. when exterior covering is nailed to sheathing
***Common, smooth, annular, spiral-thread, or galvanized box; or T-nails of the same diameter as common nails (0.113'' diameter for 6d, 0.131'' for 8d) may be used. Staples also permitted at reduced spacing.

APA panel wall sheathing specifications
Figure 6-22

The prevailing practice for installing corner bracing is to use 1/2'' thick 4' x 8' plywood sheathing at the corners (nailed according to the manufacturer's instructions) and 1/2'' thick 4' x 8' fiberboard sheathing or insulating sheathing vertically applied on the balance of the exterior walls.

When estimating corner bracing, use one of the following methods:

• If let-in bracing, estimate two 1 x 4 x 12' pieces (for 8' ceilings) for each corner.

• If sheet material is used, estimate two 4' x 8' pieces for each corner and include with the wall sheathing estimate.

Corner bracing must also be applied to the second story of two-story structures.

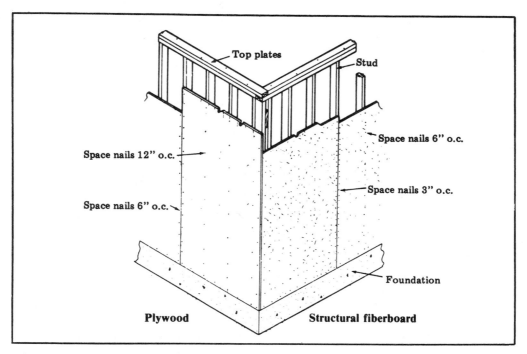

A   **Vertical application of plywood or structural fiberboard wall sheathing**

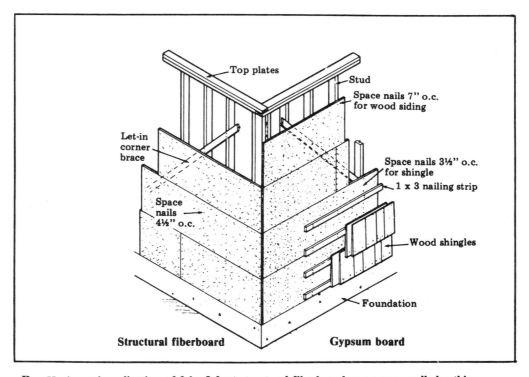

B   **Horizontal application of 2 by 8-foot structural fiberboard or gypsum wall sheathing**

Sheathing application
Figure 6-23

Estimating Wall Sheathing

Wall sheathing strengthens and adds rigidity to the exterior walls. It also adds additional insulation. You can increase the R-factor of the walls by using foam boards or panels such as polystyrene.

There are several types of wall sheathing to choose from. Let's look at them one at a time:

Wood board sheathing: T & G, square edge or shiplapped. The most common size is 3/4'' thick, 8'' wide boards. Corner bracing is not required when the boards are applied diagonally. (See Figure 6-21.)

Plywood sheathing: Minimum thickness is 5/16'' when applied on studs 16'' o.c., and 3/8'' for studs 24'' o.c. Plywood sheathing is acceptable for corner bracing if installed vertically and nailed according to the manufacturer's instructions. (See Figure 6-22).

Fiberboard sheathing: Minimum thickness is 1/2'' and corner bracing is required. Nail sheathing to studs at each bearing with 1½'' roofing nails with 3/8'' to 7/16'' heads, 4'' o.c. at edges and 8'' o.c. at intermediate supports.

Structural intermediate-density fiberboard sheathing: No corner bracing is required when using 4' x 8' panels applied vertically and nailed according to manufacturer's instructions. (See Figure 6-23 A).

Gypsum sheathing: Minimum thickness is 1/2'' and corner bracing is required. Nail sheathing to studs at each bearing with 1½'' roofing nails with 3/8'' to 7/16'' heads, according to manufacturer's instructions. (See Figure 6-23 B).

Insulating foam board sheathing: Follow manufacturer's instructions for installation.

Calculate the total linear feet of the perimeter of the exterior walls on each floor and multiply by the wall height to find the gross area in square feet. If window openings are average, I prefer to apply the sheet or panel sheathing solid and cut out the openings later when I'm ready to install the windows. This protects the interior from blowing rain.

If gables require sheathing, multiply the rise from the plate to the ridge by the width of the gable, and divide by 2. Calculate the area of each

Measured size, inches	Finished width, inches	Add for shrinkage percentage	Feet of lumber required, 100 SF surface
1 x 3	2½	25	125
1 x 4	3½	20	120
1 x 6	5½	14	114
1 x 8	7½	12	112
1 x 10	9½	10	110
1 x 12	11½	9½	109½
2 x 4	3½	20	240
2 x 6	5½	14	228
2 x 8	7½	12	225
2 x 10	9½	10	220
2 x 12	11½	9½	219
3 x 6	5½	14	343
3 x 8	7½	12	337½
3 x 10	9½	10	330
3 x 12	11½	9½	329

Note: This chart does not include waste.

Coverage of square-edge board sheathing
Figure 6-24

gable if they are different sizes, and add the totals.

Add all areas to find the total for wall sheathing. If you decide to deduct for openings, subtract the openings from the gross area for the net area in square feet to be sheathed.

For wood board sheathing, you can use Figure 6-24 to estimate the board feet required. The coverage in this chart doesn't include waste. Add 5 percent for most jobs.

For 4' x 8' sheets or panels, divide the net area in square feet by the total square feet of the sheet or panel. There are 32 square feet in a 4' x 8' panel, 36 square feet in a 4' x 9' panel.

A common method is to use plywood bracing at corners, and fiberboard or other types of panels

Panel identification index	Minimum thickness (inch)	Maximum stud spacing (inches) exterior covering nailed to:	
		Stud	Sheathing
12/0, 16/0, 20/0	5/16	16	16[2]
16/0, 20/0, 24/0	3/8	24	16 / 24[2]
24/0, 32/16	1/2	24	24

(1) When plywood sheathing is used, building paper and diagonal wall bracing can be omitted.

(2) When sidings such as shingles are nailed only to the plywood sheathing, apply plywood with face grain across studs.

Allowable spans and panel thickness
for plywood wall sheathing
Figure 6-25

Size of joist	Material Board feet required for 100 SF of surface area				Nails per 1000 BF	Labor BF Per hour
	12" o.c.	16" o.c.	20" o.c.	24" o.c.	Pounds	Board feet
2 x 4	78	59	48	42	19	60
2 x 6	115	88	72	63	13	65
2 x 8	153	117	96	84	9	65
2 x 10	194	147	121	104	7	70
2 x 12	230	176	144	126	6	70

Material and labor for ceiling joists
Figure 6-26

for the remainder of the wall. In this case, subtract the number of plywood panels from the number of fiberboard panels required for complete coverage.

Figure 6-25 gives allowable spans and panel thickness for plywood wall sheathing. When you use plywood sheathing, you can omit the building paper and diagonal wall bracing. And one more thing to remember: if you're nailing sidings such as shingles only to plywood sheathing, apply the plywood with the face grain *across* the studs.

Estimating Ceiling Joists
Figure 6-26 shows lumber, nail and labor requirements for ceiling joist jobs.

You can also estimate ceiling joists using the same procedure as for estimating floor joists. Ceiling joists are supports for the ceiling. They're not designed to support floor loads. If more than limited attic storage is used, floor joists are required.

The floor plan will show the size, spacing and direction the ceiling joists will run. Figure 6-27 can, however, be used as a guideline for ceiling joist spans. You can find the length of the joists from the floor plans by finding the dimensions of the rooms, and the direction the joists run. Normally, the joist run is the shorter span.

Most house plans have a load-bearing partition near the center of the structure. The ceiling joists rest on it or join on it. Ceiling joists seldom span the entire width of the building as trusses do. When estimating joists, estimate each section separately.

Don't forget the headers and trimmers. They're required for any openings in the ceiling such as access doors to the attic and chimneys.

Nominal Size (Inches)	Spacing (Inches o.c.)	Select Structural 1950 f	Dense Construction 1700 f	Construction 1450 f	Standard 1200 f
2 x 4	12	9'- 6"	--	8'- 2"	6'- 4"
	16	8'- 6"	--	7'- 2"	5'- 6"
	24	7'- 6"	--	5'-10"	4'- 6"
2 x 6	12	14'- 4"	14'- 4"	14'- 4"	14'- 4"
	16	13'- 0"	13'- 0"	13'- 0"	12'-10"
	24	11'- 4"	11'- 4"	11'- 4"	10'- 6"
2 x 8	12	18'- 4"	18'- 4"	18'- 4"	18'- 4"
	16	17'- 0"	17'- 0"	17'- 0"	17'- 0"
	24	15'- 4"	15'- 4"	15'- 4"	14'- 4"
2 x 10	12	21'-10"	21'-10"	21'-10"	21'-10"
	16	20'- 4"	20'- 4"	20'- 4"	20'- 4"
	24	18'- 4"	18'- 4"	18'- 4"	18'- 0"

Ceiling joist spans
Figure 6-27

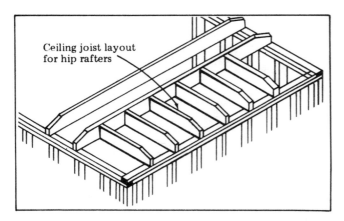

Ceiling joists for hip roof
Figure 6-28

Estimating ceiling joists is the same as for hip or gable roofs, except the layout is different. The run of the regular joists stops short of the outside wall to allow clearance for the hip and jack rafters. Short ceiling joists are installed perpendicular to the regular joists to permit the rafters to reach the outside wall plate. See Figure 6-28.

It isn't essential that all ceiling joists run the same way. Figure 6-29 illustrates how joists may run in two directions.

To estimate ceiling joists, multiply the linear feet of the wall by 0.75 and add one joist for the end for spacing 16 inches o.c. If the spacing is 24 inches o.c., multiply the linear feet of the wall by 0.50 and add one joist.

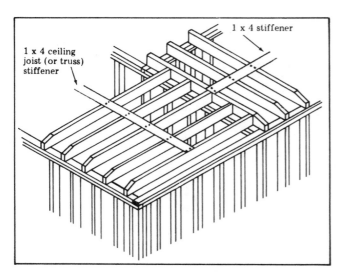

Ceiling joists running two directions
Figure 6-29

Estimating Lath Nailers

Lath or ceiling nailers form nailing support for the finish ceiling, whether it's gypsum board or gypsum lath and ceiling panels. They're commonly used on wall plates running parallel to ceiling joists or trusses. They're also sometimes used when the joists run at right angles to a wall when the joist spacing is 24 inches o.c. See Figure 6-30.

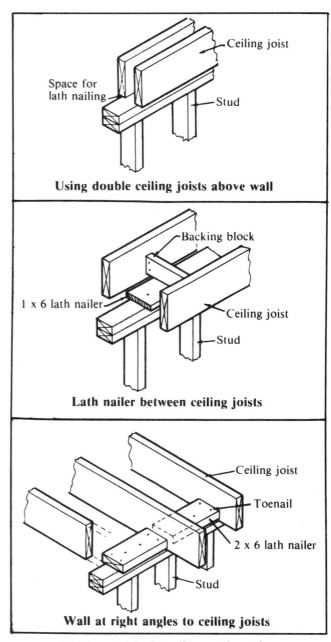

Using double ceiling joists above wall

Lath nailer between ceiling joists

Wall at right angles to ceiling joists

Horizontal lath nailers at junction of wall and ceiling framing
Figure 6-30

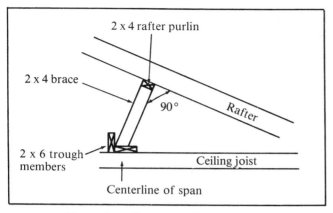

**Trough nailed to ceiling joist and
braced to rafter purlin
Figure 6-31**

There's usually enough waste material available on the site to make these nailers. Some builders, however, figure this material on the take-off, allowing 2 x 6 stock for standard 2 x 4 stud walls.

When estimating nailers, total the linear feet of all walls running parallel to the ceiling joists or trusses from the floor plans. Then divide by the length of the material to be used to find the number of pieces required.

If you use ceiling joist stiffeners or a "trough" to stiffen or hold ceiling joists in alignment, install them in the center of the span in each section.

Estimate by adding the linear feet of each section. For stiffeners, 1 x 4 material is usually sufficient. For trough work, stock the same size as the joist material is often used, up to 2 x 6 size. See Figure 6-31.

Estimating Rafters
Figure 6-32 shows the different types of rafters you'll have to estimate. Take a minute to look it over before reading on unless you have all the rafter types down pat. Figure 6-33 depicts the common roof framing terms we'll use in this chapter. Make sure you understand the difference between the span and the run, and how to arrive at the total run.

Finding the Number of Rafters
On a plain gable roof, you can find the number of common rafters the same way you found the joists. For 16-inch centers, multiply the length of the wall by 3/4 and add one. For 20-inch centers, multiply by 3/5 and add one. If extra rafters are required for the gable cornice, add them. This rule gives you the number of rafters for one side of a gable roof. Double the answer and you'll have the number of rafters needed for both sides. Most builders allow 5 percent for end waste on ordinary roof work. With modular construction, there's less waste. Use Figure 6-8, the chart for figuring floor joists, to find the number of rafters required.

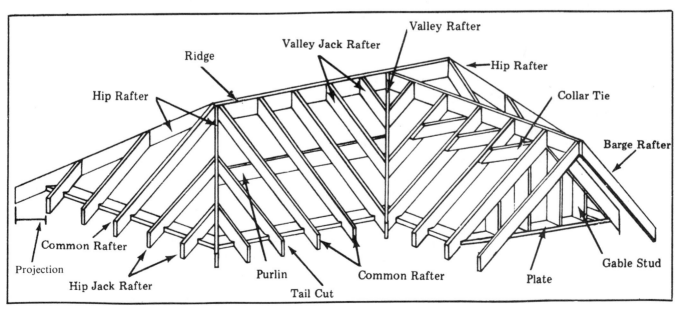

**Types of rafters
Figure 6-32**

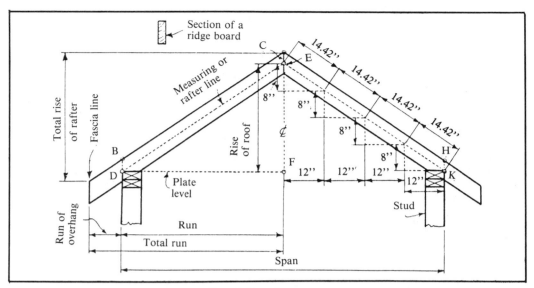

Roof framing terms
Figure 6-33

Finding the Pitch and Rise

On a pitch roof, you may know the distance between the walls rather than the length of the rafters. You have to calculate the rafter lengths based on the span and pitch. Pitch is a mathematical relationship between the slope and span of a roof. Using the pitch, you can find the height of the ridge above the plate line.

For a building 40 feet wide with a 1/4 pitch roof, the ridge is 10 feet above the plate (1/4 x 40' equals 10'). A 1/2 pitch roof with a span of 32 feet would have the ridge 16 feet above the plate (1/2 x 32' is 16').

The span multiplied by the pitch always gives the ridge height. To find the pitch of a roof, divide the rise by the span. For example, a building with a rise of 8 feet and a width or span of 24 feet has a pitch equal to 8 divided by 24, or 1/3. The most common roof pitches are 1/2, 1/3 and 1/4 pitch. The rise for 1/2 pitch equals one-half the distance of the span; for 1/3 pitch, one-third the span; and for 1/4 pitch, one-fourth the span. Figure 6-34 will help you determine the rough length of the lumber you need for rafters.

You can convert the roof pitch to the equivalent rise in inches per foot of run, and the rise into the equivalent pitch, by the following rules:

Rule 1: To find the rise of a roof per foot of run from the pitch, multiply the pitch by 24.

Let's try it. Find the rise in inches per foot of run for a 1/3 pitch roof.

$$1/3 \times 24 = 8\text{" rise}$$

Rule 2: To find the pitch of a roof from the rise per foot of run, divide the rise per foot of run by 24.

The rise per foot of run of a roof is 6 inches. Calculate the pitch.

$$6\text{"} \div 24 = 1/4 \text{ pitch}$$

Finding Rafter Length

One way to find the rafter length on a pitch roof is to use an architect's scale and read the length of the common rafter on the plan. For more accurate estimating, use Figures 6-35 and 6-36. Figure 6-35 has the factors for common rafters, and Figure 6-36 has the factors for hip and valley rafters.

You can find the length of a rafter for any given pitch by multiplying the factor (ratio) from the chart, by the amount of run for that particular rafter. The result is the rafter length. If the roof has a cornice, add the overhang length to the answer. Then round it up to the next standard length of lumber. This is a practical method for estimators, but the carpenter usually lays out the rafter lengths and cuts with his steel square.

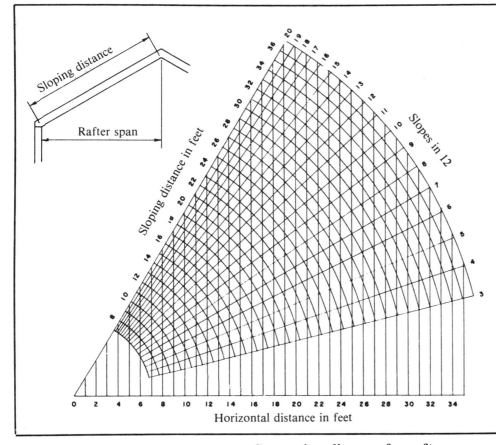

To use the diagram select the known horizontal distance and follow the vertical line to its intersection with the radial line of the specified slope, then proceed along the arc to read the sloping distance. In some cases it may be desirable to interpolate between the one foot separations. The diagram also may be used to find the horizontal distance corresponding to a given sloping distance or to find the slope when the horizontal and sloping distances are known.

Example: With a roof slope of 8 in 12 and a horizontal distance of 20 feet the sloping distance may be read as 24 feet.

Conversion diagram for rafters
Figure 6-34

Now we'll practice finding the length of a common rafter. First, find the pitch on the plans. Then go to Figure 6-35. In the first column, find the pitch (rise) of the roof. For example, assume a 1/3 pitch on your plans. This is equal to 8 inches rise per foot of run. Look under 8 inches in the first column. Reading across to the right, you'll find the ratio 1.2019. This is the length of a common rafter per foot of run.

Here's how to find the exact length of the common rafter from the outer edge of the plate to the face of the ridge board. Assume a span of 22'0", with a pitch of 1/3 and an overhang of 1'0" on each side, which makes the total width of roof 24'0". The ridge board is a 2 x 8.

One-half of 24'0" is 12'0". That's the run of the rafter including overhang. Now multiply 12'0" by the factor for a 1/3 pitch roof, 1.2019. (12'0" x

Rise	Run	Ratio	Rise	Run	Ratio
3	12	1.0308	9	12	1.2500
4	12	1.0541	10	12	1.3017
4.5	12	1.0680	11	12	1.3566
5	12	1.0833	12	12	1.4142
6	12	1.1180	13	12	1.4743
7	12	1.1577	14	12	1.5366
8	12	1.2019	15	12	1.6008

Ratios of common rafter length to run
Figure 6-35

Roof slope		Ratio	Roof slope		Ratio
Rise	Run		Rise	Run	
3	12	1.4361	9	12	1.6008
4	12	1.4530	10	12	1.6415
4.5	12	1.4631	11	12	1.6853
5	12	1.4743	12	12	1.7321
6	12	1.5000	13	12	1.7815
7	12	1.5298	14	12	1.8333
8	12	1.5635	15	12	1.8875

Ratios of hip and valley rafter length to run
Figure 6-36

Rafters

Building Width	Rise	Rafter Lengths			
		3"	4"	5"	6"
10		5'–2"	5'–3"	5'–5"	5'–7"
12		6'–2"	6'–4"	6'–6"	6'–8"
14		7'–3"	7'–5"	7'–7"	7'–10"
16		8'–3"	8'–5"	8'–8"	9'–0"
18		9'–3"	9'–6"	9'–9"	10'–1"
20		10'–4"	10'–7"	10'–10"	11'–2"
22		11'–4"	11'–7"	11'–11"	12'–4"
24		12'–4"	12'–8"	13'–0"	13'–5"
26		13'–5"	13'–8"	14'–1"	14'–6"
28		14'–5"	14'–9"	15'–2"	15'–8"
30		15'–6"	15'–10"	16'–3"	16'–9"
32		16'–6"	16'–10"	17'–4"	17'–11"

NOTE: Tables accurate only to the nearest inch.

Common rafter lengths
Figure 6-37

1.2019 is 14.42', or 14'5'', the length from the center of the ridge board to the tail end of the rafter.) A 2 x 8 ridge board has an actual size of 1½'' x 7¼''. One-half of 1½'' is 3/4''. So you subtract 3/4'' from 14'5''. That's 14'4¼'', the exact length of the common rafter. You'll need a piece of lumber 16'0'' long.

Use Figure 6-37 as a shortcut if the roof you're estimating has a span and rise that's covered on the chart. It gives rough common rafter lengths, accurate to the nearest inch. Add the length of the overhang to find the total rafter length.

If this were a hip or valley rafter, we'd go to Figure 6-36 and find that the length of a hip rafter is 1.5635 times the length per foot run of the common rafter. The length of the hip rafters for this building is 12'0'' x 1.5635. That's 18.762', or 18'9'', which requires a 20' piece.

Hip and Valley Rafters

Figure 6-38 shows the typical hip and valley roofs, including a square hip roof, in which the hip rafters run diagonally to meet at the center of the roof, and an oblong hip roof, in which the hip rafters and the common rafters are framed into the ridge.

In the gable and valley roof, two gable roofs intersect each other at the valley. In the hip and valley roof, two hip roofs intersect each other. The valley is formed the same way in both roofs.

In this brief study of hip and valley rafters, we'll only consider the roof in which these rafters form an angle of 45 degrees with the plate when looking straight down on the roof. This is known as a true hip roof.

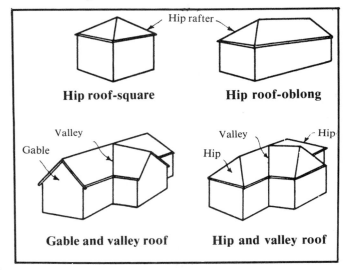

Hip roof-square   **Hip roof-oblong**

Gable and valley roof   **Hip and valley roof**

Hip and valley roofs
Figure 38

Looking at Figure 6-39, you can see that hip and valley rafters have the same rise as the common rafters on the same roof.

Figure 6-40 shows the position of the hip rafter and its relation to the common rafter in more detail. You can see that the run of the hip rafter and the rise form a right triangle. The hip rafter is the hypotenuse of that triangle. If you took high school geometry, you might still remember the Pythagorean theorem: The square of the hypotenuse of a right triangle equals the sum of the squares of the other two sides. Using this theorem, you can find the length of any side of a right

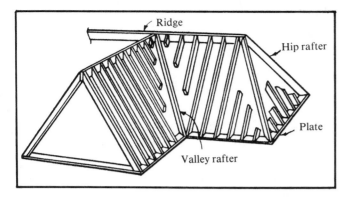

Hip and valley rafters have the same rise
Figure 6-39

triangle if you know the length of the other two sides.

But you don't need to get out your calculator and bone up on squares and square roots. Just look again at Figure 6-40. It shows the run of the common rafter divided into five 12-inch segments. The run of the hip rafter is also divided in five segments, but each segment is 16.97 inches long. Using the Pythagorean theorem, you'll find that in a right triangle with two 12-inch sides, the third side (the hypotenuse) is 16.97 inches long. Just

remember that number: while 12 inches (or 1 foot) is the basic unit of the common rafter, 16.97 inches (or 1.414 feet) is the basic unit of the hip or valley rafter. The run of the hip rafter is always 16.97 inches for every 12 inches of run of common rafter. The *total run* of a hip rafter is 16.97 inches *multiplied* by the number of feet of run of common rafter.

Let's do an example that will make this clear. Find the run of the hip rafter in Figure 6-41.

Solution: The run of the common rafter is 10'0''. To find the run of the hip rafter, just multiply that by 16.97''.

$$10' \times 16.97'' = 169.7'' = 14.14' = 14'1\tfrac{3}{4}''$$

We'll round this out to 14'2''.

Let's try one more. Find the run of the valley rafter at the wing E-E in Figure 6-41.

Solution: The run of the common rafter is 8'0''. So the run of the valley rafter is:

$$8' \times 16.97'' = 135.76'' = 11.31' = 11'3\tfrac{3}{4}''$$

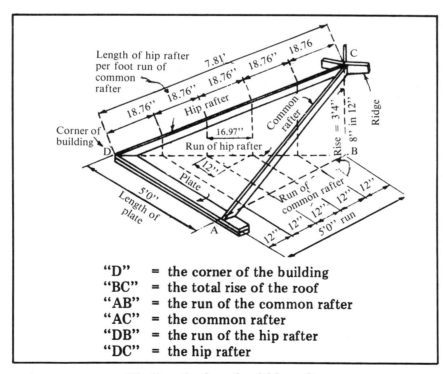

"D"   = the corner of the building
"BC" = the total rise of the roof
"AB" = the run of the common rafter
"AC" = the common rafter
"DB" = the run of the hip rafter
"DC" = the hip rafter

Finding the length of hip rafters
Figure 6-40

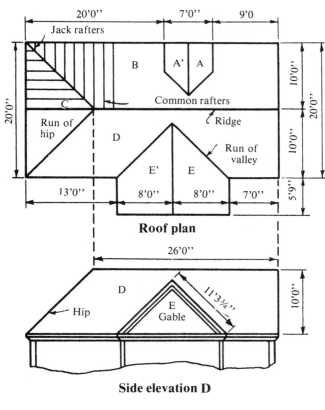

Roof plan

Side elevation D

**Sample problem
Figure 6-41**

Look back to Figure 6-36. It shows that for each foot of run of the common rafter, the length of hip rafter is 1.563', or 18.76'', on a 1/3 pitch roof. Now look at Figure 6-40 again. The common rafter in Figure 6-40 has a run of 5'0''. So the total length of the hip rafter is 5 times 1.563', which is 7.81'. You can calculate it another way, also — 5 times 18.76'' equals 93.80'', or 7.81'.

The Pythagorean theorem was used to make the chart in Figure 6-36. To find the length of any hip or valley rafter, multiply the length given in the table for the appropriate pitch by the number of feet of run of the common rafter.

Here's an example: Calculate the length of the hip rafter in Figure 6-41. The run of the common rafter is 10'0''. The pitch is 1/2.

Solution: From Figure 6-36, the length of hip rafter for each foot of run of common rafter for 1/2 pitch (12'' rise to 12'' run) is 1.7321.

$$10' \times 1.732' = 17.32' = 17'4''$$

Now find the length of the valley rafters in Figure 6-41, assuming a 1/2 pitch. The run of the common rafter is 8'0''.

$$8' \times 1.732' = 13.85' = 13'10\frac{1}{4}''$$

Jack Rafters

To find the length of jack rafters on the main roof of Figure 6-41, divide the distance between the corner of the top plate and the first common rafter into equal parts. In this case there will be seven spaces, or six jack rafters, 17 inches o.c. Make the distance between jack rafters as near to the distance between the common rafters as possible.

Find the length of the shortest jack rafter by dividing the length of the common rafter by the number of spaces there are between the jack rafters on centers. The second jack rafter will be *twice* as long as the first, the third *three* times as long, and so on.

We know the length of the common rafters in the main roof in Figure 6-41 is 14'1¾'', or 14.14'. To find the length of the jack rafter, divide that length by 7, the number of spaces between jack rafters:

$$14.14' \div 7 = 2.02' = 2'\frac{1}{4}''$$

So 2'¼'' is the length of the first jack rafter. Double that to find the length of the second:

$$2 \times 2'\frac{1}{4}'' = 4'\frac{1}{2}''$$

Triple it to find the length of the third, 6'¾'', and so on.

Calculate the length of jack rafters for a valley the same way.

In these problems, the overhang of the roof over the outside walls hasn't been included. Increase the length of rafter to provide for the cornice. When there are two or more pitches on the same roof, each pitch must be worked out separately.

Rafter Tables

Many steel squares have rafter tables stamped on them. If you don't have the manufacturer's instruction booklet that explains them, here's how to use the tables. First, look at Figure 6-42. The inch marks on the outside edge of the square indicate the rise per foot of run. For example, the figure 8

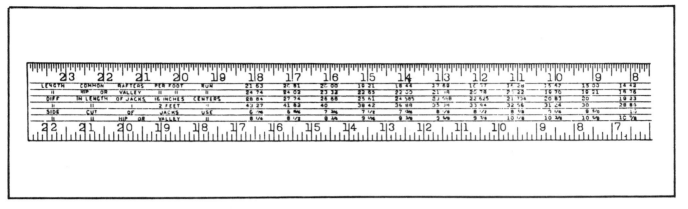

Rafter length table
Figure 6-42

means 8 inches of rise per foot of run. Directly below each of these figures is the length of the common rafters per foot of run, 14.42 in this case. If the run of the common rafter is 10 feet, multiply 14.42 inches by 10. This gives 144.2 inches, or 12.01 feet, as the length of the rafter from ridge to plate.

Find the length of the hip or valley rafter the same way. The second figure under the 8 is 18.76. This is the length of the hip or valley rafter per foot of run. If the run of this rafter is 10 feet, you'd multiply the 18.76 inches by 10. This gives 187.60 inches, or 15.63 feet, from ridge to plate.

A book titled *Rafter Length Manual,* published by Craftsman Book Company, gives the exact rafter lengths for every common roof span and rise. You can obtain one by using the order form at the back of this manual.

Listing Materials for Rafters
When listing the materials for the roof, we must first determine whether any extra plates are required. In masonry buildings we have the roof plates. In some buildings the joists are placed directly on top of the wall plates and an additional plate is placed on top of the joists. Include this plate when listing the materials for the roof.

Listing the rafters— When listing the rafters, determine the shape of the roof and the kind of framing to be used from the plans. The pitch is also indicated on the drawing or by a note on the plan.

Generally, the common rafters are listed first. Determine the pitch from the drawing, then turn to Figure 6-35 to estimate the length of the common rafters. Put down the number of common rafters,

and the length of lumber needed. There may be waste pieces you can use as lath nailers on top of the wall plates.

Next, find the length of the hip rafters, multiplying the appropriate factor from Figure 6-36 by the run of the common rafters. Write down the number of hip rafters and the length of lumber needed for them.

To find the material for the jack rafters, it isn't necessary to figure each length separately. Note, in Figure 6-43, that rafters A and H can both be cut from a piece 16 feet long. Rafters B and G can be cut from one piece. Rafters C and F can be cut from one piece, and so on. So just estimate one 16-foot piece for each two jack rafters. Figure 6-44 gives the total board feet of material and nails required for all types of rafters per square foot of surface.

Saddle boards for chimneys— Saddle boards can be measured from the roof plan or from the different elevations. Usually, each run of saddle board is shown on two elevations.

Ridge boards— Find the length of the ridge board needed by scaling on the elevation or roof plan. For a gable roof, the length of a ridge equals either the length or width dimension of the building. For a hip roof, the ridge length equals the difference between the length and width measurements.

Collar beams— Collar beams are horizontal framing members which tie rafters together to prevent roof thrust (Figure 6-45). They're installed in the upper third of the attic space below the ridge. The maximum spacing of collar beams is 48 inches o.c.

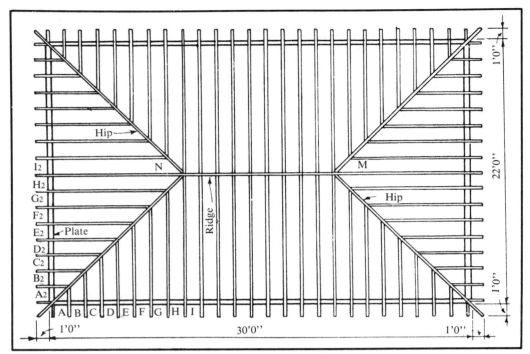

Framing detail for a hip roof
Figure 6-43

Board feet required 100 square feet surface area			
			Nails (lbs.)
	12" o.c.	16" o.c.	24" o.c.	Per 1000 BF
2 x 4	89	71	53	17
2 x 6	129	102	75	12
2 x 8	171	134	112	9
2 x 10	212	197	121	7
2 x 12	252	197	143	6

Rafter framing materials
Figure 6-44

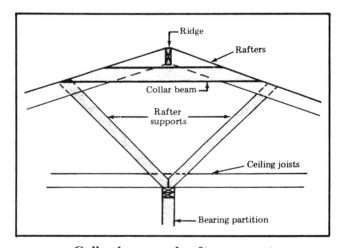

Collar beam and rafter supports
Figure 6-45

If collar beams are spaced 48 inches o.c., divide the length of the structure by 4 feet to determine the number required. Don't include the overhang. You'll usually use 1 x 6 material for collar beams.

Rafter supports— Rafter supports, or rafter braces, are used to prevent roof sag. Use 2 x 4 material, spaced the same as the rafters. They should always run from the rafter to a bearing partition, or from the rafter to a "trough" brace. The length of the rafter support depends on the roof pitch.

Lookouts— Lookouts are short framing members nailed to the sides of rafters and extending to the wall of the house. They provide the soffit for the

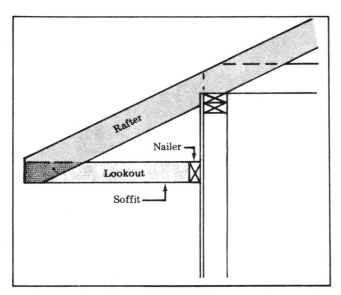

**Framing for roof overhang soffit
Figure 6-46**

roof overhang (Figure 6-46). There should be one lookout per rafter, including the gable overhang. The length of the lookouts is shown on the wall section of the plans. The material normally used for lookouts is 2 x 4.

Trusses

Many builders prefer using prefabricated roof trusses over on-site framing. Roof trusses are a framework of individual structural members fabricated into one unit and designed to span a wide distance. Interior walls aren't needed to support the roof.

Trusses are available in a variety of sizes and shapes and can be used for long or short spans, such as a roof over a porch, walk or breezeway.

Trusses are normally spaced 24 inches o.c. and save on-site labor and material costs. They eliminate the ceiling joists, rafters, ridge, collar beams, rafter supports, lookouts and gable end studs from the roof framing system.

The bottom chord of the truss serves as a ceiling joist, and the top chord serves as the rafter. The roof overhang at the eave can be fabricated with the truss as shown in Figure 6-47.

To estimate the number of roof trusses on 24-inch centers, multiply the length of the building by 0.50 and add one. Do this for each section of the house where there's a break in the roof line.

Estimating Roof Areas

To obtain the area of a plain gable roof, multiply the length of the ridge by the length of the rafters. This will give you one-half of the roof. Multiply by 2 to obtain the total square feet of roof surface.

To get the area of a hip roof, treat it as if it were a gable roof of the same pitch, and multiply the longest side of the roof by the length of the common rafter. Thus, a 48' x 28' structure having a 1'0'' overhang on all sides and common rafters 18'0'' long, would have the same roof areas whether hip or plain gable. To calculate the hip roof coverage, multiply 50' (long dimension of 48' plus 2' of overhang) by 18' (length of common rafter). There are 900 square feet in one half of the roof, or 1,800 square feet for both halves.

Roof Sheathing

Most builders use plywood roof sheathing in residential construction. A sheet of 4' x 8' plywood covers 32 square feet. A roof of 1,800 square feet would require 1,800 divided by 32, or 56.25 panels. Round it up to 57 panels and add 5% for waste.

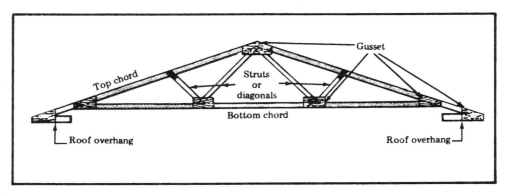

**Truss with roof overhang
Figure 6-47**

Estimated Labor Performance

Item	Size or Kind	Estimated Labor Performance	Item	Size or Kind	Estimated Labor Performance
Wall plates	2 x 6 − 2 x 8 2 x 10 − 2 x 12	65 b.f. per hour 70 b.f. per hour	Wall backing		50 b.f. per hour
			Grounds		85 lin. ft. per hour
Basement beam (girder)	2 x 8 2 x 10	30 b.f. per hour 40 b.f. per hour	Knee wall plates	2 x 4 2 x 6	40 b.f. per hour 40 b.f. per hour
Basement posts		75 b.f. per hour	Knee wall studs	2 x 4 2 x 6	40 b.f. per hou. 50 b.f. per hour
Box sills		35 b.f. per hour			
Floor joists	2 x 6 − 2 x 8 2 x 10 − 2 x 12	65 b.f. per hour 70 b.f. per hour	Sheathing (Boards)	1 x 6 diag. 1 x 8 diag. 1 x 10 diag.	65 b.f. per hour 70 b.f. per hour 75 b.f. per hour
Headers, tail joists and trimmers	2 x 6 − 2 x 8 2 x 10 − 2 x 12	65 b.f. per hour 70 b.f. per hour	(Plywood)	4' x 8' Sheets	10-12 hours per 1000 sq. ft.
Bridging		5 sets per hour	Sheathing paper	(Included in sheathing)	
Sub-flooring (Boards)	Straight Diagonal	75 b.f. per hour 65 b.f. per hour	Corner boards		25 b.f. per hour
(Plywood)	4' x 8' Sheets	10-12 hours per 1000 sq. ft.	Common Rafters		35 b.f. per hour
Ceiling joists	2 x 6 − 2 x 8 2 x 10 − 2 x 12	65 b.f. per hour 70 b.f. per hour	Hip rafters		35 b.f. per hour
			Jack rafters		35 b.f. per hour
Ceiling backing	2 x 6 − 2 x 8 2 x 10 − 2 x 12	65 b.f. per hour 70 b.f. per hour	Valley rafters		35 b.f. per hour
			Ridge pole		35 b.f. per hour
Attic floor		75 b.f. per hour	Collar beams (rafter ties)		65 b.f. per hour
Outside wall plates and shoe	2 x 4 2 x 6	40 b.f. per hour 50 b.f. per hour	Roof sheathing (Boards)	1 x 6 S4S 1 x 6 center match 1 x 8 shiplap 1 x 10 shiplap	65 b.f. per hour 55 b.f. per hour 60 b.f. per hour 75 b.f. per hour
Outside studs	2 x 4 2 x 6	40 b.f. per hour 50 b.f. per hour	(Plywood)	4' x 8' Sheets	10-12 hours per 1000 sq. ft.
Headers for wall openings	2 x 4 2 x 6	40 b.f per hour 50 b.f. per hour			
Gable-end studs		50 b.f. per hour	Roofing felt	(included in roof covering)	
Fire-stopping		50 b.f. per hour	Roof covering (Asphalt − 4-in-1 shingle)		2.5 hrs. per sq.
Corner braces		50 b.f. per hour	Cornice	2 member 3 member 4 member	20 lin. ft. per hour 12 lin. ft. per hour 10 lin. ft. per hour
Partition plates and shoe		50 b.f. per hour			
Partition studs		50 b.f. per hour			

Estimated labor performance — rough carpentry
Figure 6-48

Labor For Roof Trusses

Span in Feet	Unit	Man-Hours Assembly	Man-Hours Placement
20, placed by hand	Each	2	3
30, placed by hand	Each	4	4
40, placed by crane	Each	10	3
50, placed by crane	Each	16	3
60, placed by crane	Each	19	3
80, placed by crane	Each	25	4

Suggested Crew: Hand placement - 2 carpenters and 2 laborers.
Crane placement - 1 operator, 2 to 3 men on guylines.

CEF — Roof truss labor
Figure 6-49

Standard Nail Requirements

Description of Material	Unit of Measure	Size and Kind of Nail	Number of Nails Required	Pounds of Nails Required
Wood Shingles	1,000'	3d Common	2,560	4 lbs.
Individual Asphalt Shingles	100 sq. ft.	7/8" Roofing	848	4 lbs.
Three in One Asphalt Shingles	100 sq. ft.	7/8" Roofing	320	1 lb.
Wood Lath	1,000'	3d Fine	4,000	6 lbs.
Wood Lath	1,000'	2d Fine	4,000	4 lbs.
Bevel or Lap Siding, 1/2" x 4"	1,000'	6d Coated	2,250	*15 lbs.
Bevel or Lap Siding, 1/2" x 6"	1,000'	6d Coated	1,500	*10 lbs.
Byrkit Lath, 1" x 6"	1,000'	6d Common	2,400	15 lbs.
Drop Siding, 1" x 6"	1,000'	8d Common	3,000	25 lbs.
3/8" Hardwood Flooring	1,000'	4d Finish	9,300	16 lbs.
13/16" Hardwood Flooring	1,000'	8d Casing	9,300	64 lbs.
Softwood Flooring, 1" x 3"	1,000'	8d Casing	3,350	23 lbs.
Softwood Flooring, 1" x 4"	1,000'	8d Casing	2,500	17 lbs.
Softwood Flooring, 1" x 6"	1,000'	8d Casing	2,600	18 lbs.
Ceiling, 5/8" x 4"	1,000'	6d Casing	2,250	10 lbs.
Sheathing Boards, 1" x 4"	1,000'	8d Common	4,500	40 lbs.
Sheathing Boards, 1" x 6"	1,000'	8d Common	3,000	25 lbs.
Sheathing Boards, 1" x 8"	1,000'	8d Common	2,250	20 lbs.
Sheathing Boards, 1" x 10"	1,000'	8d Common	1,800	15 lbs.
Sheathing Boards, 1" x 12"	1,000'	8d Common	1,500	12½ lbs.
Studding, 2" x 4"	1,000'	16d Common	500	10 lbs.
Joist, 2" x 6"	1,000'	16d Common	332	7 lbs.
Joist, 2" x 8"	1,000'	16d Common	252	5 lbs.
Joist, 2" x 10"	1,000'	16d Common	200	4 lbs.
Joist, 2" x 12"	1,000'	16d Common	168	3½ lbs.
Interior Trim, 5/8" thick	1,000'	6d Finish	2,250	7 lbs.
Interior Trim, 3/4" thick	1,000'	8d Finish	3,000	14 lbs.
5/8" Trim where nailed to jamb	1,000'	4d Finish	2,250	3 lbs.
1" x 2" Furring or Bridging	1,000'	6d Common	2,400	15 lbs.
1" x 1" Grounds	1,000'	6d Common	4,800	30 lbs.

*NOTE: Cement Coated Nails sold as two-thirds of Pound equals 1 Pound of Common Nails.

NOTE: Quantities determined for 1000 board feet of material or 100 sq. ft. of Surface.

CEF — Rough carpentry nails
Figure 6-50

Rough Carpentry Labor

Figure 6-48 gives a quick rundown on the estimated hourly labor requirements for many rough carpentry items. Use your own figures, of course, if you have them. They're sure to be more accurate for your work. A sample CEF for roof truss labor is shown in Figure 6-49.

Guides to estimating nail quantities are hard to find. A good one I've found is in Figure 6-50. Use it to make this area of your estimate easier.

Rough carpentry is the largest single cost category for most residential construction. Keep accurate manhour records. No published reference can be as accurate as figures compiled from *your* jobs in *your* area with *your* crews. Keep good cost records and use them to reduce the chance of a major error.

Chapter 7

Estimating Roof Covering

Estimating roof covering involves a little more math because the shapes are irregular and nearly always have a slope. That makes the surface area greater than the apparent area when measured from the floor plan. Otherwise, the procedure is the same. Prepare the estimate step-by-step in the same order the work is done to be sure no item of material or labor is overlooked. And the first step is to find the roof area.

Estimating Surface Area for Simple Roofs

Figure 7-1 shows the common roof styles. Find the roof surface for simple roofs by multiplying the rafter length by the eave length. Of course, if you calculated the roof area when estimating the sheathing, you already know the roof surface. Just allow an extra course of shingles for the starter

strip, ridge and hip caps. Adding one extra foot to the distance between ridge and eave and one extra foot to the distance between opposite rake ends will allow enough for starter and caps. Then divide the total roof area by 100. The answer is the number of roofing squares required.

Although roofs come in many sizes, shapes and styles, nearly every roof is composed of several flat surfaces. To figure the total roof surface, just divide the roof into simple geometric planes: squares, rectangles, trapezoids and triangles.

The simplest roof has no projecting dormers or intersecting wings. See Figure 7-2. Each roof shown is several rectangles. The total roof area is equal to the sum of the rectangles.

A shed roof has only one rectangle. Find the area by multiplying the eave line (A) by the rake line (B).

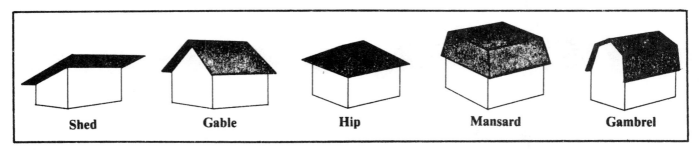

| Shed | Gable | Hip | Mansard | Gambrel |

Common roof styles
Figure 7-1

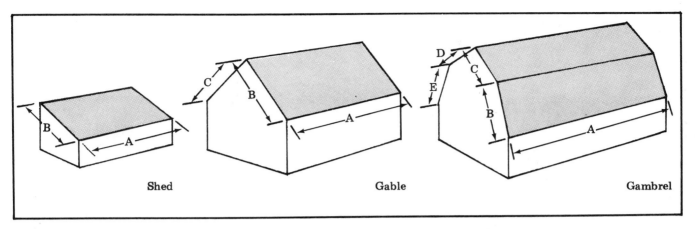

Finding the area of simple roofs
Figure 7-2

The gable roof has two rectangular planes. Find the area by multiplying the eave line by the sum of the two rake lines: A times the sum of B plus C.

A gambrel roof has four rake lines. Find the area by multiplying the eave line by the sum of the four rake lines: A times the sum of B plus C plus D plus E.

Estimating Surface Area for Complex Roofs
More complex roofs have projecting dormers or intersecting wings. Area calculations for these roofs use the same basic method as for simple roofs; the work is just a little more complicated because there are more roof surfaces to figure. Calculate each surface separately. Then add them all together to find the total roof surface. You can determine the roof surface by scaling dimensions off the plans, by measuring the distances on the roof, or indirectly, by calculating the projected horizontal area of the roof.

Calculate the projected horizontal area of the roof with a roof slope conversion table. Both the roof slope and the projected horizontal area can be determined indirectly. In the next section I'll tell you how. And I'll include tables for converting indirect measurements to true surface area.

Projected Horizontal Area
No matter how complicated a roof is, its projection onto a horizontal plane will show you the total horizontal surface covered by the roof. Figure 7-3 shows a roof complicated by valleys, dormers and ridges at different elevations. The lower half of the figure shows the projection of the roof onto a horizontal plane. In the projection, inclined surfaces appear flat, and intersecting surfaces appear as lines.

Measurements for the horizontal projection of the roof can be made from the plans, from the ground, or from inside the attic. Once the measurements are made, the horizontal area covered by the roof can be drawn to scale and calculated.

Because surface area is a function of slope, calculations must be grouped by roof slope. Don't combine calculations for different slopes until the area of each has been determined. Let's do the calculations for the roof in Figure 7-3. The horizontal area under the 9-inch slope is:

$$26 \times 30 = 780$$
$$19 \times 30 = \underline{570}$$
$$\text{Total} \quad 1,350 \text{ SF}$$

From this gross figure you have to deduct the area of the chimney and the triangular area of the ell roof that overlaps and has a different slope than the main roof:

$$\text{Chimney} = 16 \text{ SF}$$
$$\text{Ell roof: } \frac{1}{2}(16 \times 5) = \underline{40} \text{ SF (triangular area)}$$
$$56 \text{ SF}$$

The net projected area of the main roof is:

$$1,350 \text{ minus } 56 = 1,294 \text{ SF}$$

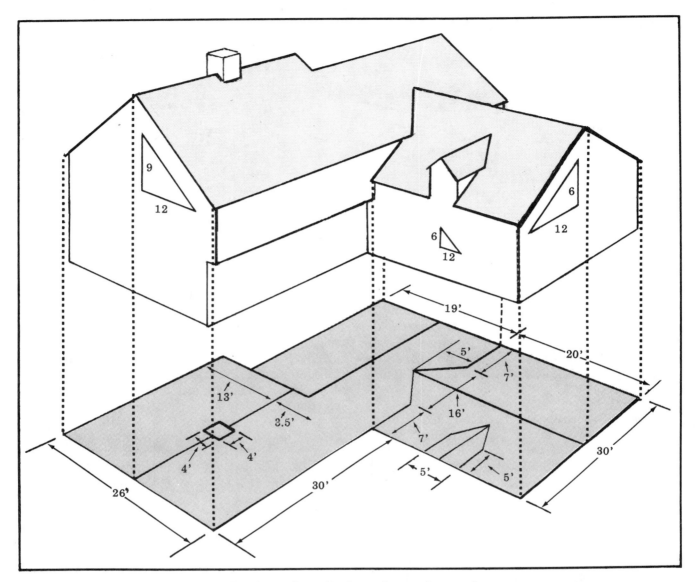

Horizontal projection of complex roof
Figure 7-3

The horizontal area under the 6-inch slope roof is:

$$20 \times 30 = 600 \text{ SF}$$
$$\tfrac{1}{2}(16 \times 5) = \underline{40 \text{ SF}}$$
$$\text{Total} \qquad 640 \text{ SF}$$

Don't forget to include the duplications. Portions of higher roof surfaces sometimes project over the roof surfaces below them. The horizontal projection doesn't show the overlap. These duplicated areas must be added to the total horizontal area. In Figure 7-3 there are three overlaps and we'll assume each one overlaps 4 inches.

1) On the 6-inch slope roof, where the two dormer eaves overhang the ell roof:

$$2(5 \times 4/12) = 3\text{-}1/3 \text{ SF}$$

Add this to the horizontal area of the 6-inch slope roof.

2) Now figure the 9-inch slope roof, where the main roof eave overhangs the ell section. The ell section is 30 feet wide. Sixteen feet of the ell roof connect with the main roof, leaving a 7-foot eave overhang on each side of the ell ridge.

$$2 (7 \times 4/12) = 4\text{-}2/3 \text{ SF}$$

Add the 4⅔ square feet to the horizontal area of the 9-inch slope roof.

3) Finally, calculate the 9-inch slope roof, where the main-roof rake overhangs the smaller section of the main roof in the rear of the building. The overhang of the 26-foot main roof covers only half of the 19-foot wide section.

$$9.5 \times 4/12 = 3\text{-}1/6 \text{ SF}$$

Add this to the horizontal area of the 9-inch slope roof.

So now you can calculate the final projected horizontal area for each roof slope. Round any fractions to the nearest square foot. For the 6-inch slope roof, the adjusted total is:

$$\begin{array}{r} 640 \\ + 3 \\ \hline 643 \text{ SF} \end{array}$$

For the 9-inch slope roof, the adjusted total is:

$$\begin{array}{r} 1{,}294 \\ + 8 \\ \hline 1{,}302 \text{ SF} \end{array}$$

Conversion to True Surface Area

Once you know the projected horizontal area for each roof slope, the next step is to convert the results to true surface area. The conversion table in Figure 7-4 handles this for you. To use the table, simply multiply the projected horizontal area by the conversion factor for the appropriate roof slope. The result is the true surface area of the roof.

For example, for the 9-inch slope roof:

Horizontal area	x Conversion factor	= Actual area
1,302 SF	x 1.250	= 1,627.5 SF

Slope (inches per foot)	Area/rake factor
4	1.054
5	1.083
6	1.118
7	1.157
8	1.202
9	1.250
10	1.302
11	1.356
12	1.414

Area/rake conversion table
Figure 7-4

For the 6-inch slope roof:

Horizontal area	x Conversion factor	= Actual area
643 SF	x 1.118	= 718.8 SF

When all horizontal areas have been converted to true surface areas, total the surface areas. The sum is the total roof area:

$$\begin{array}{r} 1{,}628 \\ + 719 \\ \hline 2{,}347 \text{ SF} \end{array}$$

You'll want to make an allowance for waste when ordering materials. In this case, assume 10% waste. Thus, the roofing material required is:

$$\begin{array}{r} 2{,}347 \\ + 235 \quad 10\% \text{ waste} \\ \hline 2{,}582 \text{ SF} \end{array}$$

The same roof slope and same horizontal area will always result in the same true surface area, regardless of roof style. In other words, if a shed roof, gable roof or hip roof each had the same slope and covered the same horizontal area, they would each require the same amount of roofing material to cover.

Estimating Asphalt Roofing

More than 80 percent of all residential roofing applied in the U.S. is asphalt, either shingles, roll

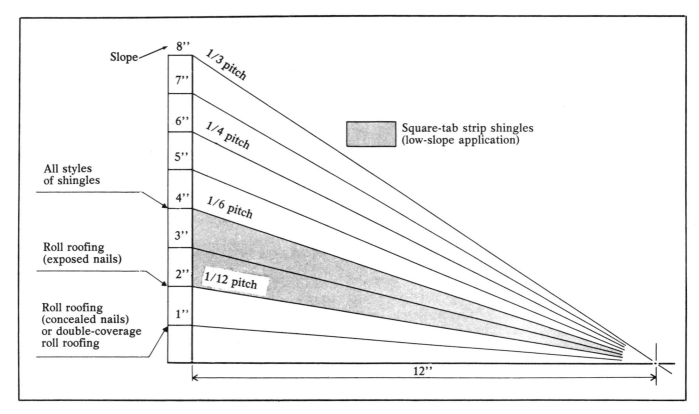

Slope limitations for asphalt roofing materials
Figure 7-5

roofing or saturated felt. Over the past dozen years, the trend in asphalt roofing has been to use fiberglass-base shingles rather than conventional organic or paper base. Figure 7-5 gives slope limitations for asphalt roofing materials.

Asphalt Shingles

Asphalt shingles are the most common roofing material used today. They're manufactured as strip shingles, interlocking shingles and giant individual shingles (16 inches long, 12 inches wide). Figure 7-6 shows the typical asphalt shingles.

Strip shingles are rectangular, measuring approximately 12 inches wide by 36 inches long, and may have as many as five cutouts. Cutouts separate the tabs of a large shingle so the finished roof looks like it's made of many smaller units. Strip shingles are also made without cutouts to produce a much different effect.

Many shingles are treated with a factory-applied, self-sealing adhesive — generally a thermoplastic material — which softens in the sun after the shingle is in place. Heat bonds each shingle securely to the one below so it won't blow off so easily in

high wind. Figure 7-7 is a production table for asphalt roofing.

The most common weight of 12 inch by 36 inch asphalt strip shingles is 235 to 240 pounds per 100 square feet (square). A 5-inch weather exposure requires 80 shingles to cover 100 square feet. They come packed three bundles to the square. One square of shingles is always the number of shingles required to cover 100 square feet of roof area at standard exposure.

Roll Roofing

Roll roofing comes in rolls approximately 36 inches wide and 36 to 38 feet long. It weighs from 40 to 90 pounds per square and is available with either a smooth surface or a surface embedded with mineral granules.

Mineral-surfaced roll roofing is available with a granule-free selvage edge in the area where each succeeding course should overlap the preceding course. The amount of overlap determines how much of the material is exposed to the weather and how much roof you can cover with each roll. Roll roofing is also used as a flashing material. Figure 7-8 gives data on typical asphalt rolls.

| PRODUCT | Configuration | Per Square | | Size | | Exposure | Underwriters Laboratories Listing |
		Approximate Shipping Weight	Shingles	Bundles	Width	Length		
Self-sealing random-tab strip shingle 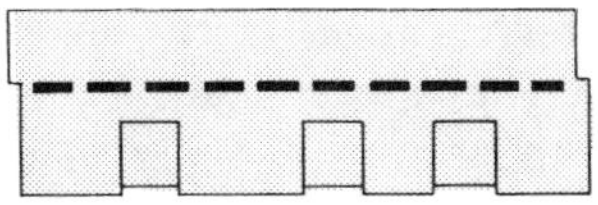 Multi-thickness	Various edge, surface texture and application treatments	285# to 390#	66 to 90	4 or 5	11½" to 14"	36" to 40"	4" to 6"	A or C - Many wind resistant
Self-sealing random-tab strip shingle 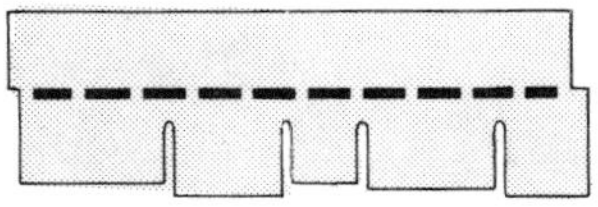 Single-thickness	Various edge, surface texture and application treatments	250# to 300#	66 to 80	3 or 4	12" to 13¼"	36" to 40"	5" to 5⅝"	A or C - Many wind resistant
Self-sealing square-tab strip shingle 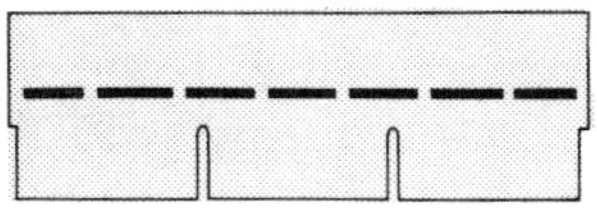 Three-tab	Two-tab or Four-tab	215# to 325#	66 to 80	3 or 4	12" to 13¼"	36" to 40"	5" to 5⅝"	A or C - All wind resistant
	Three-tab	215# to 300#	66 to 80	3 or 4	12" to 13¼"	36" to 40"	5" to 5⅝"	
Self-sealing square-tab strip shingle 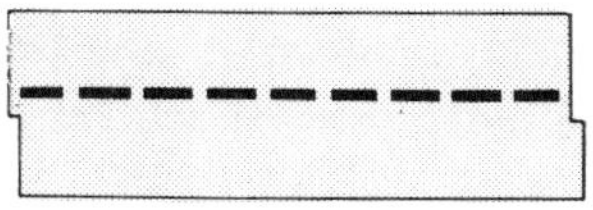 No-cutout	Various edge and surface texture treatments	215# to 290#	66 to 81	3 or 4	12" to 13¼"	36" to 40"	5" to 5⅝"	A or C - All wind resistant
Individual interlocking shingle 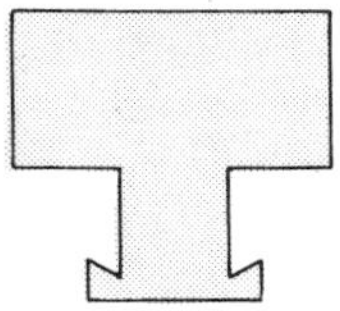 Basic design	Several design variations	180# to 250#	72 to 120	3 or 4	18" to 22¼"	20" to 22½"	—	C - Many wind resistant

Typical asphalt shingles
Figure 7-6

Labor and Materials for Asphalt Roofing

Type of Roofing	Shingles Per 100 S.F.	Nails Per Shingle	Length of Nail*	Nails Per 100 S.F.	Pounds Per 100 S.F. (Approximate)		Man-Hours Per 100 S.F.
					12 Ga. by 7/16" Head	11 Ga. by 7/16" Head	
Roll roofing on new deck	..	..	1"	252**	0.73	1.12	1.0
Roll roofing over old roofing	..	..	1¾"	252**	1.13	1.78	1.25
19" selvage over old shingles	..	..	1¾"	181	0.83	1.07	1.0
3 tab sq. butt on new deck	80	4	1¼"	336	1.22	1.44	1.5
3 tab sq. butt reroofing	80	4	1¾"	504	2.38	3.01	1.85
Hex strip on new deck	86	4	1¼"	361	1.28	1.68	1.5
Hex strip reroofing	86	4	1¾"	361	1.65	2.03	2.0
Giant American	226	2	1¼"	479	1.79	2.27	2.5
Giant Dutch lap	113	2	1¼"	236	1.07	1.39	1.5
Individual hex	82	2	1¾"	172	.79	1.03	1.5

*Length of nail should always be sufficient to penetrate at least ¾" into sound wood. Nails should show little, if any, below underside of deck.
**This is the number of nails required when spaced 2" apart.

Asphalt roofing production
Figure 7-7

Saturated Felt

This material has dry felt impregnated with asphalt. It's used primarily as an underlayment for asphalt shingles, roll roofing and other types of roofing materials, and as sheathing paper. It's available in different weights, the most common being No. 15, weighing about 15 pounds per square, and No. 30, which weighs about 30 pounds per square.

No. 15 felt comes in rolls 3 feet wide x 144 feet long (432 square feet). No. 30 comes in rolls 3 feet wide and 72 feet long (216 square feet).

To estimate the rolls of No. 15 felt required, divide the roof area by 400 and round off to the next highest number:

$$\frac{2,582 \text{ SF}}{400} = 6.45 \text{ or } 7 \text{ rolls}$$

To figure No. 30 felt, divide the roof area by 200 and round off to the next highest number.

$$\frac{2,582 \text{ SF}}{200} = 12.91 \text{ or } 13 \text{ rolls}$$

Additional Material Estimates

To complete the estimate, find the quantity of starter strip, drip edge, hip and ridge shingles, and valley flashing. The longer the eaves, ridges, rakes, hips, and valleys, the more starter strip and flashing you'll need.

Eaves and ridges are horizontal. You can scale their length directly from the horizontal projection drawing or from your plans. Rake, hips and valleys are sloped. Calculate the length the same way you calculated sloping roof surface.

To find the true length of a rake, first measure its projected horizontal length. Then use Figure 7-4 to convert projected horizontal length to true length. To use the table, multiply the rake's projected horizontal length by the conversion factor for the appropriate roof slope. The result is the true length of the rake.

For the house in Figure 7-3, the rakes at the ends of the main house have horizontal distances of 26 and 19 feet. There's another rake in the middle of the main house where the higher roof section meets the lower. Its horizontal distance is 13 plus 3.5 for the short rake, or 16.5 feet. Adding all these horizontal distances together gives a total of 61.5 feet. Then use the conversion table in Figure 7-4:

Horizontal length	x	Conversion factor	=	Actual length
61.5 feet	x	1.250	=	76.9 feet

Follow the same procedure for the ell section with its 6-inch slope roof and dormer. The horizontal length of rakes is 35 feet.

To find the quantity of drip edge required to do the job, add these rake lengths to the sum of the lengths of the eaves. The eave lengths are true horizontal distances, so no conversion is necessary.

You can take the quantity of ridge shingles required directly from the drawings, since the ridge will be horizontal.

Hips and valleys slope away from the ridge. Convert their projected horizontal lengths to true lengths with Figure 7-9.

First, measure the length of the hip or valley on the plans. Multiply that figure by the conversion factor for the appropriate roof slope. The result is the true length of the hip or valley. Total the length of all hips and valleys to find the hip shingles and valley flashing needed.

Now we'll find the total valley length for the house in Figure 7-3.

There's a valley formed on both sides of the ell-roof intersection with the main roof. The total measured distance for these valleys on the horizontal projection is 16 feet.

The fact that two different slopes are involved makes the procedure a little more complicated. If

PRODUCT	Approximate Shipping Weight		Squares Per Package	Length	Width	Side or End Lap	Top Lap	Exposure	Underwriters Laboratories Listing
	Per Roll	Per Square							
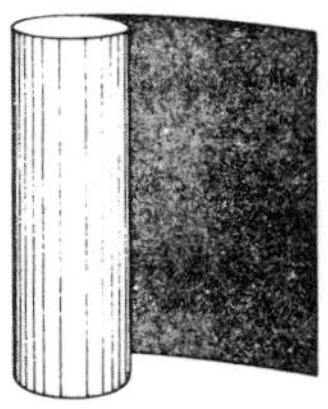 Mineral surface roll	75# to 90#	75# to 90#	1	36' to 38'	36"	6"	2" to 4"	32" to 34"	C
	Available in some areas in 9/10 or 3/4 square rolls.								
 Mineral surface roll (double coverage)	55# to 70#	110# to 140#	½	36'	36"	6"	19"	17"	C
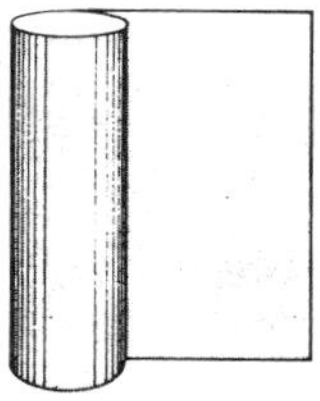 Smooth surface roll	40# to 65#	40# to 65#	1	36'	36"	6"	2"	34"	None
 Saturated felt (non-perforated)	60#	15# to 30#	2 to 4	72' to 144'	36"	4" to 6"	2" to 19"	17" to 34"	None

Typical asphalt rolls
Figure 7-8

the roof only had one slope, you could calculate the true length directly from Figure 7-9. But in this case, you need to calculate both slopes and then average them together to approximate the true length of the valleys:

Horizontal length	x	Conversion factor	=	Actual length
16 feet	x	1.600 (for 9" slope)	=	25.6 feet
16 feet	x	1.500 (for 6" slope)	=	24.0 feet
Average: (24.0 + 25.6)/2			=	24.8 feet

The approximate true length of the two valleys is 24.8 feet, or 12.4 feet each.

The projected horizontal length of the dormer valleys in Figure 7-3 is 5 feet. Since both the ell roof and the dormer roof have slopes of 6 inches, the actual length of the valleys will be 7.5 feet, using Figure 7-9. The projected horizontal length of

Slope (inches per foot)	Hip/valley factor
4	1.452
5	1.474
6	1.500
7	1.524
8	1.564
9	1.600
10	1.642
11	1.684
12	1.732

Hip/valley conversion table
Figure 7-9

valleys is the distance at the widest point between the valleys. In the case of dormer valleys, it's the width of the dormer.

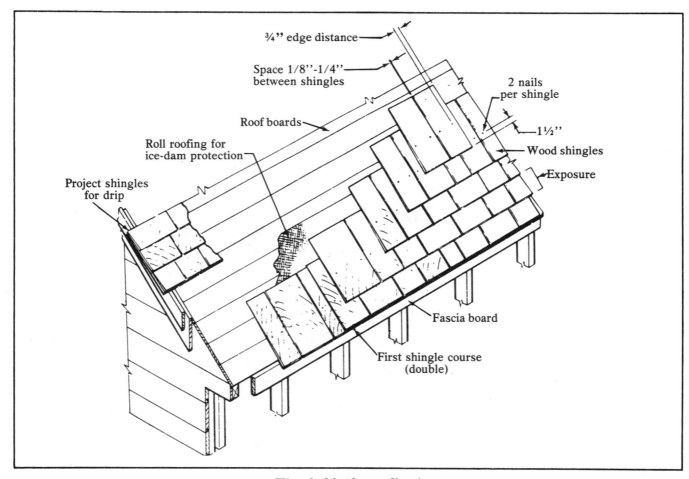

Wood shingle application
Figure 7-10

Total true valley length for the house is 25 plus 8, or 33 feet.

How to Estimate Wood Shingles

Wood shingles are sawed from several species of wood, but red cedar is the most common. The basic grades of red cedar are Number 1 (premium grade), Number 2 (a good grade for most applications), and Number 3 (a utility or economy grade).

Each grade comes in lengths of 16, 18, and 24 inches, and in random widths. The approximate weight per square for each grade is:

Length	Weight
16-inch	144 pounds
18-inch	158 pounds
24-inch	192 pounds

Figure 7-10 shows wood shingle application details.

When estimating shingles, the area exposed to the weather is the only area to consider. The lap, the part of the shingle that's covered by the piece above and isn't exposed to the weather, doesn't enter into the calculations. For instance, the face of an average wood shingle measures 4 inches by 18 inches. That's 72 square inches. When laid on a roof and lapped by the shingle above, the exposed surface is only 4 inches by about 4½ inches, or 18 square inches. That's a big difference.

There are 14,400 square inches in a square. So divide 14,400 by 18 square inches to find the number of these shingles in a square. (14,400 divided by 18 is 800.) If you add 10% for waste in cutting on hips, valleys and double courses, there are 880 shingles (4 inches x 18 inches with 4½ exposure) in each square. Eight shingles will cover one square foot of roof.

The maximum recommended weather exposure for 16-inch shingles is 5 inches; for 18-inch shingles it's 5½ inches; and for 24-inch shingles, 7½ inches. These standard exposures are recommended for all roofs with 1/4 pitch and steeper. On flatter roof slopes, the weather exposure should be reduced to 3¾ inches for 16-inch shingles, 4½ inches for 18-inch shingles, and 5¾ inches for 24-inch shingles.

Wood shingles are sold by the square but are packed in bundles of about 250 shingles. Four bundles will cover a square at standard exposure. Four bundles, then, make up one square; three bundles will cover 3/4 of a square; two bundles will cover 1/2 of a square and one bundle will cover 1/4 of a square. If you know the number of squares, it's easy to estimate the number of bundles. Figure 7-11 gives coverage figures for four bundles of each of three standard shingle sizes.

Shingle length	Minimum thickness	Approximate square foot coverage of one square (4 bundles) shingles based on these weather exposures							
		4"	4½"	5"	5½"	6"	6½"	7"	7½"
16"	5 in 2"	80	90	100*	110	120	130	140	150
18"	5 in 2¼"	72½	81½	90½	100*	109	118	127	136
24"	4 in 2"	---	---	---	---	80	86½	93	100*

Wood shingle coverage
Figure 7-11

	Material per 100 square feet of surface			Nails per 100 square feet		
Laid to weather	Shingles per 100 SF	Waste	Shingles per 100 SF w/waste	3d nails	4d nails	Labor hours per 100 SF
4"	900	10%	990	3¾ pounds	6½ pounds	3¾
5"	720	10%	792	3 pounds	5¼ pounds	3
6"	600	10%	660	2½ pounds	4¼ pounds	2½

Note: Nails based on using 2 nails per shingle. Increase time factor 25% for hip roofs.

Material and labor for estimating wood shingles
Figure 7-12

Handsplit and resawn	Approximate square foot coverage of one square of hand-split shakes based on these weather exposures						
	5½"	6½"	7"	7½"	8"	8½"	10"
18" x ½" to ¾"	55*	65	70	75**	80	85	--
18" x ¾" to 1¼"	55*	65	70	75**	80	85	--
24" x ½" to ¾"	--	65	70	75*	80	85	100**
24" x ¾" to 1¼"	--	65	70	75*	80	85	100**
32" x ¾" to 1¼"	--	--	--	--	--	--	100*

Note: *Recommended maximum weather exposure for 3-ply roof construction.
**Recommended maximum weather exposure for 2 ply roof construction.

Handsplit shake coverage
Figure 7-13

Figure 7-12 shows the material, nails and labor for estimating wood shingles. To begin the estimate, divide the roof area by the coverage for the weather exposure to be used, then add 10% for gable roofs for the starter strip and ridge. For hip roofs, add 15% for the starter strip, ridge and hips.

Example: A gable roof with a roof area of 1,250 square feet will be covered with 18-inch red cedar shingles with a weather exposure of 4½ inches. The coverage of one square in Figure 7-11 is 81.5 square feet. The calculation is:

$$\frac{1,250 \text{ SF}}{81.5 \text{ coverage}} = 15.33 \text{ squares}$$

Add 10% (1.53) = 16.86 squares

To find the number of bundles, multiply 16.86 by 4. You need 67.44 bundles. Round it up to 68 bundles.

Let's do another one. A hip roof with an area of 1,425 square feet will be covered by 16-inch red cedar shingles with a 5½-inch weather exposure. From Figure 7-11, the coverage will be 110 square feet:

$$\frac{1,425 \text{ SF}}{110 \text{ coverage}} = 12.95 \text{ squares}$$

Add 15%   12.95 squares
            1.94
           —————————————
           14.89 or 15 squares

Multiply by 4 for the number of bundles: 4 x 15 is 60 bundles of red cedar shingles.

Estimating Shakes
Handsplit or resawn shakes are split wood shingles with a rough texture. They're available only in Number 1 grade and come in lengths of 18, 24, and

Kind of shingle	Size in inches	Exposure in inches		Number per square	Weight per square	Weight of nails per square
American	12 x 24	7	x 24	86	275 lbs.	1.0 lb.
American	14 x 30	6	x 30	80	265 lbs.	2.0 lb.
Dutch lap	12 x 24	9	x 20	80	260 lbs.	1.0 lb.
Dutch lap	16 x 16	12	x 13	92	260 lbs.	0.8 lb.
Dutch lap	16 x 16	10-2/3	x 13	104	295 lbs.	1.0 lb.
Hexagonal	16 x 16	13	x 13	86	265 lbs.	1.0 lb.

Asbestos shingle data
Figure 7-14

Labor for Asbestos Cement Roofing Shingles

Work Element	Unit	Man-Hours Per Unit
14" x 30" x ⁵⁄₃₂"	SQ.	3.45
9" x 16" x ¼"	SQ.	3.20
8" x 16", 6" exposure	SQ.	4.10
16" hex, 8" exposure	SQ.	3.45
9" x 32", with underlay	SQ.	4.30
12" x 24" with underlay	SQ.	4.30

Time is man-hours per 100 square feet of roof and includes move on and off site, unloading and stacking. Shingles on sidewalls usually take 10 to 20% less time because exposures can be increased. If exposures are not increased, add 10% for shingling sidewalls.

**Shingle roofing labor
Figure 7-15**

32 inches, and random widths. They're packed five bundles per square. Shakes vary in weight from approximately 225 pounds per square for the 18" x 1/2" x 3/4" size to 450 pounds for the 32" x 3/4" x 1¼" size. Figure 7-13 gives the approximate coverage of one square of handsplit shakes for weather exposures from 5½ inches to 10 inches.

Estimate shakes the same way as we estimated wood shingles.

Labor Installing Built-up Bituminous Roofing

Work Element	Unit	Man-Hours Per Unit
Bitumen top built-up roofing with 15 lb. asphalt felts		
Three ply	SQ.	1.8
Four ply	SQ.	2.0
Five ply	SQ.	2.4
Gravel surfaced built-up roofing		
Four ply asphalt felt, 15 lb.	SQ.	2.8
Five ply asphalt felt, 15 lb.	SQ.	3.2
Gravel surfaced 90 lb. cap sheet on built-up roofing		
Three ply, 15 lb. felt	SQ.	2.3
Four ply, 15 lb. felt	SQ.	2.9
Cut and place cant strips	100 L.F.	2.0

Time includes move on and off site, unloading, stacking, cleanup and repairs as necessary. For sloped roofs over 2 in 12 pitch add 10 to 15% to time. Time does not include removal of old roofing.
Suggested crew: 2 roofers, 1 laborer.

**Built-up roofing labor
Figure 7-16**

Use rust-resistant 1½" (4d) or 2" (6d) nails. Use two pounds of 4d nails per square and three pounds of 6d nails per square.

Estimating Asbestos Shingles
Asbestos shingles are sold by the square (covering 100 square feet of surface) and are available in many styles and colors. Roof sheathing is covered with a 15 to 30 pound felt before applying the shingles. The shingles are punched at the factory so they can be fastened with roofing nails and storm anchors.

To estimate the quantity needed, divide the roof area by 100. That's the number of squares needed. The linear feet of starter shingles needed is equal to the length of the eaves. You'll need special shingles or a ridge roll for hips, valleys and ridges. To find the length of these, use the procedure we covered in the section on asphalt shingles.

Allow 5% for waste on a shed or gable roof. Add 2% more (a total of 7%) if the roof has hips or valleys. Figure 7-14 gives materials and sizes for asbestos shingle roofing. Figure 7-15 lists the manhours.

Use copper or zinc-coated nails long enough to pass through the shingle and through 3/4 of the thickness of the roof sheathing.

Sheet Metal Work

Work Element	Unit	Man-Hours Per Unit
Fabrication (galvanized steel)		
Roof gutters	100 L.F.	2.5
Downspouts	100 L.F.	2.5
Roof ridges	100 L.F.	1.5
Roof valleys	100 L.F.	1.5
Flashing	100 L.F.	2.3
Installation (galvanized steel)		
Roof ridges	100 L.F.	2.5
Roof valleys	100 L.F.	2.5
Roof flashing	100 L.F.	8
Roof gutters	100 L.F.	4
Downspouts	100 L.F.	4

Fabrication is usually performed by a sheet metal shop and includes making patterns, cutting, forming, seaming, soldering, attaching stiffeners, and loading for delivery.
Installation includes unloading, storing on site, handling into place, hanging, fastening, and soldering.
Suggested Crew: two to six steel workers depending on the weight and length of the materials

**Sheet metal labor for roofing
Figure 7-17**

Other Roof Coverings

Figure 7-16 shows labor for built-up roofing. For sheet metal work on roofs, such as flashing and gutters, consult Figure 7-17. This provides average manhours for fabrication and installation.

Generally, carpenters make slow roofers. In most cases it's more economical to subcontract roofing jobs to specialists. Roofers commonly charge by the square. The price will vary with the slope of the roof, the number of cut-ups such as valleys and hips, and the amount of flashing required.

Chapter 8

Estimating Insulation

The energy crisis of the mid-1970's permanently changed the way houses are insulated in this country. Wise builders will never go back to energy-foolish construction, even if the law permitted it. Make sure your houses are energy-efficient — they'll sell better and perform better. And be alert to energy-saving concepts so your buyers and customers get good value for their buying dollar.

Common thermal insulating materials include mineral fibers (glass, rock, and slag), vegetable fibers (wood, cane, cotton, wool, and redwood bark), expanded mineral granules (perlite and vermiculite), vegetable granules (like ground cork), foamed materials (both glass and synthetic resins) and aluminum foil.

Insulating batts, blankets, and boards are made from fibers or granules. The fibers or granules are mixed with binders, and formed into useful widths, lengths and thicknesses. The binders make insulation more water- and mildew-resistant and add strength to the finished product. Vegetable fibers are usually chemically treated to make them fire-resistant. Glass and synthetic resins are formed into blocks, sheets, or boards of various size by carefully controlling the foaming process to produce the right cell size and density.

Forms of Insulation
Thermal insulation comes in a variety of forms.

These include batts, blankets, reflective insulators, and loose fill. We'll look at them one at a time.

Insulating Batts or Blankets
Batts and blankets are placed between joists or studs. They're made in widths that fit standard joist and stud spacing. Batts or blankets may be unfaced, or they may be faced (on one or both sides) with paper, aluminum foil or plastic. If the batts or blankets are faced, the faced side has a lower vapor-permeability rate and acts as a vapor barrier.

Paper-faced batts or blankets are made with continuous paper flanges along the long edges to staple to studs and joists.

Reflective Insulators
Reflective insulators are made with aluminum foil. The foil is usually reinforced with paper backing. The insulators are generally made to provide two or more reflective surfaces with air space between the layers. Like batts or blankets, they're used between studs or joists and have flanges for nailing or stapling.

Loose Fill
Loose fill is made from mineral or vegetable fibers or granules. It's applied by blowing or hand spreading. You'll see it primarily in masonry cavity

walls, and over ceilings in attic spaces. But it can also be blown into cavities between wall studs.

Foamed-in-place insulation is two-part synthetic resin material in liquid form which can be mixed together and deposited in a cavity. Within a certain time after mixing, a foaming action takes place, filling the cavity.

Some thermal insulation is manufactured to serve more than one purpose. For example, structural insulating boards serve as sheathing, roof decks, and permanent concrete form boards. Other thermal insulating materials serve as sound insulators and decorative panels or tiles.

Vapor Barriers

Vapor barriers prevent condensation of moisture in insulated spaces. When warm, moisture-laden air comes in contact with a cold surface, little droplets of moisture collect on that surface. When moisture collects inside of insulating material, the insulation

value of the material decreases. That's why vapor barriers must be installed on the warm side of the insulation. The warm side is almost always the side facing the interior of the structure.

The vapor barrier should have a "perm" rating of one perm or less. Aluminum foil, plastic sheet material, and coated or laminated paper all qualify. Ordinary 15- or 30-pound asphalt-saturated felts are *not* acceptable. The greater the temperature difference from one side of the wall to the other, and the greater the relative humidity of the air on the warm side, the more effective the vapor barrier must be.

Many batts and blankets have vapor barriers already attached to one side. Other forms of insulation will require a separate membrane of polyethylene. A vapor barrier is essential if you're building in the condensation zone shown on the map in Figure 8-1.

When installing a separate membrane as vapor

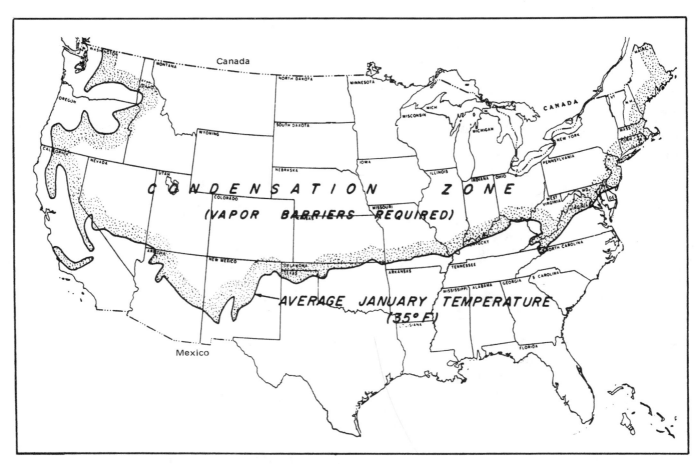

**Condensation zone: Areas where average January
temperature is 35°F or lower
Figure 8-1**

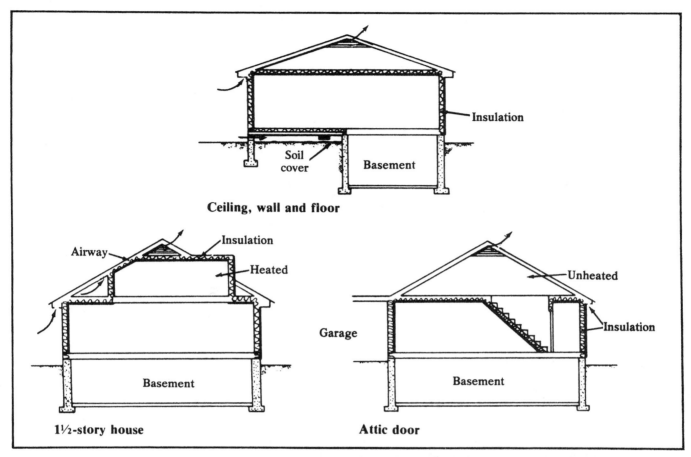

Place the insulation around heated areas of the house
Figure 8-2

barrier, estimate 2 manhours per 100 square feet of wall area and 3 manhours per 100 square feet of ceiling area.

Figure 8-2 shows how insulation should form a blanket around the heated area of a house. Insulation and vapor barriers are commonly installed in ceilings, walls, floors, second stories, and basement rooms.

Figure 8-3 shows the average winter-low-temperature zones in the United States.

Insulation Values

To get sufficient insulation values, you may have to double layers of insulation or combine layers having different insulation values. Figure 8-4 shows some of the suggested combinations.

And remember, the FTC rule "Labeling and Advertising of Home Insulation" requires that you provide certain information about the insulation to the buyer to help them evaluate competing pro-

ducts. If you're not familiar with the rule, it's covered in Volume 1 of this manual. Better yet, get a current copy of the rule on home insulation from the FTC at:

Public Reference Branch
Federal Trade Commission Headquarters
6th Street and Pennsylvania Ave., N.W.
Washington, D.C. 20580

Estimating Insulation and Labor

From the floor plan, determine the area for the ceiling and floor insulation. For the wall area, multiply the perimeter of the walls by the wall height for the gross area. Deduct any window and door openings for the net area. Divide the area to be insulated by the amount of insulation in each bag or roll, and round off to the next higher number of bags or rolls.

Here's an example to show how it works. You

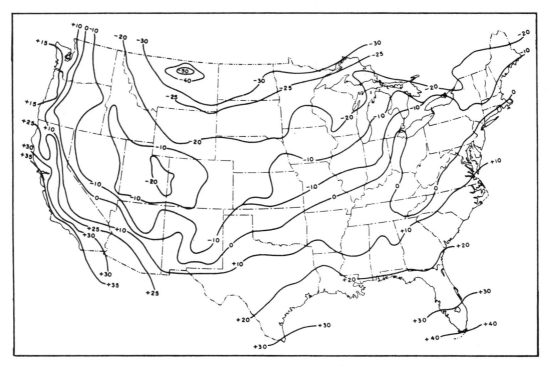

Average winter-low-temperature zones
Figure 8-3

Ceilings, double layers of batts or blankets:		**R-30**	R-19 (6'') mineral fiber and 11 bags of wool per 1,000 SF (5'')
R-38	Two layers of R-19 (6'') mineral fiber	**R-26**	R-19 (6'') mineral fiber and 8 bags of wool per 1,000 SF (3¼'')
R-33	One layer of R-22 (6½'') and one layer of R-11 (3½'') mineral fiber		
R-30	One layer of R-19 (6'') and one layer of R-11 (3½'') mineral fiber	**Walls, 2 x 4 framing:**	
R-26	Two layers of R-13 (3⅝'') mineral fiber	R-19	R-13 (3⅝'') mineral fiber batts and 1'' foam plastic sheathing
		R-11	R-11 (3½'') mineral fiber batts
Ceilings, loose fill mineral wool and batts:		**Floors:**	
R-38	R-19 (6'') mineral fiber and 20 bags of wool per 1,000 SF (8¾'')	R-22	R-22 (6½'') mineral fiber
		R-19	R-19 (6'') mineral fiber
R-33	R-22 (6½'') mineral fiber and 11 bags of wool per 1,000 SF (5'')	R-13	R-13 (3⅝'') mineral fiber
		R-11	R-11 (3½'') mineral fiber

Combining layers of insulation
Figure 8-4

need to insulate 1,645 net square feet of area with R-11 insulation (3½'' x 15''), packed 88 square feet per roll:

$$\frac{1,645}{88} = 18.69 \text{ or } 19 \text{ rolls}$$

Many estimators, when estimating batt or blanket insulation requirements, first calculate the total area of the ceiling, floor or wall and then deduct about 5% of the total area for the thickness of the framing members. So a 1,000 square foot floor area requires 950 square feet of insulation.

Size	Square feet	Material Number of batts per 100 SF	Number of SF for 100 SF wall	Staples per 100 SF
15 x 24	2.5	40	95	160
15 x 48	5.	20	95	160
19 x 24	3.7	32	96	160
19 x 48	6.33	16	95	160
23 x 24	3.84	26	95	160
23 x 48	7.67	13	100	160

Notes: Glass, mineral or rock wool batts with paper back roll insulation or strip insulation.
Studding or joist excluded.
Batts stapled @ 6'' o.c.

**Estimating materials for batt insulation
Figure 8-5**

Fill thickness	Material Number of square feet covered per cubic foot density				
	6 pounds	7 pounds	8 pounds	9 pounds	10 pounds
1''	21.1	18.0	15.9	14.1	13.0
2''	10.6	9.1	8.0	7.1	6.4
3''	7.1	6.1	5.3	4.7	4.2
4''	5.3	4.6	4.0	3.5	3.2

**Estimating materials for loose fill insulation
Figure 8-6**

Labor Installing Batt or Blanket Insulation

Work Element	Unit	Man-Hours Per Unit
Fiberglass blankets		
2'', paper backed	100 S.F.	.88
2½'', paper backed	100 S.F.	.88
4'', paper backed	100 S.F.	1.05
6'', paper backed	100 S.F.	1.40
2'', foil backed 1 side	100 S.F.	.88
3'', foil backed 1 side	100 S.F.	.96
4'', foil backed 1 side	100 S.F.	1.05
6'', foil backed 1 side	100 S.F.	1.40
2'', foil backed 2 sides	100 S.F.	.88
3'', foil backed 2 sides	100 S.F.	.96
4'', foil backed 2 sides	100 S.F.	1.05
6'', foil backed 2 sides	100 S.F.	1.40
Fiberglass batts		
2'', plain, unbacked batts	100 S.F.	.79
3'', plain, unbacked batts	100 S.F.	.96
4'', plain, unbacked batts	100 S.F.	1.14
6'', plain, unbacked batts	100 S.F.	1.49
Mineral wool batts		
2'', paper backed	100 S.F.	1.00
4'', paper backed	100 S.F.	1.30
6'', paper backed	100 S.F.	1.80

Time includes move-on and off-site, unloading and stacking, installing by staple only.
Suggested Crew: 1 installer, 1 helper

**CEF — Labor installing batt or blanket insulation
Figure 8-7**

Figure 8-5 will help you quickly estimate the materials for insulation batts. Figure 8-6 does the same thing for loose fill insulation.

You'll want your CEF to show installation labor for batt and blanket insulation. Use Figure 8-7 as a guide until you have developed your own figures based on actual work.

Figure 8-8 shows how a CEF on labor and

Labor and Materials for Poured Ceiling Insulation

Fill Thickness	Material Number of S.F. Covered By C.F. @ Density Ratings					Man-Hours Per 100 S.F. of Ceiling
	6 Lbs.	7 Lbs.	8 Lbs.	9 Lbs.	10 Lbs.	
1''	21.1	18.0	15.9	14.1	13.0	.6
2''	10.6	9.1	8.0	7.1	6.4	.6
3''	7.1	6.1	5.3	4.7	4.2	.7
4''	5.3	4.6	4.0	3.5	3.2	.8
5''	4.2	3.6	3.2	2.8	2.6	.9
6''	3.6	3.0	2.7	2.4	2.2	1.0
7''	3.1	2.6	2.3	2.0	1.9	1.1
8''	2.6	2.3	2.0	1.8	1.6	1.2

**CEF — Labor and materials for poured ceiling insulation
Figure 8-8**

Labor Installing Rigid Board Insulation

Work Element	Unit	Man-Hours Per Unit
1'' mineral fibre	100 S.F.	1.20
2'' mineral fibre	100 S.F.	1.50
1'' glass fibre	100 S.F.	1.20
1'' polystyrene	100 S.F.	1.50
¾'' urethane	100 S.F.	1.10
1'' urethane	100 S.F.	1.20
1½'' urethane	100 S.F.	1.30
2'' urethane	100 S.F.	1.50
1'' foamed glass	100 S.F.	1.10
1½'' foamed glass	100 S.F.	1.20
2'' foamed glass	100 S.F.	1.40
1'' wood fibre boards	100 S.F.	1.10
2'' wood fibre boards	100 S.F.	1.30
¾'' particle board, compressed	100 S.F.	1.10
1'' particle board, compressed	100 S.F.	1.10
2'' particle board, compressed	100 S.F.	1.40
1'' cork insulation board	100 S.F.	1.40
2'' cork insulation	100 S.F.	2.90

Time includes move on and off site, unloading, stacking, installing by nail, staple, mastic or glue.
Suggested Crew: 1 installer, 1 helper. For overhead application, add 10% to time; for deck application, deduct 10%.

**CEF —Labor installing rigid board insulation
Figure 8-9**

Insulation Value of Common Materials

Material	Thickness	"R" Value
Air film and spaces		
Air space		
Bounded by paper or wood	¾" or more	0.91
Bounded by aluminum foil	¾" or more	2.17
Exterior surface resistance	---	0.17
Interior surface resistance	---	0.68
Masonry		
Sand and gravel concrete block	8"	1.11
Sand and gravel concrete block	12"	1.28
Lightweight concrete block	8"	2.00
Lightweight concrete block	12"	2.13
Face brick	4"	0.44
Concrete cast-in-place	8"	0.64
Building materials		
Wood sheathing or subfloor	¾"	1.00
Fiberboard insulating sheathing	¾"	2.10
Plywood	5/8"	0.79
Plywood	½"	0.63
Plywood	3/8"	0.47
Bevel lapped siding	½" x 8"	0.81
Bevel lapped siding	¾" x 10"	1.05
Vertical tongue & groove board	¾"	1.00
Drop siding	¾"	0.94
Asbestos board	¼"	0.13
3/8" gypsum lath and 3/8" plaster	¾"	0.42
Gypsum board	3/8"	0.32
Interior plywood panel	¼"	0.31
Building paper	---	0.06
Vapor barrier	---	0.00
Wood shingles	---	0.87
Asphalt shingles	---	0.44
Linoleum	---	0.08
Carpet with fiber pad	---	2.08
Hardwood floor	---	0.71
Windows and doors		
Single window	---	Approximately 1.00
Double window	---	Approximately 2.00
Exterior door	---	Approximately 2.00

CEF — Insulation value of common materials
Figure 8-10

materials for poured ceiling insulation can be broken down to give you quick estimating assistance.

Labor required to install rigid board insulation is shown in Figure 8-9.

Figure 8-10 shows how to prepare a CEF card so it puts the essential information at your fingertips.

Summary

The alert builder uses every possible tool to in-crease profits and add to his or her reputation as a quality builder. Building houses with the proper in-sulation and vapor barrier is an important step. So is telling your potential buyers what kind of insula-tion is used and its R-value. Explain the potential for savings, making sure to stay within the FTC guidelines.

A successful contractor should brag a little about the quality of his buildings.

Chapter **9**

Estimating Doors and Windows

Most contractors install the exterior doors and windows as soon as the roof is sheathed and covered with felt. That completes the "dry-in" process. Work can then begin on running electric wiring, plumbing rough-in and putting in the insulation. Since estimating should follow the sequence of construction, and since we've already covered rough carpentry and roofing, doors and windows come next.

Listing doors and windows is easy estimating. It's just counting. List the number, sizes and types as shown on the plans and specs. Then add the labor and you're finished.

Our discussion here will be limited to prehung doors, since that's what most builders use.

How to List Doors

Working from the plan, start with a complete list of the number of doors required. List them by size, whether hinged left or right, and whether paint or stain grade. When listing by size, give the width first, then the height, followed by the thickness. Identify the style, number of panels, flush, solid or hollow core, and any other details.

The flush door is the most common. They're available for exterior or interior use, with lites and louvers if the designer wants them. Flush doors are usually veneered with birch, ash, mahogany, gum, cherry, oak, elm, or walnut, although other species are available. Some flush wood doors are finished with plastic or metal facings.

Flush wood doors are either hollow core or solid core. Hollow core wood doors have a core of strips of wood or cardboard. The space between these strips is hollow. The strips support the outer faces. Solid core wood doors have a solid inner core of wood blocks, composition board, wood particle board, or bonded mineral.

Wood doors are made in many stock sizes and patterns. You'll find modular sizes in all standard designs. Douglas fir and white pine are generally used for exterior doors. Fir, birch, and pine are often used for interior doors. The specifications will tell you what kind of wood door to order. Exterior doors are usually 1¾" thick and interior doors are 1⅜" thick.

Both wood and hollow metal doors (with or without insulation) are available as a single prehung unit. The door arrives on the job already installed in the frame. Exterior units are weatherstripped.

The usual door width for bathrooms is 2'4"; hallway and interior room doors 2'6"; rear or side entrance 2'8"; front entrance 3'0"; closet and individual French doors range from 2'0" to 2'6". The usual height is 6'8", but check the plans carefully for variations from these standards.

Typical door designs
Figure 9-1

Figure 9-1 illustrates a number of stock designs of exterior and interior doors. Of course there are many other kinds of interior doors available, including louvered doors, folding doors, by-pass doors and glass doors. The type of hardware is usually given in the specs. List the cost of all hardware separately in your take-off.

Estimating Labor for Installing Doors

Figure 9-2 shows the steps involved in installing the prehung door unit. But the job isn't finished until the door knob and keeper plate are installed.

A rule of thumb that's worked well for me is to allow 1.5 skilled manhours for the installation of each standard-size prehung exterior unit and 1 skilled manhour for each standard prehung interior unit. This includes the installation of the lockset.

Prehung exterior double-door units will require about 3 skilled manhours for installation.

Closet by-pass doors (with pulls and tracks) require about 2 skilled manhours.

Sliding glass doors measuring 6'0'' x 6'8'' will re-

quire about 2.5 skilled manhours for installation. Heavier doors or doors with insulating glass panels will take about an hour longer.

Look at Figure 9-3. This sample CEF gives the labor for installing folding doors in manhours per 10 square feet. A 3'0'' x 7'0'' door is 21 square feet and will take about 2.4 manhours to install.

Figure 9-4 is the information your file should include showing labor for installing steel doors and frames. Use these figures if you haven't done enough work to have your own manhour figures.

Of course, no estimating file would be complete without a card showing how much labor is needed to install sliding glass doors. Figure 9-5 might be just the information you're looking for. Remember, the figures given in this and other CEFs are averages that will apply on many jobs. But averages can be deceiving. The *average* temperature in Tucson, Arizona last May 28th was 72. Not too bad, until I explain that it was 32 degrees at dawn and 112 by 2 P.M. I've spent a full day installing one 8' x 8' unit in a remodeling job!

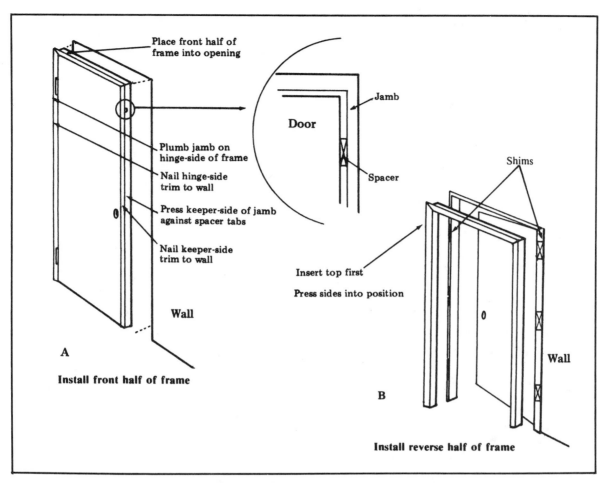

**Installing the prehung door unit
Figure 9-2**

Labor for Installing Folding Doors

Work Element	Unit	Man-Hours Per Unit
Bi-folding wood doors		
2'4'' x 6'8''	10 S.F.	1.2
2'8'' x 6'8''	10 S.F.	1.1
3'0'' x 6'8''	10 S.F.	1.3
4'0'' x 6'8''	10 S.F.	.9
Bi-folding vinyl covered wood doors		
3'0'' x 7'0''	10 S.F.	1.2
4'0'' x 7'0''	10 S.F.	1.1
8'0'' x 8'0''	10 S.F.	1.0
10'0'' x 10'0''	10 S.F.	1.3
15'0'' x 10'0''	10 S.F.	1.2
20'0'' x 10'0''	10 S.F.	1.1
Add 10% for custom grade doors.		

Time includes move on and off site, unloading, stacking, hardware installation, cleanup and repair as needed, but no finishing.

**CEF — Labor installing folding doors
Figure 9-3**

Labor Installing Steel Doors and Frames

Work Element	Unit	Man-Hours Per Unit
1¾'' thick unrated doors		
2'8'' x 6'8'', 18 gauge	Each	2.3
2'8'' x 7'0'', 18 gauge	Each	2.5
3'0'' x 6'8'', 18 gauge	Each	3.0
3'0'' x 7'0'', 18 gauge	Each	3.1
3'4'' x 7'0'', 18 gauge	Each	3.7
Hollow metal door frames		
2'8'' x 6'8'', 18 gauge	Each	1.4
2'8'' x 7'0'', 18 gauge	Each	1.5
3'0'' x 6'8'', 18 gauge	Each	1.6
3'0'' x 7'0'', 18 gauge	Each	1.6
3'4'' x 7'0'', 16 gauge	Each	1.6
6'0'' x 7'0'', 16 gauge	Each	1.8

Time includes move on and off site, unloading, stacking, installing 1 pair template butts and latchset on each door, cleanup and repair as needed.
Suggested Crew: 1 carpenter

**CEF — Labor installing steel doors and frames
Figure 9-4**

Labor Installing Sliding or Swinging Glass Patio Doors

Work Element	Unit	Man-Hours Per Unit
Standard weight doors and glass		
8' x 8'	Each	2.7
8' x 6'8"	Each	2.6
6' x 6'8"	Each	2.4
10' x 6'8"	Each	3.4
12' x 6'8"	Each	3.6
Heavier or better quality doors with insulating glass		
8' x 8'	Each	3.5
8' x 6'8"	Each	3.3
10' x 6'8"	Each	4.4
12' x 6'8"	Each	4.6
16' x 6'8"	Each	5.0

Time includes move on and off site, unloading, setup, installing in a frame opening, adjusting, cleanup and repair as needed.
Suggested Crew: 1 carpenter, 1 laborer

CEF — Labor installing glass patio doors
Figure 9-5

That job raised my average more than a little.

Figure 9-6 shows the labor required for installing overhead doors.

If you need manhour figures for installing non-prefabricated doors, look ahead to Figure 12-16 in Chapter 12. It gives labor hours for finish carpentry, including setting door frames, hanging doors, and casing or trim.

How to List Windows

Window dimensions are given by the number of lites and size in inches. Thus a "four-lite 12/28 window" has four 12" x 28" panes of glass. When giving the size of a sash, the width of the glass is always given first, then the height. If a double-hung window has a sash that isn't divided by small wood members called *muntins*, or *sash bars*, it's called a "two-lite" window.

A plan might show window size as:

9/12 - 16 Lt. D.H. WD
3'4" x 4'6"

This means it's a double-hung window that's 3'4" wide x 4'6" high, with 16 lites, each 9" wide and 12" high.

Figure 9-7 gives the window abbreviations to look for on the plans and specs you take off.

On some plans, window dimensions may be in-

Labor for Installing Overhead Doors

Work Element	Unit	Man-Hours Per Unit
Wood panel overhead doors		
8'3" x 8'7"	Each	4.3
10'3" x 10'7"	Each	6.5
12'3" x 12'7"	Each	9.3
16'3" x 16'7"	Each	16.4
22'3" x 18'11"	Each	25.4
24'3" x 18'11"	Each	26.3
Steel overhead doors, chain operated		
10' x 10' high	100 S.F.	4.5
12' x 10' high	100 S.F.	5.5
12' x 20' high	100 S.F.	6.5
Heavy duty 2" thick stock doors with standard hardware and tracks		
18' x 18' x 2" thick	Each	30.3
20' x 20' x 2" thick	Each	45.5

Deduct 10% for aluminum doors. Time includes move on and off site, unloading, stacking, cleanup and repair as required.
Suggested Crew: 1 carpenter, 1 laborer on wood door up to 10' wide. Wood doors over 10' wide: 1 carpenter, 2 laborers. Use 2 steelworkers for steel doors.

CEF — Labor installing overhead doors
Figure 9-6

dicated by the exact outside dimensions of the sash. Thus, a 2'8" x 4'8" window has a 2'8" x 4'8" sash. These figures are also the inside dimensions of the frame. Look at Figure 9-8.

Bev. Plt.	Beveled plate glass
Ck. R. Wids	Check rail windows
Bsmt.	Basement sash
C. M. C.	Crown mold cap applied to frames
D.C.	Plain drip cap applied to window and door frames
D. H.	Double hung
D. S. A.	Double strength "A" grade glass
Div.	Divided
Fr. Wd.	French window
Gla. or Glaz.	Glazed
Lt.	Light or a pane of glass
M. R.	Meeting rail
Munt	Muntin
Pl. R.	Plain rail
Pr. Blds.	Pair of blinds
P. S.	Pulley stiles
Rab.	Rabbeted, applying to door frames and windows
R. M.	Raised mold
Sh.	Sash
Shtrs.	Shutters
S. S.	Single strength glass
Wd.	Window

Common window abbreviations
Figure 9-7

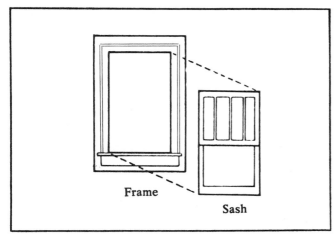

Window dimensions
Figure 9-8

The rough opening (RO) size for window frames varies slightly among manufacturers. Make the following allowances for the stiles and rails, thickness of jambs, and thickness and slope of the sill:

1) Double-hung windows (single unit): RO width equals glass width plus 6 inches. RO height: total glass height plus 10 inches.

2) Casement windows (one pair of sash): RO width equals total glass width plus 11¼ inches. RO height: total glass height plus 6⅜ inches.

Figure 9-9 shows a method of determining the RO by the glass size.

Window glass size (each sash)			Rough frame opening		
Width		Height	Width		Height
24"	x	16"	30"	x	42"
28"	x	20"	34"	x	50"
32"	x	24"	38"	x	58"
36"	x	24"	42"	x	58"

Figuring RO by glass size
Figure 9-9

Types of Windows

You should be familiar with the types of windows you're listing. Basically, windows can be classified by the type of sash-opening system. The most common types are:

- Double-hung
- Single-hung (stationary)
- Casement
- Awning
- Horizontal sliding
- Storm windows

The sash can be wood, metal, a combination of wood and metal, or wood encased in vinyl sheathing. Heat loss is greater through metal frame units.

Insulated glass is used in both stationary and movable sashes. It consists of two or more glass sheets with hermetically-sealed edges. This type of glass has more resistance to heat loss than a single thickness. Insulated glass is often used without a storm sash.

Double-Hung Windows

The double-hung window is probably the most common. It has an upper sash and a lower sash that slide vertically in separate grooves in the side jambs or in full-width metal weatherstripping. Springs, balances, or compression weatherstripping hold the sash in place. Compression weatherstripping is probably the best. It prevents air infiltration, provides tension, and acts as a counterbalance. When the window is hung, the sash should operate freely.

Hardware for double-hung windows includes sash lifts that are fastened to the bottom rail. These are not required where a finger groove is provided in the bottom rail. Other hardware includes sash locks or fasteners located at the meeting rails. They lock the window and draw the sashes together to provide a tight fit. Make sure you list all the hardware needed for the particular windows you're taking off.

Single-Hung Windows

The single-hung frame is made to hold a single sash. The sash may be fastened permanently in place, or it may swing in or out from either the side or head jamb.

Stationary windows used alone, or in combination with double-hung or casement windows, usually have a wood sash with a large single lite of insulated glass. Stationary windows are fastened permanently into the frame. Because of their size, sometimes 6 to 8 feet wide, a 1¾-inch thick sash may be needed to provide strength. Large lites of insulated glass will be heavier than other windows because insulating glass is relatively heavy.

Casement Windows

Casement windows have a side-hinged sash that swings outward. The outward-swinging sash is more weathertight than one that swings inward. Screens are installed on the inside of windows that swing outward. Some casement windows have a storm sash, but insulated glass is probably more practical in cooler climates. Unlike the double-hung unit, the entire casement window opens for ventilation.

Casement windows usually arrive on the job fully assembled, weatherstripped, and with hardware in place. Closing hardware consists of a rotary operator and sash lock. Casement sashes can be used as a pair, or in combination of two or more pairs with divided lites. Snap-in muntins provide a multiple-pane appearance.

Awning Windows

Awning windows have movable sashes that extend outward like an awning. They're often grouped into multiple units to create a window wall. Weatherstripping, storm sash and screens are usually provided. The storm sash isn't needed when insulated glass is used.

Horizontal Sliding Windows

Horizontal sliders look similar to casement windows, but the sash slides horizontally in a track located on the sill and head jambs. Again, multiple window units can be joined to create a window wall. As in most modern window units, weatherstripping, water-repellent preservative treatments, and hardware will be included in these factory-assembled units.

Prefabricated Metal Windows

Many types of complete window units are available. Most come with frame, sash and trim already assembled. Often they have the appropriate screens, weatherstripping and hardware in place or on hand for quick assembly. Some inexpensive units just aren't made to give good service. Install only quality windows intended to last the useful life of the building.

Storm Windows

Storm windows are one of the best and cheapest ways to provide insulation. They not only keep wind and rain away from the sash edges and the frame, they also provide a double thickness of glass with a dead air space. The dead air space prevents

Work Element	Unit	Man-Hours Per Unit
Casement windows and screens		
1 leaf, 1'10" x 3'2"	Each	1.4
2 leaves, 3'10" x 4'2"	Each	1.9
3 leaves, 5'11" x 5'2"	Each	2.4
Picture windows		
4'6" x 4'6"	Each	3.0
5'8" x 4'6"	Each	3.2
9' x 5'	Each	3.7
10' x 5'	Each	4.0
11' x 5'	Each	4.4
Double or single hung windows and screens		
2'0" x 3'2"	Each	1.1
2'0" x 4'6"	Each	1.6
2'8" x 3'2"	Each	2.0
2'8" x 5'2"	Each	2.1
3'4" x 5'2"	Each	2.4
5'6" x 5'2"	Each	3.4
8'4" x 5'2"	Each	5.2
Bow bay windows and screens		
8' x 5'	Each	5.1
9'9" x 6'8"	Each	7.6
7' x 5'	Each	7.1
8'9" x 5'0"	Each	7.7
7'6" x 6'0"	Each	7.6
8'9" x 6'6"	Each	8.4
One member casing on windows, ordinary work	Each	1
Hardwood, first class work	Each	1½ to 2
Two member casing on windows, ordinary work	Each	1½ to 2
Hardwood and first class work	Each	2 to 4
Window trim on brick walls, ordinary work	Each	1½ to 2
First class or difficult work	Each	2 to 4

Time is for setting factory-made assembled windows in a prepared opening and includes move on and off site, unloading, stacking, installing, repairing and cleanup as needed, but no trim or framing.
Suggested Crew: 1 carpenter, 1 laborer

Labor guidelines for installing windows
Figure 9-10

heat loss and helps to avoid condensation on the glass surface in cold weather.

Figure 9-10 shows typical labor figures for window installation. Treat awning, sliding and storm windows the same as double- or single-hung windows when estimating labor. For more detailed labor estimates, look at Figure 12-16 in Chapter 12, labor hours for finish carpentry. It covers assembling and setting window frames, fitting sash and window trim.

Window Screens

Prefabricated window frames are built to receive screens, storm sash, and shutters. Group screens of the same size and kind of mesh material together on the take-off sheet. To determine the number required, check the specifications to find which windows have screens. Check this number on the floor plans and then double-check with the elevations.

Based on the kind of frame, there are three standard types of window screens:

1) The full double-hung window screen has a horizontal bar across the center or at the meeting rails of the window. The horizontal bar is generally 5/8'' wide. The screen frame is 3/4'' wide. Both are commonly aluminum. The mesh may be aluminum or synthetic material, usually fiberglass.

2) Sliding half screens move up and down, or, in some cases, from side to side.

3) Casement screens usually have a divider bar only when special ordered. All screens that hinge in are known as "inside screens." The frames of these screens may be of wood to match the interior finish.

The width of a screen is the same as the width of the window. The length, however, is not the same as the sash or window, but longer, due to the pitch of the sill. The screen for a casement window is longer than for a double-hung window because of different measuring points. When taking off the screens, add 1 inch to the window length. Figure labor at 0.5 hour per screen.

Shutters

In some parts of the country, shutters are part of the exterior design of a home. In many, if not most cases, they're used for architectural effect only. They're nailed or screwed flat to the wall in an open position. Job specifications will usually state the kind of lumber, and the elevation drawings will show which windows require shutters, and their shape. Two shutters are required for each window. If the shutters are to be hinged, the width will be one-half the width of the window. Determine the length from the window frame detail, as the slope of the sill must be considered. Usually you'll add 1 inch to the length of the frame.

Measure the length of ornamental shutters on the elevation sheet. The width will be one-half the window width, or less. Figure shutter installation at 1 hour per first floor window and 2 hours per second floor window.

Wrapping Up Windows and Doors

Use care when taking off doors and windows. Be sure to list all hardware, screens and anchors or hinges for shutters. Double check your list against the plans and specs.

Keep accurate labor records on your jobs and adjust your CEF to reflect *actual* installation times. As I mentioned, you can spend 8 hours or more installing a sliding glass door unit. Every job is different. With accurate records, you'll soon have the "expert" working for you — your own experience.

Chapter 10

Estimating Interior Wallboard

Gypsum wallboard (sheetrock) is the most common material for walls and ceilings, so we'll cover it first. GWB has many advantages. It's cheaper than plaster, goes up quicker, can be used where a fire rating is required, has good sound resistance, is dimensionally stable, and resists cracks caused by minor frame movement.

Wallboard comes in a standard width of 4 feet and standard lengths of 8 to 16 feet. The standard thicknesses are 1/2", 5/8" and 3/4", but thicknesses of 1/4" and 3/8" are also available.

There are two types of gypsum board: wallboard, which will be exposed after installation, and backing board, which is used as backing material for gypsum board laminated assemblies, acoustical tile, ceramic tile, and other finish materials.

Figure 10-1 shows ceiling and wall application

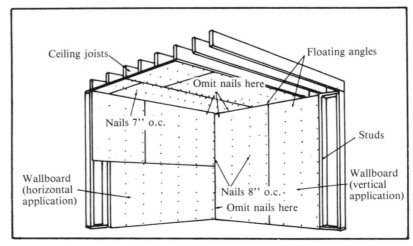

Ceiling and wall application details for wallboard
Figure 10-1

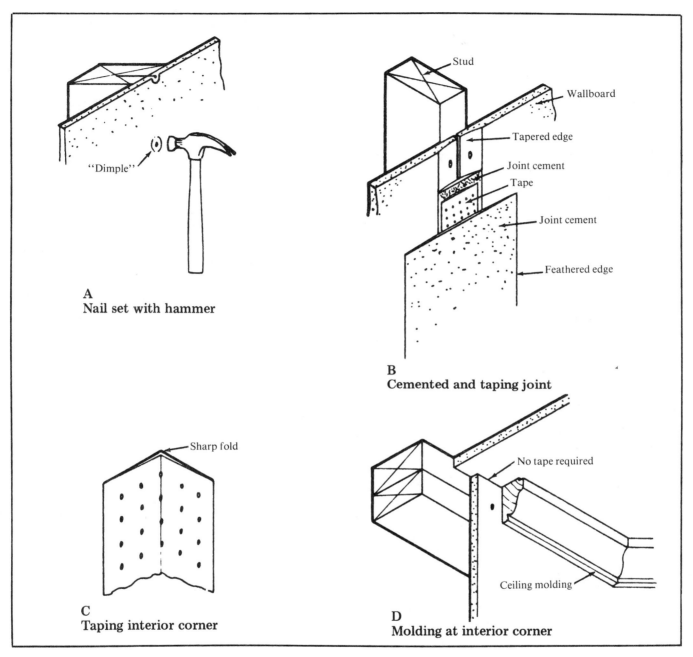

A
Nail set with hammer

B
Cemented and taping joint

C
Taping interior corner

D
Molding at interior corner

Wallboard finishing details
Figure 10-2

details. Figure 10-2 illustrates the finishing details. The nails recommended for wallboard installation are shown in Figure 10-3.

Estimating Wallboard Materials

To estimate wallboard requirements, figure the square feet of the walls and ceilings in each room to be covered. And don't forget the closets, halls and stairways. Figure 10-4 gives the square feet for walls and ceilings for rooms with 8-foot ceilings and even wall lengths.

Figure 10-5 gives the square footage for standard size wallboard panels and for bundles of 3/8'' and 1/2'' lath. Let's say you're estimating the wallboard needed for a 14'0'' x 16'0'' room with an 8-foot ceiling. According to Figure 10-4, the wall and ceiling area totals 704 square feet. Now

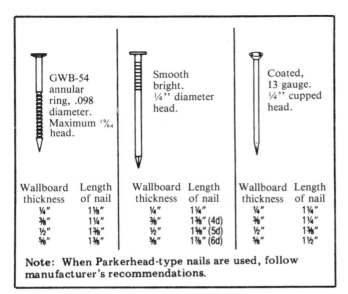

Wallboard thickness	Length of nail		Wallboard thickness	Length of nail		Wallboard thickness	Length of nail
¼"	1⅛"		¼"	1¼"		¼"	1¼"
⅜"	1¼"		⅜"	1⅜" (4d)		⅜"	1¼"
½"	1⅜"		½"	1⅝" (5d)		½"	1⅜"
⅝"	1⅜"		⅝"	1⅞" (6d)		⅝"	1½"

GWB-54 annular ring, .098 diameter. Maximum ¹⁹⁄₆₄ head.

Smooth bright. ¼" diameter head.

Coated, 13 gauge. ¼" cupped head.

Note: When Parkerhead-type nails are used, follow manufacturer's recommendations.

Nails recommended for wallboard installation
Figure 10-3

look at Figure 10-5. If you're using 4' x 8' panels, look down that column until you come to 704. Follow that line to the left until you come to the "Pieces" column. You need 22 sheets of wallboard for the room.

If you need the wallboard requirements for the walls only, use Figure 10-6 to find the number of 4' x 8' panels needed. Just add the total of the four walls to find the perimeter of the room. For the 14'0" x 16'0" room, that's 60 feet. Then look up the number of panels you need in the conversion table.

Using these figures, it's easy to estimate the amount of wallboard you need to finish the interior of a building. But go over the plans carefully to make sure you include every wall and every ceiling in the house.

Wallboard Finish Materials
Besides the gypsum board itself, you'll need these materials for wallboard application:

1) When using adhesives, allow five pounds 1⅜" annular ring nails per 1,000 square feet of wallboard.

2) Allow one tube of adhesive per 500 square feet of wallboard.

3) Allow one roll of tape (250') per 600 square feet of wallboard.

Room Areas Square Feet—4 Walls (W) and Ceilings (C)

Lineal Ft. Per Wall	6	8	10	12	14	16	18	20	22	24	26	28	30
6	C 36 / W 192	C 48 / W 224	C 60 / W 256	C 72 / W 288	C 84 / W 320	C 96 / W 352	C 108 / W 384	C 120 / W 416	C 132 / W 448	C 144 / W 480	C 156 / W 512	C 168 / W 544	C 180 / W 576
8	C 48 / W 224	C 64 / W 256	C 80 / W 288	C 96 / W 320	C 112 / W 352	C 128 / W 384	C 144 / W 416	C 160 / W 448	C 176 / W 480	C 182 / W 512	C 198 / W 544	C 224 / W 576	C 240 / W 608
10	C 60 / W 256	C 80 / W 288	C 100 / W 320	C 120 / W 352	C 140 / W 384	C 160 / W 416	C 180 / W 448	C 200 / W 480	C 220 / W 512	C 240 / W 544	C 260 / W 576	C 280 / W 608	C 300 / W 640
12	C 72 / W 288	C 96 / W 320	C 120 / W 352	C 144 / W 384	C 168 / W 416	C 192 / W 448	C 216 / W 480	C 240 / W 512	C 264 / W 544	C 288 / W 576	C 312 / W 608	C 336 / W 640	C 360 / W 672
14	C 84 / W 320	C 112 / W 352	C 140 / W 384	C 168 / W 416	C 196 / W 448	C 224 / W 480	C 252 / W 512	C 280 / W 544	C 308 / W 576	C 336 / W 608	C 364 / W 640	C 392 / W 672	C 420 / W 704
16	C 96 / W 352	C 128 / W 384	C 160 / W 416	C 192 / W 448	C 224 / W 480	C 256 / W 512	C 288 / W 544	C 320 / W 576	C 352 / W 608	C 384 / W 640	C 416 / W 672	C 448 / W 704	C 480 / W 736
18	C 108 / W 384	C 144 / W 416	C 180 / W 448	C 216 / W 480	C 252 / W 512	C 288 / W 544	C 324 / W 576	C 360 / W 608	C 396 / W 640	C 432 / W 672	C 468 / W 704	C 504 / W 736	C 540 / W 768
20	C 120 / W 416	C 160 / W 448	C 200 / W 480	C 240 / W 512	C 280 / W 544	C 320 / W 576	C 360 / W 608	C 400 / W 640	C 440 / W 672	C 480 / W 704	C 520 / W 736	C 560 / W 768	C 600 / W 800
22	C 132 / W 448	C 176 / W 480	C 220 / W 512	C 264 / W 544	C 308 / W 576	C 352 / W 608	C 396 / W 640	C 440 / W 672	C 484 / W 704	C 528 / W 736	C 572 / W 768	C 616 / W 800	C 660 / W 832
24	C 144 / W 480	C 182 / W 512	C 240 / W 544	C 288 / W 576	C 336 / W 608	C 384 / W 640	C 432 / W 672	C 480 / W 704	C 528 / W 736	C 576 / W 768	C 624 / W 800	C 672 / W 832	C 720 / W 864
26	C 156 / W 512	C 198 / W 544	C 260 / W 576	C 312 / W 608	C 364 / W 640	C 416 / W 672	C 468 / W 704	C 520 / W 736	C 572 / W 768	C 624 / W 800	C 676 / W 832	C 728 / W 864	C 780 / W 896
28	C 168 / W 544	C 224 / W 576	C 280 / W 608	C 336 / W 640	C 392 / W 672	C 448 / W 704	C 504 / W 736	C 560 / W 768	C 616 / W 800	C 672 / W 832	C 728 / W 864	C 784 / W 896	C 840 / W 928
30	C 180 / W 576	C 240 / W 608	C 300 / W 640	C 360 / W 672	C 420 / W 704	C 480 / W 736	C 540 / W 768	C 600 / W 800	C 660 / W 832	C 720 / W 864	C 780 / W 896	C 840 / W 928	C 900 / W 960

NOTE: Based on wall height of 8'-0". C — Ceiling area. W — Wall area — 4 Walls.

Room area chart
Figure 10-4

	Board products —— square feet											16'' x 48'' lath		
Pieces	4' x 6'	4' x 7'	4' x 8'	4' x 9'	4' x 10'	4' x 12'	4' x 14'	4' x 16'	2' x 8'	2' x 10'	2' x 12'	bdls.	3/8''	1/2''
2	48	56	64	72	80	96	112	128	32	40	48	1	32	21.33
4	96	112	128	144	160	192	224	256	64	80	96	2	64	42.67
6	144	168	192	216	240	288	336	384	96	120	144	3	96	64.00
8	192	224	256	288	320	384	448	512	128	160	192	4	128	85.33
10	240	280	320	360	400	480	560	640	160	200	240	5	160	106.67
12	288	336	384	432	480	576	672	768	192	240	288	6	192	128.00
14	336	392	448	504	560	672	784	896	224	280	336	7	224	149.33
16	384	448	512	576	640	768	896	1,024	256	320	384	8	256	170.67
18	432	504	576	648	720	864	1,008	1,152	288	360	432	9	288	192.00
20	480	560	640	720	800	960	1,120	1,280	320	400	480	10	320	213.33
22	528	616	704	792	880	1,056	1,232	1,408	352	440	528	11	352	234.67
24	576	672	768	864	960	1,152	1,344	1,536	384	480	576	12	384	256.00
26	624	728	832	936	1,040	1,248	1,456	1,664	416	520	624	13	416	277.33
28	672	784	896	1,008	1,120	1,344	1,568	1,792	448	560	672	14	448	298.67
30	720	840	960	1,080	1,200	1,440	1,680	1,920	480	600	720	15	480	320.00
32	768	896	1,024	1,152	1,280	1,536	1,792	2,048	512	640	768	16	512	341.33
34	816	952	1,088	1,224	1,360	1,632	1,904	2,176	544	680	816	17	544	362.67
36	864	1,008	1,152	1,296	1,440	1,728	2,016	2,304	576	720	864	18	576	384.00
38	912	1,064	1,216	1,368	1,520	1,824	2,128	2,432	608	760	912	19	608	405.33
40	960	1,120	1,280	1,440	1,600	1,920	2,240	2,560	640	800	960	20	640	426.67
42	1,008	1,176	1,344	1,512	1,680	2,016	2,352	2,688	672	840	1,008	21	672	448.00
44	1,056	1,232	1,408	1,584	1,760	2,112	2,464	2,816	704	880	1,056	22	704	469.33
46	1,104	1,288	1,472	1,656	1,840	2,208	2,576	2,944	736	920	1,104	23	736	490.67
48	1,152	1,344	1,536	1,728	1,920	2,304	2,688	3,072	768	960	1,152	24	768	512.00
50	1,200	1,400	1,600	1,800	2,000	2,400	2,800	3,200	800	1,000	1,200	25	800	533.33
52	1,248	1,456	1,664	1,872	2,080	2,496	2,912	3,328	832	1,040	1,248	26	832	554.67
54	1,296	1,512	1,728	1,944	2,160	2,592	3,024	3,456	864	1,080	1,296	27	864	576.00
56	1,344	1,568	1,792	2,016	2,240	2,688	3,136	3,584	896	1,120	1,344	28	896	597.33
58	1,392	1,624	1,856	2,088	2,320	2,784	3,248	3,712	928	1,160	1,392	29	928	618.67
60	1,440	1,680	1,920	2,160	2,400	2,880	3,360	3,840	960	1,200	1,440	30	960	640.00
62	1,488	1,736	1,984	2,232	2,480	2,976	3,472	3,968	992	1,240	1,488	31	992	661.33
64	1,536	1,792	2,048	2,304	2,560	3,072	3,584	4,096	1,024	1,280	1,536	32	1,024	682.67
66	1,584	1,848	2,112	2,376	2,640	3,168	3,696	4,224	1,056	1,320	1,584	33	1,056	704.00
68	1,632	1,904	2,176	2,448	2,720	3,264	3,808	4,352	1,088	1,360	1,632	34	1,088	725.33
70	1,680	1,960	2,240	2,520	2,800	3,360	3,920	4,480	1,120	1,400	1,680	35	1,120	746.67
72	1,728	2,016	2,304	2,592	2,880	3,456	4,032	4,608	1,152	1,440	1,728	36	1,152	767.00
74	1,776	2,072	2,368	2,664	2,960	3,552	4,144	4,736	1,184	1,480	1,776	37	1,184	789.33
76	1,824	2,128	2,432	2,736	3,040	3,648	4,256	4,864	1,216	1,520	1,824	38	1,216	810.67
78	1,872	2,184	2,496	2,808	3,120	3,744	4,368	4,992	1,248	1,560	1,872	39	1,248	832.00
80	1,920	2,240	2,560	2,880	3,200	3,840	4,480	5,120	1,280	1,600	1,920	40	1,280	853.33
82	1,968	2,296	2,624	2,952	3,280	3,936	4,592	5,248	1,312	1,640	1,968	41	1,312	874.67
84	2,016	2,352	2,688	3,024	3,360	4,032	4,704	5,376	1,344	1,680	2,016	42	1,344	896.00
86	2,064	2,408	2,752	3,096	3,440	4,128	4,816	5,504	1,376	1,720	2,064	43	1,376	917.33
88	2,112	2,464	2,816	3,168	3,520	4,224	4,928	5,632	1,408	1,760	2,112	44	1,408	938.67
90	2,160	2,520	2,880	3,240	3,600	4,320	5,040	5,760	1,440	1,800	2,160	45	1,440	960.00
92	2,208	2,576	2,944	3,312	3,680	4,416	5,152	5,888	1,472	1,840	2,208	46	1,472	981.33
94	2,256	2,632	3,008	3,384	3,760	4,512	5,264	6,016	1,504	1,880	2,256	47	1,504	1,002.67
96	2,304	2,688	3,072	3,456	3,840	4,608	5,376	6,144	1,536	1,920	2,304	48	1,536	1,024.00
98	2,352	2,744	3,136	3,528	3,920	4,704	5,488	6,272	1,568	1,960	2,352	49	1,568	1,045.33
100	2,400	2,800	3,200	3,600	4,000	4,800	5,600	6,400	1,600	2,000	2,400	50	1,600	1,066.67
200	4,800	5,600	6,400	7,200	8,000	9,600	11,200	12,800	3,200	4,000	4,800	60	1,920	1,280.00
300	7,200	8,400	9,600	10,800	12,000	14,400	16,800	19,200	4,800	6,000	7,200	70	2,240	1,493.33
400	9,600	11,200	12,800	14,400	16,000	19,200	22,400	25,600	6,400	8,000	9,600	80	2,560	1,706.67
500	12,000	14,000	16,000	18,000	20,000	24,000	28,000	32,000	8,000	10,000	12,000	90	2,880	1,920.00
600	14,400	16,800	19,200	21,600	24,000	28,800	33,600	38,400	9,600	12,000	14,400	100	3,200	2,133.33
700	16,800	19,600	22,400	25,200	28,000	33,600	39,200	44,800	11,200	14,000	16,800	200	6,400	4,266.67
800	19,200	22,400	25,600	28,800	32,000	38,400	44,800	51,200	12,800	16,000	19,200	300	9,600	6,400.00
900	21,600	25,200	28,800	32,400	36,000	43,200	50,400	57,600	14,400	18,000	21,600	400	12,800	8,533.33
1,000	24,000	28,000	32,000	36,000	40,000	48,000	56,000	64,000	16,000	20,000	24,000	500	16,000	10,666.67

Wallboard area table
Figure 10-5

How to Figure a Room

Determine the perimeter. This is merely the total of the widths of each wall in the room. Use the below conversion table to figure the number of panels needed.

Perimeter	No. of 4' x 8' Panels Needed
36'	9
40'	10
44'	11
48'	12
52'	13
56'	14
60'	15
64'	16
68'	17
72'	18
92'	23

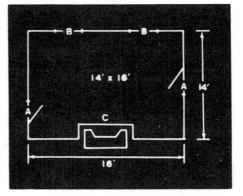

For example, if your room walls measured 14' + 14' + 16' + 16', this would equal 60' or 15 panels required. To allow for areas such as windows, doors, fireplaces, etc., use these deductions listed below:

Deductions:
Door   1/2 panel (A)
Window   1/4 panel (B)
Fireplace   1/2 panel (C)

Thus, the actual number of panels for this room would be 13 pieces (15 pieces minus 2 total deductions). If the perimeter of the room falls in between the figures in the above table, use the next highest number to determine panels required. These figures are for rooms with 8' ceiling heights or less.

Figuring wallboard panels by room perimeter
Figure 10-6

4) Allow five gallons (one can) of joint compound per 1,000 square feet of wallboard.

5) For texture-finished ceilings using joint compound, allow five gallons per 400 square feet.

6) Allow one 8-foot metal corner bead for each outside corner.

Furring
Furring is sometimes needed to provide an even base for the drywall. Figure 10-7 shows the board feet of furring strips and pounds of nails you'll use. A carpenter should cut and install 350 to 400 linear feet of furring in eight hours.

Labor for Wallboard Installation
Many builders subcontract sheetrock hanging and taping at a set cost per square foot. Many drywall subs also install spray-on ceiling texture finish. The cost is usually less than other finishes when applied

at the same time as the board is finished. But if your crew hangs the wallboard, it will probably be cheaper to paint the ceiling than to get a separate bid for spray-on ceiling texture.

Size of Strips	O.C. Spacing of Furring	BF Per SF of Wall	Lbs. Nails Per 1000 BF
1" x 2"	12"	.18	
	16"	.14	
	20"	.11	55
	24"	.10	
1" x 3"	12"	.28	
	16"	.21	
	20"	.17	37
	24"	.14	
1" x 4"	12"	.36	
	16"	.28	
	20"	.22	30
	24"	.20	

Materials required for furring
Figure 10-7

Labor Installing Gypsum Drywall

Work Element	Unit	Man-Hours Per Unit	Gypsum Wallboard Types
Drywall on one face of metal or wood studs or furring			**Regular** is available in several thicknesses for both new and remodeling construction.
1 layer, ⅜"	100 S F	1.8	**Fire rated** is designed especially for fire resistance. Major additives are vermiculite and fiberglass.
1 layer, ½"	100 S F	1.9	
1 layer, ⅝"	100 S F	2.1	**Sound deadening board** is usually applied in combination with other wallboard products to achieve higher sound and fire ratings.
2 layers, ⅜" (mastic)	100 S F	2.7	
2 layers, ½" (mastic)	100 S F	3.0	**Tile backer board** is recommended as a base for adhesive
2 layers, ⅝" (mastic	100 S F	3.4	application of ceramic, metal or plastic tile for interior areas
Drywall for columns, pipe chases or fire partitions			where moisture and humidity are a problem. (Direct, continuous contact with moisture should be avoided.)
1 layer, ⅜", nailed	100 S F	4.4	**Sheathing** is for exterior applications. Used as a substrate for
1 layer, ½", nailed	100 S F	4.5	siding, masonry, brick veneer and stucco.
1 layer, ⅝", nailed	100 S F	4.6	**Backerboard** is recommended for backing paneling and other
2 layers, ½", mastic	100 S F	8.5	multi-layered applications. Adds strength and fire protection.
2 layers, ⅝", mastic	100 S F	8.9	Also can be used effectively with ceiling tile.
3 layers, ½", mastic	100 S F	12.5	
3 layers, ⅝", mastic	100 S F	13.0	**Vinyl-surfaced wallboard** resists scuffs, cracks and chips. Ideal
1 layer, 1½", coreboard	100 S F	4.0	for commercial and institutional use.
Drywall for beams and soffits			**Tapered edge** inclines into the board from the long edge. With
1 layer, ½"	100 S F	4.0	joint finishing results in a smooth, monolithic wall.
1 layer, ⅝"	100 S F	3.9	**Square edge** is used where an exposed joint is desired.
2 layers, ½"	100 S F	7.3	**Tapered, round edge** is for the same applications as tapered edge
2 layers, ⅝"	100 S F	8.0	board. Designed to reduce beading and ridging problems often
Drywall, glued			associated with poorly finished joints.
1 layer, ½"	100 S F	2.0	**Beveled edge** is used where a "panel" effect is desired. In this
1 layer, ⅝"	100 S F	1.9	application the joints are left exposed.
Screwed drywall			
1 layer, ½"	100 S F	1.9	**Tongue and groove** is available on 24" wide sheathing and backer
1 layer, ⅝"	100 S F	2.2	boards.
Additional time requirements			**Modified beveled edge** needs no special joint finishing, though
Add for ceiling work	100 S F	.6	matching batten strips may be used if desired.
Add for walls over 9' high	100 S F	.5	Thickness: ¼", 3/8", ½", 5/8". (Not all products are available
Add for resilient clip application	100 S F	.4	in all thicknesses.) Width: 4'. Length: 6' through 16'.
Add for vinyl covered drywall	100 S F	.4	
Add for thincoat plaster finish	100 S F	1.4	
Deduct for no taping, finish or sanding	100 S F	.9	

Time includes move on and off site, unloading, stacking, installing drywall, repair and cleanup as needed. Taping, joint finishing and sanding are included.   **Suggested Crew: 1 applicator and 1 laborer**

Labor installing gypsum wallboard
Figure 10-8

Once again, I want to emphasize that only your own carefully-compiled records can give you accurate labor figures. But Figure 10-8 has some average manhour figures for installing gypsum drywall.

Estimating Lath and Plaster
Lath and plaster work is almost always figured by the square yard. Unit costs are based on the square yard or 100 square yards. Linear foot measure is only used when estimating moldings or continuous trim.

The Plaster
Plaster is a mixture of cement binders, inert aggregate fillers, and water. Other materials may be added for color or workability, and to make the cured material harder or more fire resistant. It's generally applied by trowel in one or more coats to form a durable wall or ceiling finish material, a backing material for tile, or as fireproofing for structural framing.

There are two types of plaster; one made from portland cement and the other from gypsum. Portland cement plaster is suitable for interior and exterior use, but don't use it over gypsum plaster, gypsum masonry, or gypsum lath. It's less stable than gypsum plaster and more susceptible to cracking. It can't be finished as smooth as gypsum plaster, or with the same intricate detail. And it's less fire resistant than gypsum plaster. In some areas, portland cement plaster with integral color is called *stucco*.

Gypsum plaster isn't suitable for exterior uses except in well-protected areas. Don't use it in interior areas subject to wetting or extremely high humidity conditions. It can be applied over concrete masonry, cement plaster and gypsum and metal laths.

Both cement and gypsum plasters are made with sand, various lightweight aggregates, organic and inorganic fibers, and a wide variety of admixtures. Each coat in a multiple coat cement or gypsum plaster job is likely to have different ingredients or different proportions of the same ingredients.

Plaster can be applied directly to concrete or masonry. Or it can be supported by lath attached to concrete, masonry, sheathing, or wood or metal furring strips. Another method of support is lath attached to a metal grid system suspended on wires from the structure.

Plaster Aggregate

Aggregates in plaster extend coverage, reduce shrinkage, increase strength, and lower cost. Aggregates for plaster include wood fiber, sand, perlite and vermiculite.

Sand is the most common aggregate. It's dense, strong and a good barrier against sound transmission. Vermiculite and perlite are used to improve fire resistance, insulation value, sound absorption, and to reduce weight.

Fiber aggregate, such as shredded wood or other fiber, is generally used with particle aggregates. Fibers increase plaster strength and reduce weight. They also aid placement in gun-applied plaster.

Lath

Rough (unplaned) wood strips nailed to wood framing members used to be the common lathing method. Wood lath is still available, but it's seldom used in today's construction. Today lath is either gypsum or metal.

Gypsum lath is made of paper bonded to a gypsum core, like gypsum wallboard, except that the long edges are generally rounded. It may be plain or perforated with 3/4" diameter holes spaced about 4 inches apart. It's available in 3/8", 1/2", 5/8" and 1" thicknesses, 16", 24", and 48" widths, and in 48" and 96" lengths. Plain (unperforated) lath is available with aluminum foil backing for thermal insulation. Type X gypsum lath has special additives in the core to make it more fire resistant. Perforated lath allows a mechanical as well as a natural chemical bond with gypsum plaster.

Expanded metal lath is available in several forms, in galvanized steel and factory painted copper-bearing steel. Flat diamond mesh is a uniformly expanded flat sheet that's produced in two weights. Self-furring diamond mesh is similar to diamond mesh except that it's indented at regular intervals to hold the body of the lath away from the sheathing.

Flat rib lath has 1/8"-deep ribs evenly spaced and parallel to the long dimension of the lath. There are deeper rib laths available, with 3/8" or 3/4" ribs. Each variety is made in two weights.

Galvanized woven wire fabric with a hexagonal mesh pattern (sometimes called stucco netting, or poultry netting) is generally used for exterior work. It comes in sizes from 1" to 2¼" mesh with a maximum open area of 4 square inches per mesh. It may be dimpled at regular intervals to hold it away from the sheathing, or you can use special nails with fibered spacer washers on the shank to hold it out. For unsheathed framing, choose woven wire fabric with a paper backing.

Welded wire fabric lath is made from 16 gauge wires, spaced not over 2 inches in either direction and stiffened longitudinally with heavier wire. All wires are welded at all intersections. Most welded wire lath in use today is galvanized and provided with paper backing. Some welded wire lath is also crimped at regular intervals to hold the body of the lath away from the contact surface.

Wire cloth is sometimes used as a plaster base. It's generally 19 gauge wires woven in a straight grid pattern and galvanized after weaving.

Fasteners

Many types of fasteners are used in the lathing and plastering industry. Some are common types found in a variety of assemblies. Others have been developed for specific purposes in the lathing and plastering process. We'll cover some of the common fasteners:

- Annealed galvanized steel wire is probably the most common fastener. It's used to support horizontal grid frames, to tie vertical and horizontal framing and furring members together, and to secure lath to supporting members.

- Nails and staples of various wire gauges, lengths, shank and head designs and finishes are used to secure metal and gypsum lath to wood and metal supports. Choose from galvanized, zinc plated, blued, cement coated, or bright finishes. Shanks are smooth, barbed, ringed, or threaded design. Hardened stub nails are used to secure metal or wood to concrete or masonry.

• Explosive-powder driven fasteners are used to secure metal runner tracks to concrete.

• Self-drilling and thread-cutting or thread-forming fasteners attach both metal and gypsum lath to metal supports, and connect metal framing members. Bolts are also used for some framing member connections and to secure framing members to concrete or masonry.

• Clips of various designs, made of spring wire or sheet metal, are used to attach furring channels to runner channels, gypsum lath to metal supports, and resilient furring to rigid supports. Clips are generally zinc plated.

Multiple Coat Application

Gypsum base coats are applied in one, or more often, two coats. In two-coat work, the base coat is usually a factory mix that only needs mixing with water. Additional aggregate is sometimes added to the second (brown) coat. Sand is also sometimes added to the first (scratch) coat over gypsum lath or masonry bases.

Job-mixed base coats contain neat gypsum plaster mixed with graded sand or vermiculite or perlite.

Gypsum Finish Coats

Lime putty in a plaster mix provides whiteness, plasticity, and bulk. Lime, used alone, doesn't set hard, and shrinks as it dries. Quicklime requires a long slaking (soaking) period at the job site to develop the plasticity needed for plastering. For these reasons, lime is blended with gypsum gauging plaster or Keene's cement to provide the desired properties.

Hydrated lime is quicklime which has been partially slaked at the factory to reduce the field slaking period. There are two basic types of hydrated lime: normal (Type N, ASTM C-6) which requires 16 to 24 hours of soaking, and special (Type S, ASTM C-206) which doesn't require soaking. It can be used immediately.

Gauging plasters are specially ground to provide controlled set, strength, and minimum shrinkage when mixed with lime. Gauging plaster is available

in slow and quick-set types. This allows control of the hardening process without the use of retarders or accelerators.

Keene's cement provides a dense, harder-than-average finish with more resistance to moisture. It's available in Type I regular slow setting (3 to 6 hours), and Type II quick setting (1 to 2 hours).

Molding plaster is a finely-ground gypsum mix which produces a smooth, workable material suitable for intricate ornamental and decorative work.

Gypsum finish coats can be made from several possible mixes for trowel, spray or float finish. They include gauging plaster, Keene's cement, and gypsum white coat. There are also special finishes that include integral coloring and special texturing, and veneer or skin coat plaster, a thin finish coat applied directly over a suitable base. It can provide a smooth finish to an otherwise porous surface, add texture, or conceal joints or irregularities.

Plaster used to fireproof the material it covers is a specially-formulated spray-applied material with mineral fibers or expanded mineral aggregates. It's applied in one or more coats to structural members.

Figuring Surface Areas

When figuring the wall surface, there's some controversy over which openings to ignore and which openings to consider. For example, you would almost certainly ignore a small window in a wall. The 4 or 5 square feet of wallboard or lath that's cut out for the window opening will almost certainly be wasted. It's better to figure the quantity needed as though the window weren't there. Doors are larger, but still won't offer much saving. You'll probably want to ignore most door openings when calculating wall surface. Some estimators deduct openings only if they exceed 100 square feet. Other estimators deduct one-half of all openings. The plasterer's union in your area may require certain rules of measuring. Use the method that works for you. For the rest of this chapter we'll deduct one-half of all openings.

There are two distinct methods of getting the area of plaster work from the plans. One method is to find the inside dimensions of the rooms and use these to calculate the area, as we did earlier for figuring wallboard. This gives you the actual surface area. This is the method a lath and plaster subcontractor will use to figure the contract price.

When using this method, note whether dimensions given on the plan are to the face of the wall or

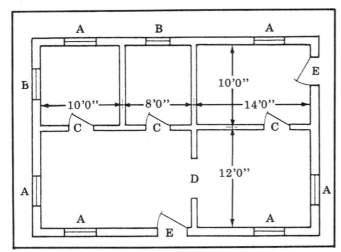

Sample area problem
Figure 10-9

to the center line of partitions. If they're to the center line, deduct half the thickness of the partitions on each side of the room. Figure partitions as 6 inches thick if they're built of 2 x 4 studs with lath and plaster.

There's a second, faster way, of figuring the area of a surface in estimating plaster work. Count the feet of partitions and outside wall areas, not deducting anything for the thickness of the partitions. This method isn't as accurate. You'll get an answer that's a little too high. But it saves time in calculating the area and is good enough when estimating jobs that won't be subbed out. Here are the steps to follow:

1) Count the linear feet of outside wall surface.

2) Find the linear feet of partitions and multiply by two.

3) Add the totals of the outside wall and the partitions, then multiply by the height of the wall to get the areas of the walls in square feet.

4) Add to this total area the area of the ceiling. (On your estimate sheet, identify the ceiling area, since labor for ceiling plaster work is generally more costly.)

5) Subtract the number of square feet allowed for openings.

6) Divide by 9 to convert square feet to square yards.

Look at Figure 10-9 and we'll work through an example to see how this method works. Assume the ceiling height in Figure 10-9 is 9'0''.

Size of openings:
A) 3'6'' x 4'0''
B) 2'6'' x 4'0''
C) 2'8'' x 7'0''
D) 5'0'' x 7'0''
E) 3'0'' x 7'0''

Outside walls: 32' + 22' x 2 = 108 linear feet
Center partition (lengthways): 32'
Center partition (crossways): 22'
Short partition: 10'

$$\frac{64}{\times\ 2}$$
128 linear wall feet

Total linear feet:
$$\begin{array}{r} 108 \\ +128 \\ \hline 236 \\ \times\ 9' \text{ ceiling heights} \\ \hline 2,124 \text{ SF wall area} \end{array}$$

Area of ceiling (22' x 32') = 704 SF

Total area of walls and ceiling:
$$\begin{array}{r} 2,124 \\ 704 \\ \hline 2,828 \text{ SF} \end{array}$$

Deduct openings:

A)	3.5 x 4 x 6 =	84	SF
B)	2.5 x 4 x 2 =	20	SF
C)	2.66 x 7 x 6 =	111.72	SF
D)	5 x 7 x 2 =	70	SF
E)	3 x 7 x 2 =	42	SF
	Total	327.72 or 328	SF

We said earlier we'd deduct half of the opening area in our example, so divide 328 square feet by 2. Now deduct 164 square feet from the total area:

$$\begin{array}{r} 2,828 \text{ total area} \\ -\ 164 \text{ ½ of openings} \\ \hline 2,664 \text{ SF} \end{array}$$

Divide the 2,664 net square feet of area by 9 to convert to square yards. That's 296 square yards of surface area to plaster.

If you want to be even more accurate, you can deduct 100% instead of 50% of openings C and D, since they're interior doors that appear on both sides of the walls.

Kind of construction	Lath	Other materials	Labor
Metal lath, wood studs	105 square yards	8 lbs. nails or 15 lbs. staples	8 hours
Metal lath on steel studs for 2" solid partition	1000 lin. ft. of ¾" channels 105 sq. yds. of lath	10 lbs. tie wire	24 hours including studs
Plasterboard on wood studs	900 sq. ft.	7 lbs. nails	8 hours
Cornice furring	110 sq. yds. of lathing	8 lbs. nails	--
Beam and cornice furring	1800 lin. ft. of ¾" channel 108 sq. yds. of lath	27 lbs. tie wire	144 hours
Suspended ceiling with cross channels 12" o.c.	285 lin. ft. of 1½" channel 1000 lin. ft. of ¾" channel, 105 sq. yds. of lath	85, ¼" hanger rods and 18 lbs. wire	30 hours

Lath and furring for 100 square yards of wall or ceiling
Figure 10-10

Estimating the Costs

To estimate the costs of lath and plaster work, first figure the quantities of material or the cost per hundred square yards. It's usually best to figure different unit costs per hundred square yards or per square yard for each kind of work. If your bid is successful, then you can make out a bill of materials in detail. A detailed bill of materials isn't necessary to prepare a bid.

Materials for Lathing and Plastering

Figure 10-10 gives material and labor figures for lathing and furring 100 square yards of wall or ceiling. The first column lists the kind of construction. Look in the second column for the square yards of metal lath and the linear feet of furring that you'll need. The third column gives the pounds of nails or wire required to fasten the lath. The last column estimates labor requirements.

Figure 10-11 shows the cubic feet of plaster or mortar required to cover 100 square yards of surface for various thicknesses. Use this table as a guide when no other figures are available. It shows, for example, that it takes 20 cubic feet of plaster to put a 1/4"-thick surface on 100 square yards.

Then, using Figure 10-12, you can translate that 20 cubic feet of plaster into quantities of material for the proportions of the particular plaster mix you'll be using. For instance, assume you're using gypsum cement plaster with 1 to 2 proportions. For one cubic foot, you need 33 pounds of plaster and 0.66 cubic feet of sand. Multiply both by 20 to find out how much you need for the whole job: 660

pounds of plaster and 13.2 cubic feet of sand. You'll need to order seven 100-pound bags of plaster and 1/2 cubic yard of sand. Notice that the quantity of plaster is listed by weight, and the sand by volume.

Figure 10-13 can save you a step. It gives the quantity of plaster and sand you need for 100 square yards of scratch and brown coats of gypsum cement plaster. Figure 10-14 covers gypsum wood fiber plasters while Figure 10-15 gives you the material for gypsum sanded plasters.

The wide selection of finishing coats makes it impossible to give a complete chart of materials in the limited space of this chapter. However, Figure 10-16 will give you an idea of the material required for six different finishes.

Figure 10-17 is similar to Figure 10-11, except that it gives the quantity of cement required for various thicknesses of portland cement plaster over cement, tile or stone. For other thicknesses, the amounts will be in proportion to the thicknesses

Thickness	Quantity
1/8" thick	10 cubic feet
1/4" thick	20 cubic feet
3/8" thick	30 cubic feet
1/2" thick	40 cubic feet

Plaster needed to cover 100 square yards
Figure 10-11

Kind of mortar	Proportion	Lime or cement in pounds	Sand in cubic feet
Pure lime putty	--	44 lbs. hydrated lime	No sand
Lime mortar	1 to 2	19.6 lbs. hydrated lime	0.87
Lime mortar	1 to 3	15 lbs. hydrated lime	1.0
Portland cement mortar	1 to 2	41 lbs. cement	0.87
Portland cement mortar	1 to 3	32 lbs. cement	1.0
Gypsum cement plaster	1 to 2	33 lbs. plaster	0.66
Gypsum cement plaster	1 to 3	20 lbs. plaster	0.6
Gypsum wood fiber plaster	No sand	58 lbs. plaster	No sand
Gypsum sanded plaster	--	79 lbs. plaster	No sand

Materials for 1 cubic foot of mortar
Figure 10-12

Construction	Proportion by weight	100 lb. sacks plaster	Cubic yards sand
Over metal lath	1 to 2	17 to 20	1.3 to 1.5
Over plasterboard	1 to 2	8 to 9	.6 to .7
Over brick or clay tile	1 to 3	14 to 17	1.5 to 1.9
Over gypsum tile	1 to 3	10 to 12	1.1 to 1.3

Gypsum cement plaster for 100 square yards of
scratch and brown coat
Figure 10-13

Construction	Proportion by weight	100 lb. sacks plaster	Cubic yards sand
Over metal lath	no sand	22 to 27	--
Over plasterboard or gypsum lath	no sand	13 to 16	--
Over brick and clay tile	1 to 1	18 to 20	.7 to .8
Over gypsum tile	1 to 1	14 to 16	.5 to .6

(Note: Proportions for gypsum plasters are by weight).

Gypsum wool fiber plaster for 100 square yards of
scratch and brown coat
Figure 10-14

listed. For 3/8" thickness, for example, you could multiply the 1/8" figures by three, or add the 1/8" and 1/4" figures to find the total.

You can also use Figure 10-17 for figuring materials for portland cement plaster over metal lath. In this case, determine the size of ground and subtract the thickness of the metal lath from the ground. For example, if the grounds are 3/4" and the metal lath 1/8", it takes a 5/8" thickness of plaster. That will require as much material as twice the 1/4" thickness plus the 1/8" thickness from the table.

For lime plaster, the proportions are generally based on the volume of putty and the volume of sand. You need to know how much putty you'll get from a certain quantity of hydrated lime. Figure 10-18 will help you do the calculations. The first column gives the hydrated lime in pounds and sacks. The second column gives cubic feet of putty obtained from the different quantities of hydrated lime.

Construction	100 lb. sacks plaster
Over metal lath	45 to 50
Over plasterboard	20 to 22
Over brick or clay tile	35 to 40
Over gypsum tile	25 to 28

Gypsum sanded plaster for 100 square yards of scratch and brown coat
Figure 10-15

Kind of finish	50 lb. sack hydrated lime	Cubic yards sand	100 lb. sacks plaster
Sand finish using lime and plaster	6 to 7	1/4	--
White finish using lime and plaster	6 to 9	--	1 to 1¼
Hard cement finish using Keene's cement	--	--	4 to 5
Smooth cement finish using Keene's cement and hydrated lime	2	--	3 to 4
Sand finish using special finishing plaster	--	.12 to .25	2 to 2.5
Lime putty	6 to 9	--	--

Materials for 100 square yards of finishing coat
Figure 10-16

Thickness	Proportion by volume	Sacks of cement	Cubic yards sand
1/8" thick	1 to 1	6.75	0.25
1/8" thick	1 to 1½	5.36	0.30
1/8" thick	1 to 2	4.48	0.33
1/8" thick	1 to 2½	3.84	0.35
1/8" thick	1 to 3	3.32	0.37
1/4" thick	1 to 1	13.52	0.50
1/4" thick	1 to 2	8.96	0.66
1/4" thick	1 to 3	6.64	0.74

Portland cement plaster over concrete, tile or stone, 100 square yards
Figure 10-17

Hydrated lime	Putty in cubic feet
44 lbs.	1
50 lbs. (1 paper bag)	1.2
100 lbs. (1 cloth bag)	2.3
308 lbs. (6 sacks)	7
456 lbs. (9 sacks)	10.5

Yields of hydrated lime
Figure 10-18

Bases	Grounds	Actual plaster thickness	Coats	Proportions	Amount per 100 square yards	
					Bags plaster (100 pounds)	Cubic feet aggregate
Metal Lath	3/4"	1/4"	Scratch	1-2	10	20
		7/16"	Brown	1-3	12	36
Gypsum Lath	7/8"	3/16"	Scratch	1-2	5	10
		1/4"	Brown	1-3	6	18
Unit Masonry	5/8"	3/16"	Scratch	1-3	4	12
		3/8"	Brown	1-3	8	24

Gypsum lightweight plaster
Figure 10-19

Figure 10-19 gives approximate quantities of gypsum plaster and vermiculite aggregate for 100 square yards of various mixtures. The proportion "1-2" means 100 pounds of gypsum plaster to 2 cubic feet of vermiculite or any other lightweight aggregate.

Gypsum ready-mix plasters with lightweight aggregates already in them are also available. They're packed in 80-pound sacks for lath bases and in 67-pound sacks for masonry bases. For 100 square yards on a metal lath base, you'll need 34 sacks. For a gypsum lath base, 19 sacks are needed. For a masonry base, allow 26 sacks.

Labor for Lathing and Plastering
Figure 10-20 gives typical manhour figures for labor required per 100 square yards for lath, plastering and finish plaster. Use them for figuring unit costs for a square of lathing and plastering if you don't have your own manhour figures.

Estimating Corner Bead
Corner beads are used around openings and on the corners of plastered walls to prevent chipping. Standard lengths are 8, 9, 10 and 12 feet. To estimate corner beads, find the combined length of the outside corners in all rooms to be plastered. If any two rooms are connected by a plastered archway, multiply the distance around the archway by 2 and add this number of linear feet to the corner sum.

If you're ordering corner beads in stock lengths, select any suitable length and divide the combined length of bead required by that stock length, counting any fractional part of a piece as one additional length.

Labor Plastering

Work Element	Unit	Man-Hours Per Unit
Solid plaster partitions, including lath and ¾" channel studs		
2" thick partitions	100 SY	115.0
3" thick partitions	100 SY	128.0
2 coat plaster, with finish indicated, on walls		
Gypsum or lime finish	100 SY	42.0
Keene's cement finish	100 SY	54.0
Wood fiber finish	100 SY	47.0
Vermiculite finish	100 SY	40.0
Portland cement finish	100 SY	45.0
Acoustical finish	100 SY	58.0
3 coat plaster, with finish indicated, on walls		
Gypsum or lime finish	100 SY	55.0
Keene's cement finish	100 SY	69.0
Wood fiber finish	100 SY	68.0
Vermiculite finish	100 SY	55.0
Portland cement finish	100 SY	66.0
Acoustical finish	100 SY	72.0
Scratch coat	100 SY	12.0
Brown coat	100 SY	13.0
3 coat stucco on metal lath		
¾", float finish	100 SY	70.0
¾", trowel finish	100 SY	74.0
Additional plastering requirements		
Add for ceiling work	100 SY	4.5
Add for column work	100 SY	23.0
Add for chases, fascia, recesses or soffits	100 SY	33.5
Add for beams	100 SY	26.0
Add for irregular surfaces	100 SY	21.0
Deduct for plaster over gypsum lath	100 SY	6.0

Plastering includes mixing plaster, installing and finishing plaster, scaffolding, curing and drying plaster.
Suggested Crew: 2 plasterers, 1 or 2 tenders

Labor plastering, 100 square yards
Figure 10-20

Labor Installing Wall Paneling

Work Element	Unit	Man-Hours Per Unit
Plastic faced hardboard, including moulding and trim		
1/8"	100 S F	2.8
1/4"	100 S F	2.9
Plywood, 4' x 8' panels, including trim		
¼"	100 S F	3.5
½"	100 S F	4.4
Plank paneling		
¼"	100 S F	3.9
¾"	100 S F	5.0
¾", random width	100 S F	5.4
Cedar closet lining		
1" x 4" plank	100 S F	5.9
¼" plywood	100 S F	4.5

Allow about 25% more time for ceiling installation. Deduct 5 to 15% when 9' or 12' high plywood panels can be used. If installtion is on metal studs, add 10% to the times listed.

CEF — Labor installing wall paneling
Figure 10-21

Estimating Ornamental Plaster Work

Ornamental plaster work is hard to estimate. I'll try to give you a little guidance though, just in case you come across any. There's not much ornamental plaster work done in modern buildings.

Narrow cornices or moldings are generally estimated by the linear foot. Wider cornices are estimated by the square foot. In estimating this kind of work, consider the size of the room and the number of miters that have to be made. Very often the cost of making a miter in an internal or an external corner will be almost as much as running a sketch of stretch mold or cornice.

Portland cement base 12 inches or less in height will require about 16 hours of plasterer's time and four hours of helper time per 100 linear feet. If the base has a cove at the floor, it will require about 24 hours for the plasterer and 6 hours for the helper per 100 linear feet. Any cornice or beam that has a girth of 12 inches or less will require approximately 18 hours per 100 linear feet for the plasterer, and about 6 hours for the helper.

It will require 1 hour extra labor for each internal corner or miter, and about 1/2 hour extra plasterer's time for each external miter.

Estimate cornices 12" or more in girth by the 100 square feet. It will require about 17 hours plasterer's time and 6 hours helper time per 100 square feet. Figure from 1½ to 2 hours plasterer's time for each internal miter and 3/4 to 1 hour for each external miter. If the cornices are run in Keene's cement or portland cement, the labor may run as much as 25% more than these estimates.

Figure beams like cornices unless they have plain surfaces, in which case they may be figured like ordinary plastering surfaces or reveals. Usually they require about twice as much labor as a large flat surface.

For running small molds to form panels, figure about 15 hours plasterer's time and 5 hours helper's time per 100 linear feet. For small coves and rounded corners, estimate about 10 hours of plasterer's time per 100 linear feet, and 2 to 3 hours helper's time.

These labor hours don't include the time for applying the scratch and brown coat. Include the time for scratch and brown coats under the ornamental work at the same rate as under regular plaster work.

Conclusion

The more information you have on the materials, labor and equipment needed, the easier it will be to estimate costs accurately. Look at Figure 10-21. If your CEF includes this information, you have a starting point for most estimates.

The labor figures used throughout this book are for quality work. Accept no less from your crew or subs. Shoddy work will ruin you.

Chapter 11

Estimating
Exterior Finish Carpentry

Exterior finish carpentry includes siding and trim. We'll start with the siding, then go on to the water table, cornices, soffits and porch ceilings.

Estimating Siding

Plywood, hardboard and wood boards are the most common siding materials in many parts of the U.S. Woods appropriate for outside finish include Douglas fir, redwood, Western cedar, and white pine. Most manufactured wood materials used for exterior finish carry the manufacturer's recommendations for installation, and often include instructions for applying the finish. Of course, you have to know how a material will be applied to estimate the installed cost.

Plywood Siding
Plywood siding is a popular exterior covering. In many cases it can be applied directly to the studs, omitting any exterior sheathing. Plywood siding includes APA Sturd-I-Wall, plywood panel siding and plywood lap siding.

APA Sturd-I-Wall— The American Plywood Association's (APA) Sturd-I-Wall system uses APA 303 plywood panel siding. The siding can be applied directly to studs or over nonstructural sheathing materials such as fiberboard, gypsum or rigid foam insulated sheathing. *Nonstructural*

means that the sheathing material does not meet the bending and racking strength requirements specified by building codes.

Sturd-I-Wall is accepted by HUD and most building codes. A single layer of Sturd-I-Wall is strong and rack-resistant enough to eliminate the need for separate structural sheathing and diagonal wall bracing.

Sturd-I-Wall siding is normally installed vertically, as shown in Figure 11-1. It can be applied horizontally (Figure 11-2) with 2 x 4 blocking at the horizontal joints. Use 6d nails for panels 1/2'' thick or less, 8d nails for thicker panels. Nail spacing is 6 inches along panel edges and 12 inches for intermediate spaces.

Plywood Panel Siding and Lap Siding— Most plywood siding is installed over sheathing. Figures 11-3 and 11-4 show how plywood panel and lap siding are installed.

Lap siding must be applied with the face grain across the supports. Siding joints should be staggered and don't have to fall on a stud if nailable panel or lumber sheathing is used under the siding. Nailable sheathing includes:

1) Nominal 1-inch boards with studs spaced 16 or 24 inches o.c.

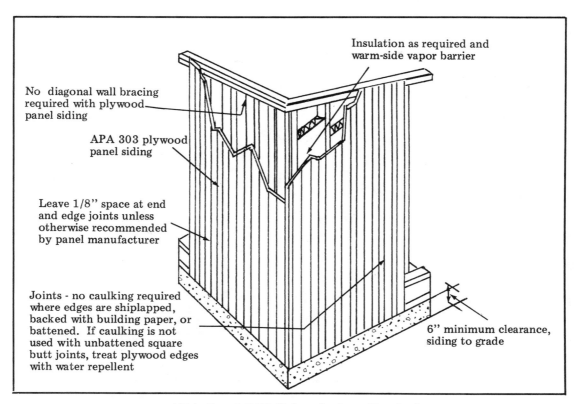

No diagonal wall bracing required with plywood panel siding

APA 303 plywood panel siding

Leave 1/8" space at end and edge joints unless otherwise recommended by panel manufacturer

Joints - no caulking required where edges are shiplapped, backed with building paper, or battened. If caulking is not used with unbattened square butt joints, treat plywood edges with water repellent

Insulation as required and warm-side vapor barrier

6" minimum clearance, siding to grade

APA Sturd-I-Wall (vertical application)
Figure 11-1

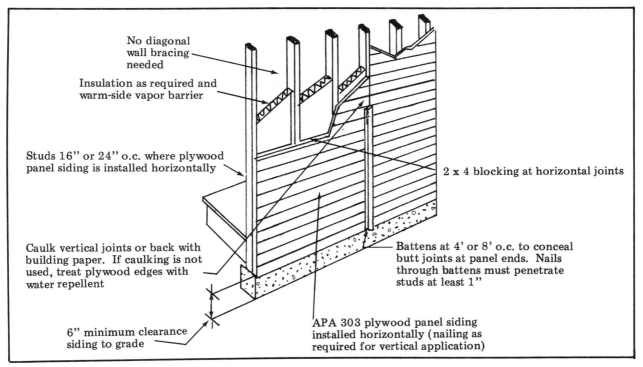

No diagonal wall bracing needed

Insulation as required and warm-side vapor barrier

Studs 16" or 24" o.c. where plywood panel siding is installed horizontally

Caulk vertical joints or back with building paper. If caulking is not used, treat plywood edges with water repellent

6" minimum clearance siding to grade

2 x 4 blocking at horizontal joints

Battens at 4' or 8' o.c. to conceal butt joints at panel ends. Nails through battens must penetrate studs at least 1"

APA 303 plywood panel siding installed horizontally (nailing as required for vertical application)

APA Sturd-I-Wall (horizontal application)
Figure 11-2

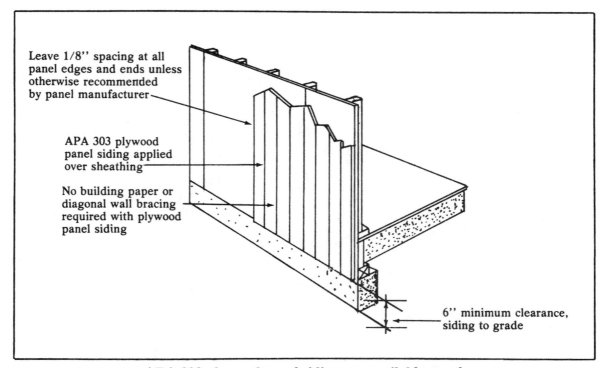

Leave 1/8" spacing at all panel edges and ends unless otherwise recommended by panel manufacturer

APA 303 plywood panel siding applied over sheathing

No building paper or diagonal wall bracing required with plywood panel siding

6" minimum clearance, siding to grade

APA 303 plywood panel siding over nailable panel
or lumber sheathing
Figure 11-3

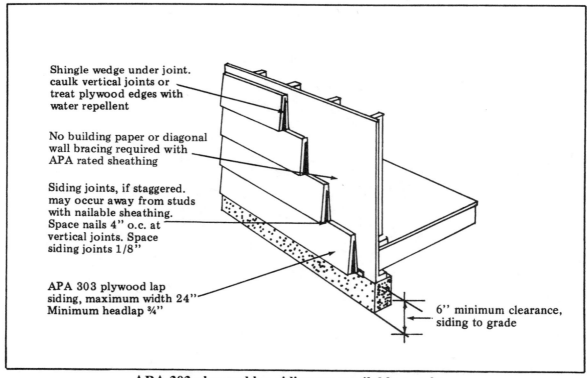

Shingle wedge under joint. caulk vertical joints or treat plywood edges with water repellent

No building paper or diagonal wall bracing required with APA rated sheathing

Siding joints, if staggered. may occur away from studs with nailable sheathing. Space nails 4" o.c. at vertical joints. Space siding joints 1/8"

APA 303 plywood lap siding, maximum width 24" Minimum headlap ¾"

6" minimum clearance, siding to grade

APA 303 plywood lap siding over nailable panel or
lumber sheathing
Figure 11-4

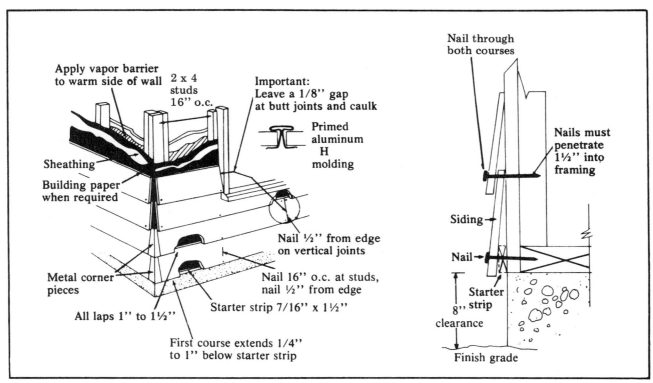

Hardboard lap siding application
Figure 11-5

2) Rated sheathing panels with a span rating of 24/0 or greater, and the long dimension either parallel to or perpendicular to studs spaced 16 or 24 inches o.c. But note that three-ply plywood panels must be applied with the long dimension across studs when studs are spaced 24 inches o.c. Check your local code on this point.

Hardboard Siding

Hardboard is made from wood chips and sawmill by-products. It's a popular exterior siding because it comes in many designs and patterns. Also, it's one of the least expensive siding materials and is fairly easy to install without special tools. Hardboard is free of knots and is uniform in thickness, density and appearance. Like all wood, it has to be maintained and kept free of termites. But there's no grain to rise or cause splitting.

The two most common types of hardboard siding are lap (Figure 11-5) and panel siding (Figure 11-6).

Estimating Panel Siding

Calculate the area of the walls, including the gables. To find the area of the gables, multiply the rise from the plate to the ridge by the width of the gable, then divide by 2. Look at Figure 11-7. Deduct for openings 4 feet wide or more. Add the areas of all the gables to the net wall area for the rest of the house. Then divide the net wall area by the area covered by one panel. Round your answer up for the number of pieces required. For example, 1,460 square feet divided by 32 (area of one 4' x 8' panel) equals 45.62, or 46 panels. Use rust-resistant 8d nails. Estimate 50 nails to a 4' x 8' panel.

For manhour estimates for panel siding, look at Figure 11-8. But use these figures only until you've compiled good manhour data for your own crews.

Estimating Lap Siding

To estimate lap siding, first determine the area of the walls, including gables. Refer to the wall sections and floor plans for the dimensions of the exterior walls. Disregard openings less than 10 square feet; this reduces the waste factor. When you have the net wall area, multiply it by the factor shown in Figure 11-9 for the appropriate siding width. The result is the number of board feet of siding you'll need, including the allowance for lapping and waste. There's a table of standard nail requirements at the end of Chapter 6. Refer to it for the nails you'll need for lap siding.

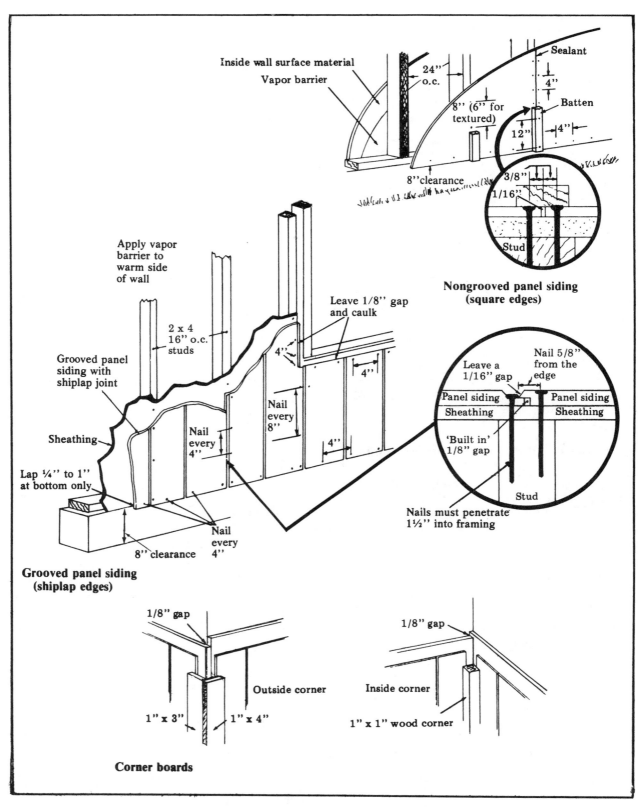

Hardboard panel siding application details
Figure 11-6

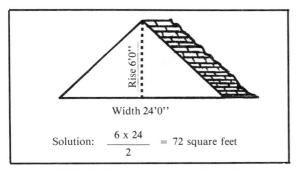

Calculating the area of gables
Figure 11-7

Item	Size or kind	Estimated labor performance
Boards	½ x 6 bevel	40 BF per hour
Boards	½ x 8 bevel	50 BF per hour
Boards	¾ x 10 bevel	60 BF per hour
Plywood	4' x 8' sheets	10-12 hours per 1,000 SF

Manhours for installing panel siding
Figure 11-8

Hardboard Shakes

Hardboard shakes are another popular exterior covering. You can usually apply shakes either over sheathing or directly to studs. No building paper is needed over vertical sheathing unless it's required by your code.

Shakes can be applied to walls that slope as much as 15 degrees from vertical, provided the wall is sheathed and covered with building paper.

Corner bracing is required when hardboard shakes are installed directly over studs, or when nonstructural sheathing is used. When shakes are applied over sheathing, the type of sheathing determines the corner bracing requirements. Figure 11-10 shows hardboard shake application details.

To estimate hardboard shake requirements, compute the wall area, deducting door and window areas. Add 20% to the net area. To cover 1,000 square feet with 10½'' wide x 48'' long hardboard shakes with a 9-inch exposure, you need 333 pieces, allowing for waste and cutting. Estimate 2.5 manhours per 100 SF. Use 6d galvanized nails. It'll take 10 nails to each 48-inch piece, or 1 pound to 100 SF.

Nominal width (inches)	Multiply net wall area by:
8	1.34*
10	1.26*
12	1.21*

*Allows for lapping and waste

Estimating lap siding
Figure 11-9

Wood Siding

Wood siding is more expensive than hardboard siding, but it's still used on quality homes in many parts of the country.

Wood siding should meet these requirements: It should be easy to work with and should not warp. If the siding is to be painted, the wood should be reasonably free of knots, pitch pockets and tapered edges. Cedar, Eastern white pine, sugar pine, Western white pine, cypress and redwood have all of these qualities. Other species used for siding include Western hemlock, pine, spruce, yellow poplar, Douglas fir, Western larch, and Southern pine.

Vertical-grain lumber makes the best siding because it doesn't expand and contract much with changes in temperature and humidity. Redwood and Western red cedar are usually available in vertical grain and mixed-grain grades.

Horizontal Wood Siding— Some wood siding patterns are used only horizontally. Others are applied only vertically. Some can be used either way if adequate nailing area is available. Check the detail to find out what corner treatment is used with horizontal siding. Your estimate will include corner posts for inside corners, and metal or wood corner boards for outside corners. Figure 11-11 shows typical corner treatments for horizontal siding. A little later we'll talk about estimating the corner boards.

Figure 11-12 gives a breakdown of materials and labor for bevel and drop siding. And don't forget to include the furring strip or drip cap in your estimate for bevel siding. Figure 11-13 shows the first course of bevel siding installed two ways: over a furring strip, and with a drip cap and water table.

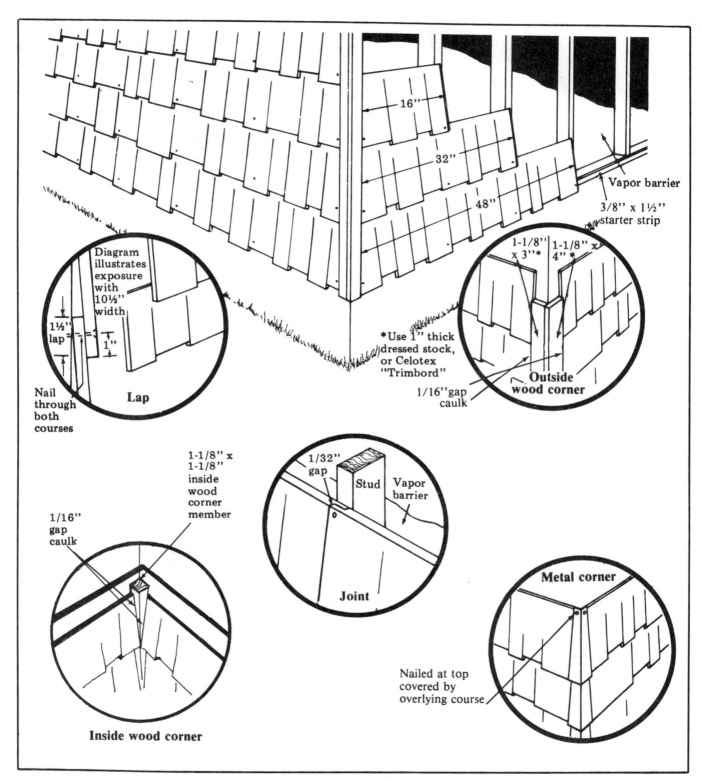

Hardboard shake application details
Figure 11-10

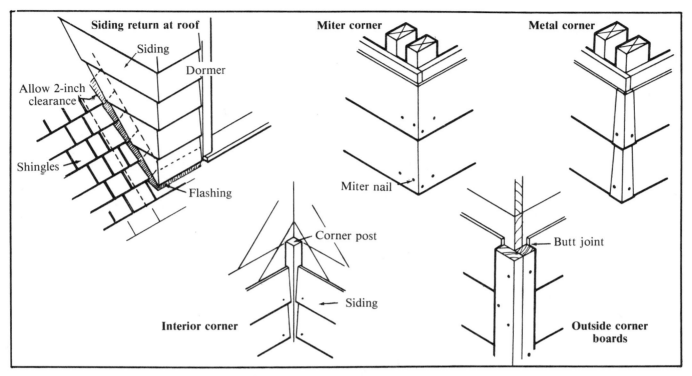

Corner treatments for horizontal siding
Figure 11-11

| Size | Bevel Siding Material Siding for 100 square foot wall | | | Nails per 100 SF | Labor board feet per hour |
	Exposed to weather	Add for lap	BM per 100 SF		
½ x 4	2¾	46%	151	1½ pounds	30
½ x 5	3¾	33%	138	1½ pounds	40
½ x 6	4¾	26%	131	1 pound	45
½ x 8	6¾	18%	123	¾ pound	50
⅝ x 8	6¾	18%	123	¾ pound	50
¾ x 8	6¾	18%	123	¾ pound	50
⅝ x 10	8¾	14%	119	½ pound	55
¾ x 10	8¾	14%	119	½ pound	55
¾ x 12	10¾	12%	117	½ pound	55

| Size | Drop Siding Material Siding for 100 square foot wall | | | Nails per 100 SF | Labor board feet per hour |
	Exposed to weather	Add for lap	BM per 100 SF		
1 x 6	5¼	14%	119	2½ pounds	50
1 x 8	7¼	10%	115	2 pounds	55

Note: Quantities include 5% for end cutting and waste. Deduct for all openings over ten square feet.

Materials and labor for bevel and drop siding
Figure 11-12

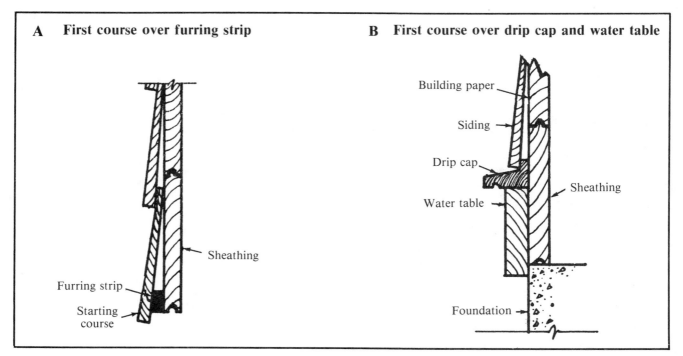

A First course over furring strip

Sheathing

Furring strip

Starting
course

B First course over drip cap and water table

Building paper

Siding

Drip cap

Water table

Sheathing

Foundation

Bevel siding application details
Figure 11-13

Vertical Wood Siding— Vertical siding is appropriate for some architectural styles. It's usually made of rough-sawn boards and battens. The boards and battens can be arranged in several ways: boards and batten, batten and boards, or board and board. See Figure 11-14.

Vertical siding should be applied over 1-inch sheathing boards or over plywood sheathing that's 1/2-inch or 5/8-inch thick. Sheathing provides the required nailing surface.

If you use composition or foam sheathing, or thinner plywood sheathing, 1 x 4 nailing blocks should be installed horizontally between the studs. Allow vertical spacing of 16 to 24 inches between nailing blocks. And it's good practice to apply building paper over the sheathing before applying vertical siding. List nailers and building paper on your take-off. Calculate nailer requirements by taking the linear feet of the wall times the number of runs. Estimate building paper by the square feet of wall area, as we did for the panel siding.

To estimate vertical siding requirements, first compute the area of the walls, including the gables. Disregard openings less than 10 square feet. Use Figure 11-15 to estimate the material for vertical siding.

Estimate labor at 50 board feet per hour. Use 6d galvanized or aluminum nails. Figure 575 nails to 500 board feet of siding.

Aluminum Siding
Aluminum siding is available in horizontal and vertical styles. The most common siding pattern currently in use is the horizontal type with an 8-inch exposure. Double 4-inch and double 5-inch are also available. Vertical siding is similar in appearance to board and batten or V-groove siding. It's used as a design break, and sometimes used for gable ends.

Many builders prefer to subcontract out aluminum siding jobs. The siding is sold by the square, which covers 100 square feet on the wall. To estimate, divide the net wall area by 100 and add the waste factor suggested by the manufacturer. Take off trim and accessories separately.

Vinyl Siding
Vinyl siding is made for both vertical and horizontal applications. A vinyl soffit system is also available. All the necessary equipment, such as posts for inside and outside corners, starter strips and dividers, is available. Horizontal vinyl siding usually comes 12'6'' long, with a single-board, 8-inch exposure. Vertical vinyl siding is usually 10 feet long with 12-inch-wide exposure.

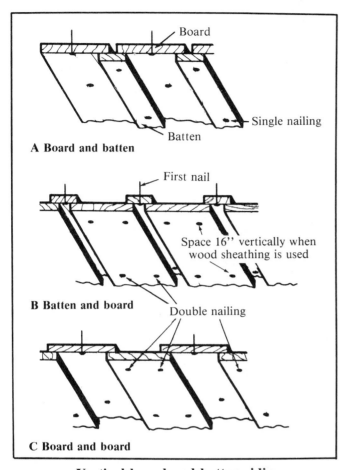

A Board and batten

B Batten and board

Space 16" vertically when wood sheathing is used

Double nailing

C Board and board

Vertical board and batten siding
Figure 11-14

Nominal width (inches)	Multiply net wall area by:
8	1.10
10	1.08
12	1.07

Estimating vertical board siding
Figure 11-15

should provide good drainage. For example, if vertical board-and-batten siding is used at the gable ends, and horizontal siding is used on the walls below, a drip cap and flashing are required. Another way to make the material transition is to extend the plate and studs of the gable end out from the wall. Your plans will indicate the distance of the extension. The gable siding will project beyond the wall siding and provide good drainage. Figure 11-16 shows both methods.

Estimating Drip Cap and Water Table
In Figure 11-13 we saw bevel siding installed over drip cap and water table. To estimate them, measure from the first floor plan instead of from the elevations. Begin at one corner and work around the building until you arrive at the starting point. Record each length taken from the plan and add them for the total linear feet. For every outside corner which is mitered, allow 12 inches extra. This is the amount of plain stock or molding you'll need.

Estimating Corner Boards
The size and kind are given on the building plans or in the specifications. To find the linear footage, take the first floor plan and one or more of the elevations. Locate one of the corners on the first floor plan, then find it on the elevation and scale the height. Write the height down on your estimate sheet. Do each corner this way.

Add up the figures that you put down on your pad. Find the total linear feet of corner board and enter this item on the estimate sheet with the total width, thickness and kind of lumber. A corner board should always be one piece. If the height of the corner is 9 feet, it will be cut from a 10-foot piece if an 18-foot piece isn't available for making two 9-foot corner boards.

To estimate metal corners, divide the wall height by the siding exposure (6 inches, 9 inches, etc.). Multiply the answer by the number of outside corners.

Vinyl siding is sold by the square, which covers 100 SF on the wall. You divide the net wall area by 100 and allow for waste as recommended by the manufacturer. Take off trim and accessories separately.

Vinyl siding expands and contracts with temperature changes and must be free to move back and forth on the nail shank. Nails shouldn't bind the vinyl and must be near the center of the slot so the siding can move.

Subs specializing in the installation of vinyl siding usually do a professional job at a reasonable cost.

Material Transitions
Some houses use two types of siding. Others use both vertical and horizontal siding of the same type. The siding used for gable ends may be different from the siding used for walls.

The joint between the two types of sidings

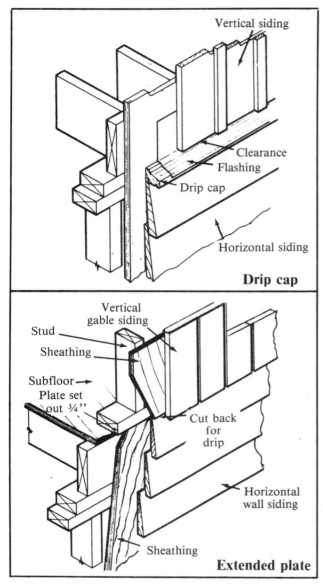

Drip cap

Extended plate

Providing drainage at material transition
Figure 11-16

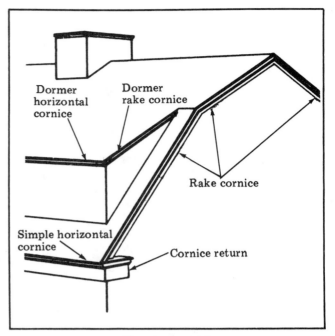

Cornice positions
Figure 11-17

Figure 11-18 shows the popular wide box cornice and a cornice return. Figure 11-19 details cornice treatment on a brick veneer house. Double check that you're including each element of the cornice on your estimate sheet: soffit, frieze, fascia and trim.

Estimating Cornice Manhours

The time required for cornice trim work depends on the type of horizontal cornice and rake cornice treatment. For example, the box cornice in Figure 11-18 A has a frieze board on the face of the house, a soffit underneath the cornice and a fascia on the cornice face. The soffit and fascia are nailed to a nailing header. Additional trim may be used on the cornice face and on the frieze board. Examine the plans and specs carefully to ensure you've included all the material required.

When a fascia is used to finish off the gable end of a building, as shown in Figure 11-18 B, allow 14 hours per 100 linear feet.

Manhours required to install 100 linear feet of cornice trim work are shown in Figure 11-20.

A two-member closed cornice (including molding and frieze board) on a 40 x 24-foot house will take about 18 skilled manhours. Approximately 23 skilled hours will be required for a three-member box cornice.

Estimating Exterior Cornices

On a box cornice, if the soffit is 12 inches, if the frieze is 8 inches and if the fascia is 4 inches, the combined width is 24 inches. That yields 2 board feet per linear foot of cornice. The lengths of the cornice are shown in the elevation drawings. Compute the lengths from the measurements shown on the plan or by scaling off the dimension. For each outside corner, allow an extra 12 inches. Figure 11-17 shows the various cornice positions. Check the plans to make sure you don't miss any. It's easy to omit cornice on buildings with irregular shapes.

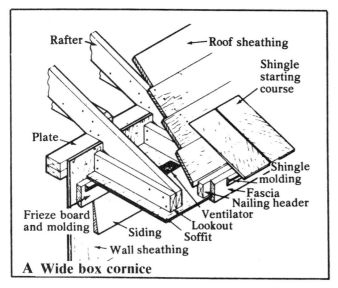

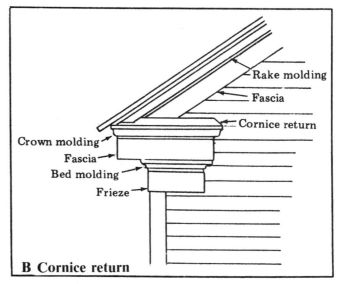

Popular cornice treatments
Figure 11-18

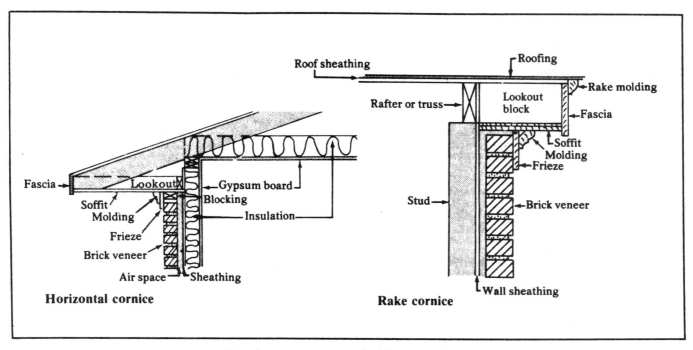

Cornice treatment on brick veneer house
Figure 11-19

Cornice trim work must be done from ladders or scaffolding. Take this into consideration when estimating manhour requirements.

Soffit and Porch Ceilings

A-C exterior grade plywood 3/8'' x 4' x 8' is com-monly used for soffit and porch ceilings. It's easy to handle and doesn't warp easily.

When estimating plywood for soffits and porch ceilings, determine the square footage of the area to be covered and divide this total by the number of square feet in each panel of plywood used (32 SF

Trim	Skilled manhours per 100 LF
2-member closed cornice	8
3-member boxed cornice	12
Molding	3
Barge board (verge) or separate frieze board	4

**Manhours for installing cornice trim
Figure 11-20**

for a 4' x 8' panel). A cornice 16'' wide will require cutting a plywood sheet lengthwise into three pieces, giving enough soffit ceiling for 24 linear feet. A 12'' wide cornice will let you get four 12'' x 96'' pieces from a 4' x 8' panel. That's 32 linear feet of soffit.

Estimate soffit labor at 4 hours per 100 linear feet. Use 4d common or finishing nails. Figure 1 pound to 100 square feet for porch ceilings. Estimate labor at 10 manhours per 100 square feet.

Other Labor Estimates

For the water table shown in Figure 11-13, allow 5 hours per 100 linear feet.

Allow 1 hour for installing a pair of shutters on first floor windows, 2 hours per pair for second floor windows.

Allow 2.5 hours for each screen door and ½ hour for each window screen.

These figures assume that the crew is already on site and has tools and materials readily at hand. Allow extra time for moving onto the site, setting up and clean up.

Caution

If you keep accurate records, you'll find that 2-story finish work takes more labor hours than single-story work. More time is needed to erect and remove scaffolding or to shift ladders. Work up your Construction Estimate File to show this difference. It could be 25 percent or more.

Exterior finish is a key indicator of quality. Make sure your houses will pass a close inspection of all exterior trim. I've know some excellent carpenters to become careless when trimming out the rear of the house or finishing closets.

The labor required to complete the exterior trim or finish of identical houses may vary as much as 10 to 25 percent even when the same crew is involved. Here's why — no two jobs are exactly alike. And even if they were, the people, materials, weather, and working conditions would always be different. Estimating is an art, not a science. But there are good estimates and bad estimates; good estimators and bad estimators. Good estimators learn to anticipate most problems, and produce consistently reliable estimates for most of their jobs. This book is intended to help you do this.

Chapter **12**

Estimating Interior Finish Carpentry

Interior finish carpentry includes all exposed woodwork on the inside of a building. In this chapter we'll estimate finish carpentry in the order it's usually done on a job, starting with stairs.

Estimating Stairs

Stair construction is specialty work, usually done by the most experienced carpenter on the job. More intricate and decorative stairs are often subbed out to stair specialty shops. But this section will help you estimate nearly any job-built stairs you come across.

Main Stairs and Basement Stairs

The two most common types of stairs are main stairs and basement (or service) stairs. Main stairs lead up to the second floor. Besides providing easy ascent and descent, they must look good, since a staircase is often a key feature in the interior design of the home. Basement stairs lead to a basement or garage area. They can be somewhat steeper than main stairs, and are usually built with less expensive materials.

Most main stairs and basement stairs are constructed in place. The main stairs may be assembled from prefabricated parts, including housed stringers, treads and risers. Basement stairs are usually made of plank treads and 2 x 12 carriages.

Stair construction can include both hardwood and softwood lumber species. Use oak, maple or birch for main stairway components, such as treads, risers, handrails and balusters. Consider Douglas fir or southern pine for basement stair treads and risers. For an economical, wear-resistant stairway, combine a hardwood tread with a softwood riser.

The terms *stringers, strings, horses,* and *carriages* refer to the supporting members of the stairs. They all mean the same thing.

Figure 12-1 gives standard dimensions for straight stairs and stairs with landings.

Since estimating stairs is fairly complex, I've included Figure 12-2. It's a stair checklist to help you do the take-off without forgetting anything.

Attic Stairs

If the attic is intended primarily for storage, or if there's no space for a fixed stairway, the plans will usually call for a prefabricated folding unit. This type of stairway provides attic access through an opening in a hall ceiling. When the stairs aren't in use, they fold up into the attic space. Your estimate will include the cost of the unit plus the labor to install it.

Dimensions for straight stairs

Height floor-to-floor H	Number of risers	Height of risers R	Width of treads T	Total run L	Minimum headroom Y	Well opening U
8'0''	12	8''	9''	8' 3''	6'6''	8' 1''
8'0''	13	7 3/8'' +	9½''	9' 6''	6'6''	9' 2½''
8'0''	13	7 3/8'' +	10''	10' 0''	6'6''	9' 8½''
8'6''	13	7 7/8''—	9''	9' 0''	6'6''	8' 3''
8'6''	14	7 5/16''—	9½''	10' 3½''	6'6''	9' 4''
8'6''	14	7 5/16''—	10''	10'10''	6'6''	9'10''
9'0''	14	7 11/16''+	9''	9' 9''	6'6''	8' 5''
9'0''	15	7 3/16''+	9½''	11' 1''	6'6''	9' 6½''
9'0''	15	7 3/16''+	10''	11' 8''	6'6''	9'11½''
9'6''	15	7 5/8''—	9''	10' 6''	6'6''	8' 6½''
9'6''	16	7 1/8''	9½''	11'10½''	6'6''	9' 7''
9'6''	16	7 1/8''	10''	12' 6''	6'6''	10' 1''

Dimensions shown under well opening ''U'' are based on 6'6'' minimum headroom. If headroom is increased well opening also increases.

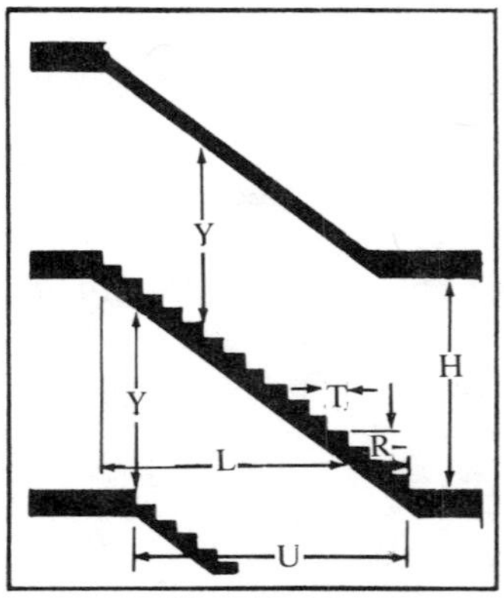

Straight stair

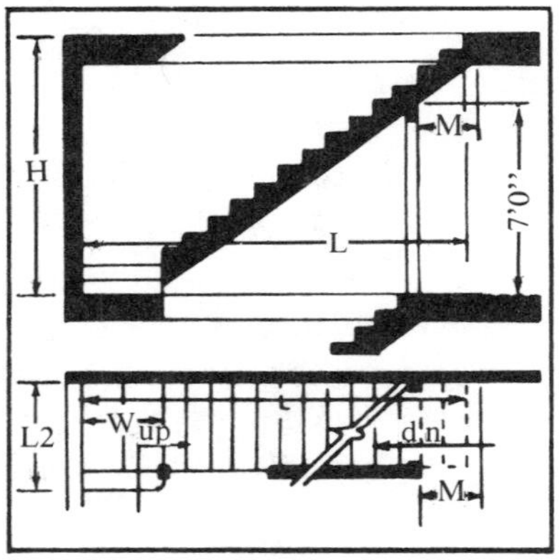

Stair with landing

Dimensions for stairs with landings

Height floor to floor H	Number of risers	Height of risers R	Width of tread T	Run		Run	
				Number of risers	L	Number of risers	L2
8'0''	13	7 3/8'' +	10''	11	8' 4'' + W	2	0'10'' + W
8'6''	14	7 5/16''—	10''	12	9' 2'' + W	2	0'10'' + W
9'0''	15	7 3/16''+	10''	13	10' 0'' + W	2	0'10'' + W
9'6''	16	7 1/8''	10''	14	10'10'' + W	2	0'10'' + W

Stairs with landings are safer and reduce the required stair space. The landing provides a resting point and a logical place for a right angle turn.

Stair dimensions
Figure 12-1

Stair Checklist

1) The house is: ☐1-story ☐1½-story ☐2-story

2) Stairs required: ☐basement ☐main ☐attic ☐attic folding ☐exterior

3) Stair type: ☐long "L" ☐narrow "U" ☐landing ☐winders

4) Stair dimensions:

   ☐Floor-to-floor: _____   ☐Total run: _____

   ☐Number of risers: _____   ☐Minimum headroom: _____

   ☐Height of risers: _____   ☐Well opening: _____

   ☐Width of treads: _____

5) Number of carriages or stringers: ☐two ☐ three

6) Number of handrails: ☐one ☐two

7) Main stairway is: ☐open ☐closed

8) Stairway has: ☐railing ☐balusters ☐newel post

9) Material costs:

Carriages: (Size _____)	_____ LF	@ $ _____ LF	= $ _____
Stringers: (Size _____)	_____ LF	@ $ _____ LF	= $ _____
Treads: (Size _____)	_____ LF	@ $ _____ LF	= $ _____
Risers: (Size _____)	_____ LF	@ $ _____ LF	= $ _____
Handrail:	_____ LF	@ $ _____ LF	= $ _____
Newel posts:	_____ ea	@ $_____ ea	= $ _____
Balusters:	_____ ea	@ $_____ ea	= $ _____
Molding:	_____ LF	@ $ _____ LF	= $ _____
Nails:	_____ lbs	@ $_____ lb	= $ _____
Framing and bracing: (Size _____)	_____ LF	@ $ _____ LF	= $ _____
Other materials: _____ (not listed above)		@ $_____	= $ _____

10) Labor costs:

Skilled:	_____ hours	@ $_____ hour	= $ _____
Unskilled:	_____ hours	@ $_____ hour	= $ _____
		Total	$ _____

Stair checklist
Figure 12-2

Labor Erecting Stairs

Work element	Unit	Manhours per unit
Erecting stairwork, hours per 9' rise		
Building ordinary plain box stairs on the job	Each	8 to 16
Rails, balusters and newel post for above	Each	4 to 8
Erecting plain flight of stairs built-up in shop	Each	6 to 8
Erecting two short flights	Each	10 to 12
Erecting open stairs	Each	10 to 12
Erecting open stairs with two flights	Each	12 to 16
Newels, balusters and hand rail for the above	Each	6 to 8
Erecting prefabricated wood stairs, hours per 9' rise		
Circular, 6' diameter, oak	Each	23.0
Circular, 9' diameter, oak	Each	31.0
Straight, 3' wide, assembled	Each	3.0
Straight, 4' wide, assembled	Each	3.2
Installing folding stairs (attic)	Each	2.5

CEF — Labor erecting stairs
Figure 12-3

Estimating Labor for Stairs

Unless you have good records of your own labor costs, use Figure 12-3 in your CEF as a reference for manhours for building stairs. It includes job-built and prefabricated main or basement stairs and folding attic stairs.

Estimating Cabinets

Cabinets in a quality home should have the appearance of fine furniture. Savvy homebuyers have come to expect more than just plywood doors attached to boxes. Manufacturers make meeting these expectations easy by designing complete units and sections ready to install. For high-quality custom homes, there are cabinet shops in most areas that make and install custom-built cabinets.

Stock cabinets are made to standard dimensions, but they're available in enough different sizes to fit almost any space. Figure 12-4 shows the common cabinets and sizes. Stock cabinets come in widths from 12 to 48 inches (in 3-inch increments). Narrow filler strips are available to fill in spaces smaller than 3 inches wide.

When ordering cabinets, there are three main dimensions required: width, height and depth. Look for a sketch or detail drawing on the house plans or get one from the architect. When you're taking off kitchen cabinets, keep in mind the minimum and maximum dimensions in Figure 12-5. If there's a problem, now's the time to find it.

When you have the dimensions down, look for this information:

1) Kind of lumber. Look in the specifications.

2) Kind of drawers, whether lip or flush. Size of stock for the stiles and rails of the doors if they're paneled

3) Whether the doors have wood or glass panels. If glass, note the kind of glass, such as clear or frosted, and the grade

4) Type and style of hardware

5) If the shelves are adjustable

6) Kind of material to be used for the back of the cabinets

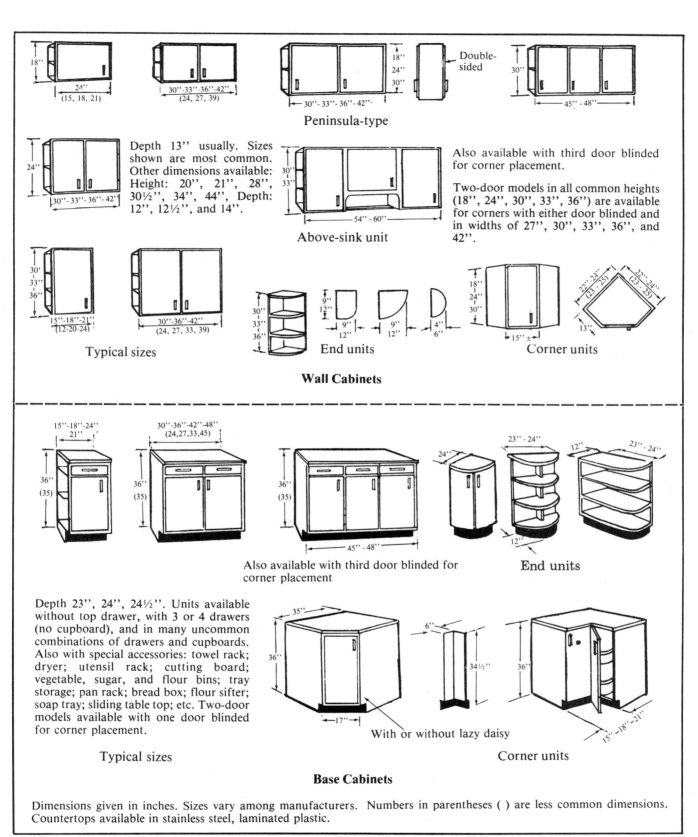

Wall Cabinets

Depth 13" usually. Sizes shown are most common. Other dimensions available: Height: 20", 21", 28", 30½", 34", 44", Depth: 12", 12½", and 14".

Also available with third door blinded for corner placement.

Two-door models in all common heights (18", 24", 30", 33", 36") are available for corners with either door blinded and in widths of 27", 30", 33", 36", and 42".

Base Cabinets

Depth 23", 24", 24½". Units available without top drawer, with 3 or 4 drawers (no cupboard), and in many uncommon combinations of drawers and cupboards. Also with special accessories: towel rack; dryer; utensil rack; cutting board; vegetable, sugar, and flour bins; tray storage; pan rack; bread box; flour sifter; soap tray; sliding table top; etc. Two-door models available with one door blinded for corner placement.

Dimensions given in inches. Sizes vary among manufacturers. Numbers in parentheses ( ) are less common dimensions. Countertops available in stainless steel, laminated plastic.

Wood kitchen cabinets
Figure 12-4

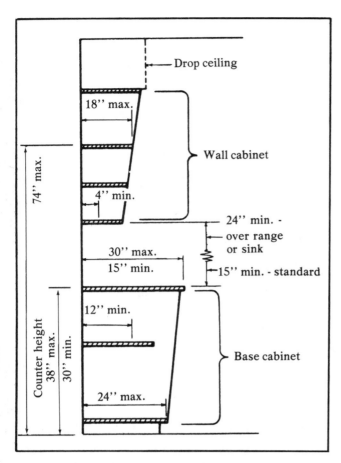

Kitchen cabinet dimensions
Figure 12-5

7) Type of countertops. Are they inexpensive laminated plastic or custom-formed with an integral front edge and backsplash?

Labor Installing Cabinets
Figure 12-6 belongs in your CEF unless you have accurate labor figures of your own for cabinet work. It includes the most common types of cabinets, and the manhours needed to install them.

Estimating Wood Flooring
Common hardwoods used in flooring are white oak, red oak, hard (sugar) maple, birch and beech. Pecan, walnut, cherry, ash, hickory and teak are expensive woods used only occasionally for flooring. The most common softwoods are Douglas fir and southern yellow pine.

The single most popular wood used for flooring is oak. It's a hard, durable and abundant wood that makes an attractive floor. Maple is also popular. Some don't think it's as attractive as oak, but it takes a better finish. Beech and birch are similar to maple, but they're not as easy to get.

Flooring lumber is produced plain sawed or quarter sawed. Plain sawing produces boards with the wood's annual rings at an angle to the face of the board. It's the fastest way to cut lumber and reduces waste at the mill. Quarter-sawed lumber, however, has better wear and warp resistance. Oak is the only wood generally available quarter sawed.

Common Types of Wood Flooring
You'll probably see four types of hardwood flooring:

- Strip flooring
- Plank flooring
- Parquet flooring
- Laminated block flooring

Strip flooring— This is the most common type. It consists of narrow strips 1½ to 3½ inches wide in various thicknesses. Most strip flooring is tongue and groove, although the smaller sizes are square-edged. End-matched strip flooring (Figure 12-7) is also available. Strip flooring is generally applied over plywood subfloor with nails.

Plank flooring— Plank floors simulate the original Colonial floors of hand-sawn planks secured to joists with wood pegs. These planks are now machine planed, tongue and grooved. They are blind nailed and secured with countersunk screws covered by walnut plugs to simulate the pegs.

Parquet flooring— This is usually the most expensive hardwood flooring. Parquet has variations in species or shades of the same species that create intricate patterns. Each tile is factory cut to exact size and should fit adjoining pieces perfectly. Each piece is laid separately, either by nailing or setting in mastic. The most popular patterns are square, rectangular, and herringbone. Most parquet flooring is oak, although it's also made in maple, birch, beech, mahogany, walnut and teak.

Laminated block flooring— Laminated block is a three-ply assembly. It comes in many patterns and species, in 5/16, 3/8, and 13/16-inch thickness. The cross-ply lamination process prevents shrinking and expansion so you don't need to allow for dimension changes, as you do with strip flooring.

Labor Installing Cabinets and Tops

Work element	Unit	Manhours per unit
Base cabinets, 36" high		
24" wide	Each	1.0
36" wide	Each	1.2
Base corner cabinets, 36" wide	Each	3.2
Wall cabinets, 12" deep		
12" x 18"	Each	0.8
18" x 18"	Each	0.9
18" x 36"	Each	1.0
24" x 36"	Each	1.3
Cabinet stain finish	100 S F	1.0
Cabinet paint finish	100 S F	1.2
Cabinet vinyl finish	100 S F	1.4
Factory formed tops, 4" backsplash		
24" wide	L F	0.20
32" wide	L F	0.21
Backsplash only, 4" high	L F	0.08
Cutting blocks, custom sizes	S F	0.42
Broom closets, 7' high	Each	1.8

Time includes layout, unloading and all necessary trim work, cleanup and repairs as needed. Time does not include any cleanup prior to installing cabinets or demolition work.
Suggested Crew: 1 carpenter and 1 laborer, 1 painter for finishing.

CEF — Labor installing cabinets and tops
Figure 12-6

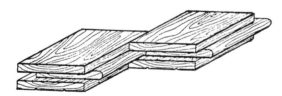

End matched flooring
Figure 12-7

Flooring sizes
Figure 12-8

Material and Labor for Wood Flooring
Finish wood flooring is sold by the original size of the rough stock before it's milled. Look at Figure 12-8. The width of finished strip flooring will be about 3/4-inch less than the rough measurement. The term "face" means the net surface width of the stock after it's been milled.

To make an accurate estimate, first find the actual floor area of each room that is to have wood flooring. Then add a certain percentage to the floor area to make up for the wood lost in milling. For 2¼-inch stock, add 1/3 to the actual floor area to find the number of board feet of milled wood required. For 1 x 4 fir or pine flooring, add 1/4 to the floor area. Add 1/6 to the floor area for 6-inch stock. For other widths of strip flooring, check the materials column in Figure 12-9. It lists both the materials and labor required for various sizes of

Size	Material BF per 100 SF	Material 1000 BF will lay SF	Nails per 100 SF	Labor Hours per 100 SF Laying	Labor Hours per 100 SF Sanding	Labor Hours per 100 SF Finishing
$^{25}/_{32}$ x 1½	155.0	645.0	3.7 pounds	3.7 hours	--	--
$^{25}/_{32}$ x 2	142.5	701.8	3.0 pounds	3.4 hours	--	--
$^{25}/_{32}$ x 2¼	138.3	723.0	3.0 pounds	3.0 hours	Average	Average
$^{25}/_{32}$ x 3¼	129.0	775.2	2.3 pounds	2.6 hours	1.3	2.6
⅜ x 1½	138.3	723.0	3.7 pounds	3.7 hours	hours	hours
⅜ x 2	130.0	769.2	3.0 pounds	3.4 hours	--	--
½ x 1½	138.3	723.0	3.7 pounds	3.7 hours	--	--
½ x 2	130.0	769.2	3.0 pounds	3.4 hours	--	--

Material and labor for strip flooring
Figure 12-9

Size flooring	Type and size of nails	Spacing
Tongue and groove flooring must always be blind-nailed, square edge flooring face-nailed.		
(Tongue & groove) $^{25}/_{32}$ x 3¼	7d or 8d screw type, cut steel nails or 2" barbed fasteners*	10-12" apart
(Tongue & groove) $^{25}/_{32}$ x 2¼, $^{25}/_{32}$ x 1½	Same as above	Same as above
(Tongue & groove) ½ x 2, ½ x 1½	5d screw, cut or wire nail. Or 1½" barbed fasteners*	8-10" apart
Following flooring must be laid on wood subfloor.		
(Tongue and groove) ⅜ x 2, ⅜ x 1½	4d bright casing, wire, cut, screw nail or 1¼" barbed fasteners*	6-8" apart
(Square-edge) $^{5}/_{16}$ x 2, $^{5}/_{16}$ x 1½	1-in. 15 gauge fully barbed flooring brad, preferably cement coated.	2 nails every 7"

*If steel wire flooring nails are used they should be 8d, preferably cement coated. Machine-driven barbed fasteners, used as recommended by the manufacturer, are acceptable.

Nail schedule for strip flooring
Figure 12-10

strip flooring. The labor hours in Figure 12-19 are based on mechanical nailing. A mechanical nailer greatly improves the production rate over manual nailing. Figure 12-10 is a nail schedule for strip flooring.

If you don't have good manhour figures of your own, use Figure 12-11 to estimate the labor for wood block flooring

Estimating Fireplace Mantels

A fireplace mantel can be almost any design, size and height, depending on the shape and size of the chimney and the fireplace opening. The mantel may be left plain, or decorated with a casing, frieze, cap, or carved molding. Pilasters can be used to support the mantel.

Labor for Installing Wood Block Flooring

Work element	Unit	Manhours per unit
Parquet, prefinished		
5/16″ thick	100 SF	6.0
1/2″ thick	100 SF	6.1
Parquet, prefinished, top quality		
5/16″ thick	100 SF	6.2
1/2″ thick	100 SF	6.4
Simulated parquet and tile flooring, 5 ply, 5/8″ thick plywood	100 SF	3.5
Wood block flooring (factory type)		
Creosoted 2″ thick	100 SF	3.7
Creosoted 2½″ thick	100 SF	4.0
Creosoted 3″ thick	100 SF	4.3
Natural finish end grain 1½″ thick	100 SF	3.8
Expansion strip 1″ thick	100 LF	2.7
Gym floor (on shims, sleepers and screeds, with sub-floor included)		
25/32″ thick maple	100 SF	13.0
25/32″ thick maple plus extra sub-floor (2 ply)	100 SF	15.3
25/32″ thick maple with steel springs and 2 ply sub-floor	100 SF	16.4

Time includes move on and off site, unloading, stacking, cleanup and repair as needed, but no sanding or finishing.
Suggested Crew: 1 carpenter, 1 laborer

CEF — Labor installing wood block flooring
Figure 12-11

Material and Labor for Fireplace Mantels
List each item of material separately on your take-off sheet. Estimate in linear feet, as shown on the plans or as scaled from the section drawings. If you're ordering the mantel millwork as a complete unit, or if you're having a subcontractor install it for you, just list the total cost.

Figure 12-12 shows labor requirements for fireplace mantels. Your CEF should include these estimates until you can replace them with your own manhour figures.

Estimating Molding or Trim
Look on the plan's detail sheet for the design of all interior trim. Find the kind of lumber in the specifications. Figure 12-13 shows common molding patterns and locations. Two types of joints are generally used; the miter and the butt joint. For the miter joint, add 6 inches of length for each corner.

Standard trim is available in various lengths by the linear feet. Some estimators, however, list the material for a certain part of a building as a unit, listing it as a separate item on the estimate. For example, instead of listing the number of linear feet of casing around the door and window openings, some builders will figure quantities by so many *sides of trim*. These sides of trim are pieces which have been cut roughly to the proper length and tied in bundles at the mill. Finish carpenters will fit them accurately at the job site. These "specified lengths" often are more expensive than random lengths. Before starting a take-off, make sure you know which way to list the trim. You can order precut door and window casing for a specific window size. The pieces are mitered and ready to in-

Labor Installing Fireplace Mantels

Work element	Unit	Manhours per unit
Prefabricated milled decorative unit, 42" high, 6' wide	Each	3.2
Bracket mounted 10" wide, 3" thick hardwood beam		
6' long	Each	1.7
8' long	Each	2.0
Rough sawn oak or pine beam		
4" x 8"	L F	0.25
4" x 10"	L F	0.27
4" x 12"	L F	0.31

Time includes layout, cutting, drilling and placing of shields where required, repairs and clean-up.
Suggested Crew: 1 carpenter

CEF — Labor installing fireplace mantels
Figure 12-12

stall. Most finish carpenters, however, prefer to do their own cutting. They have little tricks they use in cutting molding which guarantees a good fit where there might otherwise be a "bad joint." Order the appropriate type for the crew doing the finish carpentry.

Finger-jointed facing is also available in specified lengths. Finger jointing consists of fitting short pieces together and gluing them to make one long piece. You can save money this way in areas where the trim will be painted.

One of the advantages of estimating trim by the side is that it saves time and effort, because you don't have to figure the exact lengths. You should know how to figure the lengths, however, as it's often more economical to buy large quantities of some kinds of interior trim to be cut to length on the job. Tract builders can save money by buying the stock material in random lengths from the lumberyard and cutting it to rough lengths before sending it to the building sites.

Rules for Estimating Lengths of Trim

Head casings: Add to the width of the door or width of window sash *twice* the width of the side casing stock plus 2 inches. This equals one length of head casing. If the window frame has a mullion casing, add the width of each mullion to the answer. (Prehung interior doors have the casing already installed. Prehung exterior doors have the molding installed on the exterior side.)

Side casing: Where side casing is butted to the head casing, as in diagram A in Figure 12-14, add 2 inches to the length of the door. For miter joint construction (Figure 12-14 B), the length equals the door or window height, plus the width of casing stock, plus 2 inches. Allow two pieces for each window frame.

Back band: For head back bands, add 12 inches to the width of the door or sash opening. For side back bands, add 6 inches to the length of the door or sash opening. You can see the back bands in C in Figure 12-14.

Cap trim: Where a special cap trim is used, this is usually the length of the head casing (Figure 12-14 D).

Window stools: The length equals the width of the window plus 12 inches. Add the width of all mullion center casings, if any, to the answer. Combine into standard lengths of lumber. See Figure 12-14 E.

Apron trim: The length equals the width of the frame plus twice the width of casing stock, plus 2 inches.

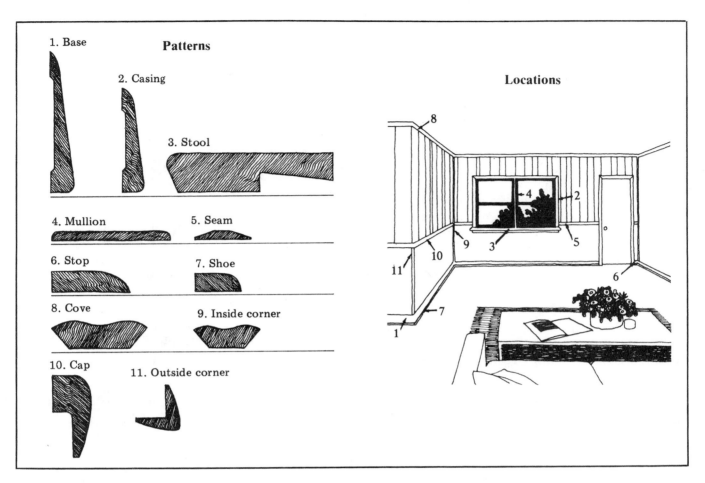

Molding patterns and locations
Figure 12-13

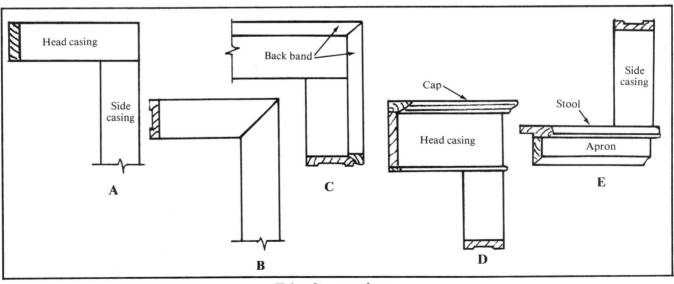

Trim for openings
Figure 12-14

Estimating Baseboards and Cove or Crown Molding

Where baseboards are used, they're nailed on the wall at the floor line to form a finish between wall and floor. In most houses, the baseboard is a plain pattern, using a single strip of wood. In more expensive construction, look for a built-up type with a molding fitted to the top of a plain board. Figure 12-15 shows several designs. The backs should be hollowed out to avoid warping and to make it easier to fit them to the wall.

Sometimes the joint between baseboard and floor is covered with a quarter round or a base shoe.

To estimate the linear feet of baseboard and shoe for one room, find on the plans the distance around the room, to the nearest foot, less each door or wall opening. Repeat this for all rooms, halls and closets, and add the results.

To find the linear feet of crown or cove molding, find the distance around a room and add two feet. Repeat for all the rooms requiring molding. Increase the results to a standard length of stock.

Finish Carpentry Labor

Figure 12-16 lists typical installation times for most finish carpentry items, including manhours for setting door and window frames for units that aren't

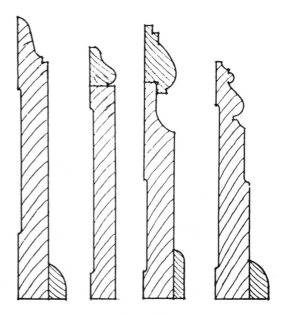

Base trim
Figure 12-15

prehung or prefabricated. Highly experienced crews with specialized equipment can reduce these installation times considerably. Once again, your own records will allow more accurate manhour estimates for your jobs.

Exterior trim			
Corner boards, verge, fascia, hours per 100 LF	4		
Cornice, 3 member, hours per 100 LF	12		
Porch post, plain, hours each	1.0		
Porch post, built-up, hours each	2.0		
Clothes closets, hours per each			
1 shelf, hookstrip, hook and pole	2.0		
Open shelving and cleats	.5		
Linen closet, shelving and cleats	3.0		
Baseboard, hours per 100 LF			
Two member, ordinary work	4	to	6
Two member, hardwood and first class or difficult work	6	to	8
Three member, ordinary work	5	to	7
Three member, hardwood and first class or difficult work	7	to	9
Crown molding, hours per 100 LF			
Ordinary work	2¾	to	3
Hardwood, first class or difficult work	4	to	5

Chair rail, hours per 100 LF			
Ordinary work	2½	to	3
Hardwood, first class or difficult work	3	to	4
Plate rail, hours per 100 LF			
Two member, ordinary work	10	to	12
Two member, first class work	12	to	15
Interior cornice			
Ordinary work	10	to	16
First class or difficult work	18	to	22
Setting door frames, hours per frame			
Exterior door frames			
Ordinary size and ordinary workmanship	¾ to	1¼	
Ordinary size, hardwood and first class workmanship	1	to	2
Interior door frames			
Ordinary size and ordinary workmanship	½ to	1	
Ordinary size, hardwood, first class work	1	to	1½
Ordinary size with transom and ordinary workmanship	1	to	1¼
Ordinary size with transom and first class workmanship	1	to	2

Labor hours for finish carpentry
Figure 12-16

Assembling window frames, hours per frame		Window trim, hours per opening		
Single frame	¾ to 1¼	One member casing on windows, ordinary work	1	
Double frame	1¼ to 2	Hardwood first class work	1½ to 2	
Setting window frames, hours per frame		Two member casing on windows, ordinary work	1½ to 2	
Single frame in frame wall	1 to 1¼	Hardwood and first class work	2 to 4	
Double frame in frame wall	1½ to 2	Window trim on brick walls, ordinary work	1½ to 2	
Single frame in brick wall	1 to 1½	First class or difficult work	2 to 4	
Double frame in brick wall	1½ to 2	Hanging doors, hours per unit		
Fitting sash, hours per opening		Single, ordinary work	1	
Fitting and hanging ordinary sash for double-hung window,	1 to 1½	Single, first class work	2 to 4	
Fitting and hanging casement sash	1½ to 3	Double or sliding, ordinary work	3 to 6	
Casing or trim, hours per side		Double or sliding, first class work	4 to 10	
One member casing on door, ordinary work	½ to ¾	Fitting locks, hours per unit		
Hardwood and first class work	¾ to 1	Ordinary work	½ to 1¼	
Two member casing on doors, ordinary work	¾ to 1	First class or difficult work	1¼ to 2	
Hardwood and first class work	1 to 1½	Fitting window locks and lifts	¼ to ½	

Labor hours for finish carpentry
Figure 12-16 (continued)

Chapter **13**

Estimating Specialty Finishes

This chapter will hit a few high spots on what is necessarily a grab-bag of interior specialties: ceramic tile, quarry tile, terrazzo, resilient floor covering, and painting. Most builders subcontract work like this to specialists who do nothing else. But sometimes it may be easier to handle the work yourself, especially if the job is small and if you have someone on the payroll who's experienced in the trade. Even if you subcontract all specialty finish work, know enough about each trade to check the sub's estimates. We'll start with one of the most common interior finishes — ceramic tile.

Ceramic Tile

Ceramic tile is made from natural clays and finely-ground ceramic materials such as silica, quartz, feldspar, marble, fluxes, cements, pigments or acetylene black. The powders are either pressed to the desired shape or mixed with water and formed. Heat fuses the particles into a solid mass.

Ceramic tile can be divided into two general types, glazed and unglazed. Unglazed tiles are dense, hard, vitrified units that are available in a variety of integral solid and mottled colors. They don't have slippery faces and are manufactured primarily for interior and exterior flooring or paving. Unglazed tile is available in a number of sizes.

Unglazed mosaic tile has a face area of less than 6 square inches and a 1/4 inch nominal thickness. It's usually factory mounted in sheets of about 2 square feet to make setting easier. Mosaic tile is available in porcelain and natural clay tiles, and in slip-resistant and conductive grades.

Unglazed quarry tile is made by the extrusion process from selected natural clays or shale. It's dense, coarse grained, and available in several solid colors. Quarry tile is generally 6 square inches or more and from 1/2 to 3/4 inch thick.

Glazed tile is available in impervious porcelain, vitrified natural clay, quarry tile, nonvitreous white bodies, and many sizes. Ordinary glazed wall tile has an impervious glaze over a white nonvitreous body. It's intended for interior use, although it can be used to cover exterior locations not subject to freezing.

Figure 13-1 suggests manhour estimates for installing ceramic tile base. Figure 13-2 gives labor figures for installing quarry tile. Use these to begin your CEF for ceramic tile, and add to it figures you compile from your experience and other sources.

In most residential construction, use a multi-purpose adhesive to install ceramic tile to a drywall or masonry backing.

For 100 SF of wall or floor, estimate:

- 100 SF of tile
- 2 gallons wall or floor adhesive
- 15 pounds of grout

Allow 10 percent for waste in tile, adhesive and grout.

Work element	Unit	Manhours per unit
Portland cement bed with grout		
3" x 3" cove	10 LF	2.5
4¼" x 4¼" cove	10 LF	2.3
6" x 2" cove	10 LF	1.8
6" x 6" cove	10 LF	2.4
4¼" x 4¼" sanitary base	10 LF	2.3
6" x 6" sanitary base	10 LF	2.4
Adhesive bed with grout		
6" x 4¼" sanitary base	10 LF	1.7
6" x 6" sanitary base	10 LF	1.7

Time includes move on and off site, unloading, cleanup and repair as needed.
Suggested crew: 1 tile setter, 1 helper.

Labor installing ceramic tile base
Figure 13-1

Terrazzo

Terrazzo is a hard mosaic material composed of marble or other aggregates held together in a binder. Other materials may be added to create special properties. It's commonly used as a cast-in-place finish flooring material, but can be used in precast form in tiles, panels, or special shapes. There are three principal methods of installing cast-in-place terrazzo:

1) *Monolithic,* often referred to as thinset or one-course, in which the terrazzo topping is bonded directly to the concrete slab. Thinset terrazzo ranges in thickness from 3/16 inch for synthetic resin type to 1/2 inch or more for portland cement type.

2) *Bonded* or two-course terrazzo uses a topping bonded to a portland cement mortar underbed. The underbed in turn is bonded to a concrete slab, laid directly over steel deck or laid over waterproof membrane on wood. These installations vary in thickness from 1¾ inches when bonded to concrete to 3 inches or more for installation over wood or metal.

3) *Sand cushion* or three-course terrazzo lay-up includes a sand bed about 1/4 inch thick over the concrete base slab and a membrane over the sand. The underbed and topping are similar to bonded or two-course terrazzo, except that reinforcing steel is always placed in the underbed. This is the most costly of the three installation methods, but is the least vulnerable to cracks caused by horizontal shifting in the concrete base slab.

For residential work, you're most likely to be estimating precast terrazzo. Figure 13-3 gives labor estimates for laying precast terrazzo in a variety of situations. If you do run into cast-in-place terrazzo, use Figure 13-4 as your guide for manhour calculations.

Resilient Flooring

Resilient floor covering includes asphalt tile, vinyl-asbestos tile, vinyl tile and vinyl sheet, rubber tile and rubber sheet, cork tile, and linoleum sheet goods. Accessories used with resilient flooring include rubber and vinyl base, stair treads, feature strips, edge nosings (reducer strips), thresholds (saddles), and stringer or stair skirting material.

Resilient floor covering ranges in thickness from less than 1/16 inch through 1/8 inch. The most popular tiles are 9-inch and 12-inch squares. Sheet goods come in rolls up to 12 feet wide.

All types of resilient floor coverings, except cork, are available in a wide variety of colors, patterns, and textured finishes.

Work element	Unit	Manhours per unit
Floor quarry tile		
4" x 4" x ½", 1/8" straight joints	100 SF	24.0
6" x 6" x ½", 1/4" straight joints	100 SF	21.6
6" x 6" x ¾", 1/4" straight joints	100 SF	22.7
9" x 9" x ¾", 3/8" straight joints	100 SF	20.3
6" x 6" x ½", 1/4" hexagon joints	100 SF	24.0
Wall quarry tile		
4" x 4" x ½", 1/8" straight joints	100 SF	26.9
6" x 6" x ¾", 1/4" straight joints	100 SF	26.3
Quarry tile trim, cove base		
5" x 6" x ½", straight top	100 LF	18.0
6" x 6" x ¾", round top	100 LF	18.6
Stair treads, 6" x 6" x ¾", tile 12" wide tread	100 LF	26.4
Window sills, 6" x 6" x ¾", tile 6" wide	100 LF	16.7

All tile is assumed to be set in a ¾" portland cement bed. Deduct 33% from manhours if thin set epoxy bed and grout are used. Time includes move on and off site, unloading, mixing and placing mortar or epoxy bed, setting and grouting tiles, cleanup and repairs as needed.
Suggested crew: 1 tile setter and 1 helper.

Labor installing quarry tile
Figure 13-2

Work element	Unit	Manhours per unit	Work element	Unit	Manhours per unit
Wall tile			¾" thick, curved	100 LF	12.6
½" x 6" x 6"	100 SF	11.5	1" thick, straight	100 LF	12.6
½" x 9" x 9"	100 SF	10.3	1" thick, curved	100 LF	13.7
½" x 12" x 12"	100 SF	9.2	1½" thick, straight	100 LF	13.7
1" x 9" x 9"	100 SF	14.3	1½" thick, curved	100 LF	15.5
1" x 12" x 12"	100 SF	12.6	Tread & riser combinations,		
1" x 18" x 18"	100 SF	11.5	straight stairs		
1½" x 12" x 12"	100 SF	13.8	1½" tread, ¾" riser	100 LF	29.0
1½" x 18" x 18"	100 SF	12.0	2" tread, 1" riser	100 LF	31.0
1½" x 24" x 24"	100 SF	10.4	3" tread, 1" riser	100 LF	33.0
Base, coved or straight face			Tread & riser combinations,		
4" high	100 LF	7.0	curved stairs		
6" high	100 LF	7.5	1½" tread, ¾" riser	100 LF	31.0
8" high	100 LF	7.9	2" tread, 1" riser	100 LF	34.0
Curbing			3" tread, 1" riser	100 LF	36.0
4" x 4"	100 LF	18.0	Structural notched stair		
6" x 6"	100 LF	23.0	stringers, LF of carriage length		
8" x 8"	100 LF	27.0	1" thick	10 LF	2.6
Wainscot			1½" thick	10 LF	3.0
1" x 12" x 12"	100 SF	14.4	2" thick	10 LF	3.3
1½" x 18" x 18"	100 SF	15.4	2½" thick	10 LF	4.0
Stair treads, to 12" wide			3" thick	10 LF	4.6
1½" thick, straight	100 LF	20.6	Landings, structural		
1½" thick, curved	100 LF	21.7	1½" thick	10 SF	1.7
2" thick, straight	100 LF	21.7	2" thick	10 SF	1.8
2" thick, curved	100 LF	23.0	2½" thick	10 SF	2.0
3" thick, straight	100 LF	22.9	3" thick	10 SF	2.1
3" thick, curved	100 LF	24.6			
Stair risers, 6" high			Add for metal dividers, 4' x 4'		
¾" thick, straight	100 LF	11.5	grid, SF of floor	100 SF	1.6

Time includes moving materials into place, layout, setting, finishing, repair and cleanup as needed.
Suggested crew: 1 mason and 1 helper.

Labor installing precast terrazzo
Figure 13-3

Work element	Unit	Manhours per unit	Work element	Unit	Manhours per unit
Flooring, 4' x 4' squares with metal divider strip			Epoxy thinset terrazzo, ½" thick	100 SF	15.8
Sand cushion, 2½" to 3" thick	100 SF	18.0	Base, 6" high with divider strips 16" o.c.		
Bonded to concrete, 1¾" thick	100 SF	12.5	Flush type	100 LF	20.0
Epoxy thinset, terrazzo, ¼" to ½" thick	100 SF	9.5	Projecting, ¼" to ½"	100 LF	20.0
Monolithic ½" terrazzo with 3½" concrete base panels, approx. 25 SF each	100 SF	7.8	Splay edge type	100 LF	21.5
Bonded conductive, 1¾" thick	100 SF	13.7	Stairs, 1½" to 2" on concrete or metal with dividers		
Sand cushion conductive, 3", regular	100 SF	23.0	Treads, to 12" wide	100 LF	33.0
Sand cushion conductive, 3", mosaic textured	100 SF	32.0	Treads and risers combined	100 LF	75.0
Add for heavy-duty abrasive	100 SF	2.5	Stringers, curb and fascia	100 SF	31.0
Add for venetian type topping	100 SF	3.4	Landings	100 SF	14.5
Wainscot			Add for heavy-duty homogenous abrasive surface	100 SF	1.0
Bonded to concrete or masonry, 1½" thick	100 SF	19.4	Add for light-duty embedded abrasive surface	100 SF	2.0
			Add for surface abrasive strips	100 LF	1.0
			Add for abrasive metal nosing	100 LF	2.6

Time includes move on and off site, placing, grinding, cleanup and repair as needed.
Suggested crew: 2 terrazzo workers

Labor installing cast-in-place terrazzo
Figure 13-4

Type and use	Approximate coverage in SF per gallon
Primer — For treating on or below grade concrete subfloors before installing asphalt tile	250 to 350
Asphalt cement — For installing asphalt tile over primed concrete subfloors in direct contact with the ground	200
Emulsion adhesive — For installing asphalt tile over lining felt	130 to 150
Lining paste — For cementing lining felt to wood subfloor	160
Floor and wall size — For priming chalky or dusty suspended concrete subfloors before installing resilient tile other than asphalt	200 to 300
Waterproof cement — Recommended for installing linoleum tile, rubber and cork tile over any type of suspended subfloor in areas where surface moisture is a problem	130 to 150

Adhesive for resilient floor tile
Figure 13-5

Figure 13-5 gives coverage figures for adhesives for floor tile. Figure 13-6 covers the tile itself, and Figure 13-7 suggests the waste factor to use for different size floors. Figures 13-8 and 13-9 give labor manhours for installing resilient tile and sheet flooring.

Wall-to-Wall Carpet

Wall-to-wall carpet is used in most houses built today. The most popular materials are nylon and polyester fibers. Carpet and pad are frequently laid directly over a plywood subfloor, eliminating the need for any other underlayment.

The pad or cushion is made of animal hair, fibers, rubberized fibers or cellular rubber. It may be bonded on the underside of the carpet or laid separately. Bonded backing is normally used when carpet is installed over concrete floors. On wood floors the backing is usually installed separately.

Carpet is available in widths of 9, 12, 15 and 18 feet. The 9- and 12-foot widths are the most common in residential construction.

Estimate carpet and pad by the square yard. For example, a 12' x 16' room has 192 square feet, or 21.33 square yards.

Paints

Most paints are based on a binder which is either dissolved in a solvent or emulsified in water. When applied in a thin film, the paint will dry or cure to form a dry, tough coating. Solutions of such binders in a solvent have various names: clear

SF of floor	Number of tiles 9" x 9"	12" x 12"	6" x 6"	9" x 18"
1	2	1	4	1
2	4	2	8	2
3	6	3	12	3
4	8	4	16	4
5	9	5	20	5
6	11	6	24	6
7	13	7	28	7
8	15	8	32	8
9	16	9	36	8
10	18	10	40	9
20	36	20	80	18
30	54	30	120	27
40	72	40	160	36
50	89	50	200	45
60	107	60	240	54
70	125	70	280	63
80	143	80	320	72
90	160	90	360	80
100	178	100	400	90
200	356	200	800	178
300	534	300	1200	267
400	712	400	1600	356
500	890	500	2000	445

To find the number of tile required for an area not shown in this table, such as the number of 9" x 9" tile required for an area of 550 SF, add the number of tile needed for 50 SF to the number of tile needed for 500 SF. The result will be 979 tile, to which must be added 5% for waste. The total number of tile required will be 1,028.

Estimating material for floor tile
Figure 13-6

1	to	50 SF	14%
50	to	100 SF	10%
100	to	200 SF	8%
200	to	300 SF	7%
300	to	1000 SF	5%
Over		1000 SF	3%

Waste factors for floor tile
Figure 13-7

Work element	Unit	Manhours per unit
9" x 9" x 1/8", most colors, vinyl	100 SF	1.6
9" x 9" x 1/8", color group D, vinyl	100 SF	1.7
12" x 12" x 1/8", most colors and grades, vinyl	100 SF	1.5
9" x 9" x 3/16", cork tile	100 SF	2.2
9" x 9" x 5/16", cork tile	100 SF	2.3
12" x 12" x 3/16", cork tile	100 SF	1.9
12" x 12" x 5/16", cork tile	100 SF	2.0
9" x 9" x 1/8", rubber tile	100 SF	1.6
9" x 9" x 3/16", rubber tile	100 SF	1.6
12" x 12" x 1/8", rubber tile	100 SF	1.5
12" x 12" x 3/16", rubber tile	100 SF	1.5
Vinyl or rubber top set cove base	100 SF	2.8

Time includes move on and off site, typical area prep, unloading, stacking, installation, cleanup and repair as needed.
Add 15% if less than 500 SF, deduct 5% if over 5,000 SF.
Suggested crew: 1 tile setter

Labor installing resilient floor tile
Figure 13-8

Work element	Unit	Manhours per unit
Sheet vinyl or linoleum		
.070" thick	100 SF	1.8
.090" thick	100 SF	2.1
.125" thick	100 SF	2.3
.140" thick	100 SF	2.5
Natural cork sheets, vinyl faced		
1/8"	100 SF	3.4
1/4"	100 SF	3.5
Add for wall application	100 SF	1.3

Labor installing resilient sheet flooring
Figure 13-9

finishes, varnishes (if they dry by oxidation), or lacquers (if they dry by evaporation). If opaque pigments of colors are dispersed in the binder, the result is paint, which will produce a white or colored film.

Pigment concentration can also be varied to produce a high gloss, a semigloss or a lusterless (flat) finish. Special pigments such as red lead and zinc chromate can be used to provide corrosion resistance to primers. Metallic pigment can be added to varnishes to produce metallic coatings such as aluminum paints. The way a paint performs depends on the type of binder used.

Figure 13-10 shows the important properties of the major types of paint binders. Some combination binders not covered in this figure are:

- Oil-alkyd: properties similar to oil and alkyd paints

- Cleoresinous: similar to alkyds but with less color retention

- Phenolic-alkyd: similar to phenolic and alkyd paints

- Oil-modified urethane: similar to phenolic and alkyd paints

- Vinyl-alkyd: similar to vinyl and alkyd paints

Figure 13-11 gives approximate paint requirements for typical interior and exterior spaces. For irregular shapes, use Figure 13-12 to help you calculate the number of square feet to be covered.

	Alkyd	Cement	Epoxy	Latex	Oil	Phenolic	Rubber	Moisture Curing Urethane	Vinyl
Ready for use	Yes	No	No[3]	Yes	Yes	Yes	Yes	Yes	Yes
Brushability	A	A	A	+	+	A	A	A	—
Odor	+[1]	+	—	+	A	A	A	—	—
Cure normal temp.	A	A	A	+	—	A	+	+	+
Cure low temp.	A	A	—	—	—	A	+	+	+
Film build/coat	A	+	+	A	+	A	A	+	—
Safety	A	+	—	+	A	A	A	—	—
Use on wood	A	—	A	A	A	A	—	A	—
Use on fresh conc.	—	+	+	+	—	—	+	A	+
Use on metal	+	—	+	—	+	+	A	A	+
Corrosive service	A	—	+	—	—	A	A	A	+
Gloss - choice	+	—	+	—	A	+	+	A	A
Gloss - retention	+	X	—	X	—	+	A	A	+
Color - initial	+	A	A	+	A	—	+	+	+
Color - retention	+	—	A	+	A	—	A	—	+
Hardness	A	+	+	A	—	+	+	+	A
Adhesion	A	—	+	A	+	A	A	+	—
Flexibility	A	—	+	+	+	A	A	+	+
Resistance to:									
Abrasion	A	A	+	A	—	+	A	+	+
Water	A	A	A	A	A	+	+	+	+
Acid	A	—	A	A	—	+	+	+	+
Alkali	A	+	+	A	—	A	+	+	+
Strong solvent	—	+	+	A	—	A	—	+	A
Heat	A	A	A	A	A	A	+[2]	A	—
Moisture permeability	Mod.	V.High	Low	High	Mod.	Low	Low	Low	Low

\+ = Among the best for this property
— = Among the poorest for this property
A = Average
X = Not applicable

[1] Odorless type [2] Special types [3] Two component type

Comparison of paint binders' principal properties
Figure 13-10

Distance around the room	Ceiling height 8 feet	Ceiling height 8½ feet	Ceiling height 9 feet	Ceiling height 9½ feet	Paint for ceiling	Finish for floors	For each door or window
30 feet	⅝ gallon	⅝ gallon	¾ gallon	¾ gallon	1 pint	1 pint	Each window
35 feet	¾ gallon	¾ gallon	¾ gallon	⅞ gallon	1 quart	1 pint	and frame
40 feet	⅞ gallon	⅞ gallon	⅞ gallon	1 gallon	1 quart	1 quart	requires
45 feet	⅞ gallon	1 gallon	1 gallon	1⅛ gallons	3 pints	1 quart	¼ pint
50 feet	1 gallon	1⅛ gallons	1⅛ gallons	1¼ gallons	3 pints	1 quart	Each door
55 feet	1⅛ gallons	1⅛ gallons	1¼ gallons	1¼ gallons	2 quarts	3 pints	and frame
60 feet	1¼ gallons	1¼ gallons	1⅜ gallons	1⅜ gallons	2 quarts	3 pints	requires
70 feet	1⅜ gallons	1½ gallons	1½ gallons	1⅝ gallons	3 quarts	2 quarts	½ pint
80 feet	1½ gallons	1⅝ gallons	1¾ gallons	1⅞ gallons	1 gallon	5 pints	

Distance around the house	Average height 12 feet	Average height 15 feet	Average height 18 feet	Average height 21 feet	Average height 24 feet
60 feet	1 gallon	1¼ gallons	1½ gallons	1¾ gallons	2 gallons
76 feet	1¼ gallons	1½ gallons	2 gallons	2¼ gallons	2½ gallons
92 feet	1½ gallons	2 gallons	2½ gallons	2¾ gallons	3 gallons
108 feet	1¾ gallons	2¼ gallons	2¾ gallons	3¼ gallons	3¾ gallons
124 feet	2 gallons	2½ gallons	3¼ gallons	3¾ gallons	4¼ gallons
140 feet	2½ gallons	3 gallons	3½ gallons	4 gallons	4½ gallons
156 feet	2¾ gallons	3¼ gallons	4 gallons	4½ gallons	5¼ gallons
172 feet	3 gallons	3¾ gallons	4½ gallons	5 gallons	5¾ gallons

On interior work, for rough, sand-finished walls or unpainted gypsum board, add 50% to quantities; for each door or window deduct ½ pint of materials for walls. For trim, add ⅛ to ⅕ of the amount required for the body. For exterior blinds, ½ gallon will cover 12 to 14 blinds, one coat.

Approximate paint requirements for interiors and exteriors
Figure 13-11

Triangle

To find the number of square feet in any shape triangle or 3 sided surface, multiply the height by the width and divide the total by 2.

Square

Multiply the base measurement in feet times the height in feet.

Cylinder

When circumference (distance around cylinder) is known, multiply height by circumference.

When diameter (distance across) is known, multiply diameter by 3.1416. This gives circumference. Then multiply by height.

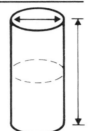

Circle

To find the number of square feet in a circle multiply the diameter (distance across) by itself and then multiply this total by .7854.

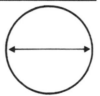

Arch Roof

Multiply length (B) by width (A) and add one-half the total.

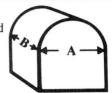

Gambrel Roof

Multiply length (B) by width (A) and add one-third of the total.

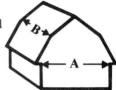

Cone

Determine area of base by multiplying 3.1416 times radius (A) in feet.

Determine the surface area of a cone by multiplying circumference of base (in feet) times one-half of the slant height (B) in feet.

Add the square foot area of the base to the square foot area of the cone side for total square foot area.

Rectangle

Multiply the base measurement in feet times the height in feet.

Calculating square feet in different shapes
Figure 13-12

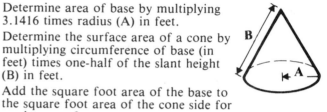

	Aluminum	Cement Base Paint	Exterior Clear Finish	House Paint	Metal Roof Paint	Porch-and-Deck Paint	Primer or Undercoater	Rubber Base Paint	Spar Varnish	Transparent Sealer	Trim-and-Trellis Paint	Wood Stain	Metal Primer
Wood													
Natural finish	-	-	P	-	-	-	-	-	P	-	-	P	-
Porch floor	-	-	-	-	-	P	-	-	-	-	-	-	-
Shingle roof	-	-	-	-	-	-	-	-	-	-	-	-	-
Shutters and trim	-	-	-	P+	-	-	P	-	-	-	P+	-	-
Siding	P	-	-	P+	-	-	P	-	-	-	-	-	-
Windows	P	-	-	P+	-	-	P	-	-	-	P+	-	-
Masonry													
Asbestos cement	-	-	-	P+	-	-	P	P	-	-	-	-	-
Brick	P	P	-	P+	-	-	P	P	-	P	-	-	-
Cement & Cinder block	P	P	-	P+	-	-	P	P	-	P	-	-	-
Cement porch floor	-	-	-	-	-	P	-	P	-	-	-	-	-
Stucco	P	P	-	P+	-	-	P	P	-	P	-	-	-
Metal													
Copper	-	-	-	-	-	-	-	-	P	-	-	-	-
Galvanized	P+	-	-	P+	-	-	P	-	P	-	P+	-	P
Iron	P+	-	-	P+	-	-	-	-	-	-	P+	-	P
Roofing	-	-	-	-	P+	-	-	-	-	-	-	-	P
Siding	P+	-	-	P+	-	-	-	-	-	-	P+	-	P
Windows, alum.	P	-	-	P+	-	-	-	-	-	-	P+	-	P
Windows, steel	P+	-	-	P+	-	-	-	-	-	-	P+	-	P

"P" indicates preferred coating for this surface. "P + " indicates that a primer or sealer may be necessary before the finishing coat or coats (unless the surface has been previously finished).

Exterior paint selection chart
Figure 13-13

Work element	Unit	Manhours per unit
Brush painting, per coat		
Wood flat work	1000 SF	8.5
Doors and windows, area of opening	1000 SF	9.0
Trim	1000 SF	8.0
Plaster, sand finish	1000 SF	7.0
Plaster, smooth finish	1000 SF	6.0
Wallboard	1000 SF	5.5
Metal	1000 SF	8.5
Masonry	1000 SF	7.0
Varnish flat work	1000 SF	8.5
Enamel flat work	1000 SF	6.5
Enamel trim	1000 SF	8.0
Roller painting, per coat		
Wood flat work	1000 SF	6.0
Doors	1000 SF	8.5
Plaster, sand finish	1000 SF	2.5
Plaster, smooth finish	1000 SF	3.0
Wallboard	1000 SF	3.0
Metal	1000 SF	5.5
Masonry	1000 SF	3.0
Spray painting, per coat		
Wood flat work	1000 SF	2.0
Doors	1000 SF	3.0
Plaster, wallboard	1000 SF	2.5
Metal	1000 SF	3.5

The painting of interior surfaces includes minimum surface preparation, mixing paint materials, and application of paint to surface.

Labor for interior painting
Figure 13-14

Work element	Unit	Manhours per unit
Exterior wood trim, 3 coats	100 LF	1.7
Interior wood trim, 3 coats	100 LF	1.7
Kitchen cabinets, 3 coats	100 SF	2.7
Wood casework, 3 coats	100 SF	2.7
Metal casework, 2 coats	100 SF	1.7
Wardrobes, 3 coats	100 SF	2.7
Bookcases, 3 coats	100 SF	2.7

Time includes move on and off site, set up, surface preparation, masking and taping, light sanding between coats, remove masking and tape, cleanup and touchup as required. These figures will apply on most jobs.

Labor painting millwork
Figure 13-16

Work element	Unit	Manhours per unit
Brush paint, per coat		
Wood siding	1000 SF	7.5
Wood doors and windows, area of opening	1000 SF	9.5
Trim	1000 SF	8.5
Steel sash, area of opening	1000 SF	5.0
Flat metal	1000 SF	7.0
Metal roofing and siding	1000 SF	7.5
Masonry	1000 SF	7.5
Roller painting, per coat		
Masonry	1000 SF	5.5
Flat metal	1000 SF	4.5
Doors	1000 SF	7.0
Spray painting, per coat		
Wood siding	1000 SF	4.0
Doors	1000 SF	5.0
Masonry	1000 SF	6.0
Flat metal	1000 SF	5.0
Metal roofing and siding	1000 SF	6.0
Highway or airfield lines and symbols, including glass beads	1000 SF	8.5
Cementitious paint, including curing	1000 SF	10.0
Sandblasting steel	1000 SF	55.0
Wire brush cleaning of steel	1000 SF	17.5
Clean and spray waterproofing on masonry	1000 SF	10.0

Surface preparation for exterior painting includes removing mill scale from metal surfaces with wire brushes or by sandblasting, removing dust with brush or cloth, removing oil and grease, masking and taping adjacent surfaces, removing masking and taping. Sometimes it is necessary to lightly sand between coats or size and fill porous materials before painting, all of which is surface preparation.

Suggested crew: One or two men spraying, one or two men tending (one man is used to mix and prepare paint for larger crews).

Labor for exterior painting
Figure 13-15

Use Figure 13-13 as a guide for selecting exterior paints. Figures 13-14 and 13-15 give typical manhour figures for interior and exterior painting under normal conditions, while Figure 13-16 is a guide for painting millwork. Remember, however, that painting wood trim is subject to wide variables. The location, height, surface conditions, amount of masking and covering, quality specifica-

Material	Touch	Recoat	Rub
Lacquer	1-10 min.	1½-3 hrs.	16-24 hrs.
Lacquer sealer	1-10 min.	30-45 min.	1 hr. (sand)
Paste wood filler	--	24-48 hrs.	--
Paste wood filler (Q.D.)	--	3-4 hrs.	--
Water stain	1 hr.	12 hrs.	--
Oil stain	1 hr.	24 hrs.	--
Spirit stain	Zero	10 min.	--
Shading stain	Zero	Zero	--
Non-grain raising stain	15 min.	3 hrs.	--
NGR stain (quick-dry)	2 min.	15 min.	--
Pigment oil stain	1 hr.	12 hrs.	--
Pigment oil stain (Q.D)	1 hr.	3 hrs.	--
Shellac	15 min.	2 hrs.	12-18 hrs.
Shellac (wash coat)	2 min.	30 min.	--
Varnish	1½ hrs.	18-24 hrs.	24-48 hrs.
Varnish (Q.D. synthetic)	½ hr.	4 hrs.	12-24 hrs.

Average times. Different products will vary.

Drying times of coatings
Figure 13-17

Work element	Unit	Manhours per unit
Wallpaper		
Light to medium weight, butt joint	100 SF	1.4
Heavy weight, butt joint	100 SF	1.5
Vinyl wall covering		
Light to medium weight, butt joint	100 SF	1.9
Heavy weight, butt joint	100 SF	2.1
Special wall coatings	100 SF	3.4
Flexwood	100 SF	7.1
Flexi-wall	100 SF	8.1

Time includes move on and off site, unloading, limited surface and material preparation, cleanup and repair as needed.
Suggested crew: 1 paper hanger, 1 laborer

Labor installing wall coverings
Figure 13-18

Size of room	Height of ceiling 8'	9'	10'	Yards of border	Rolls for ceiling
4 x 8	6	7	8	9	2
4 x 10	7	8	9	11	2
4 x 12	8	9	10	12	2
6 x 10	8	9	10	12	2
6 x 12	9	10	11	13	3
8 x 12	10	11	13	15	4
8 x 14	11	12	14	16	4
10 x 14	12	14	15	18	5
10 x 16	13	15	16	19	6
12 x 16	14	16	17	20	7
12 x 18	15	17	19	22	8
14 x 18	16	18	20	23	8
14 x 22	18	20	22	26	10
15 x 16	15	17	19	23	8
15 x 18	16	18	20	24	9
15 x 20	17	20	22	25	10
15 x 23	19	21	23	28	11
16 x 18	17	19	21	25	10
16 x 20	18	20	22	26	10
16 x 22	19	21	23	28	11
16 x 24	20	22	25	29	12
16 x 26	21	23	26	31	13
17 x 22	19	22	24	23	12
17 x 25	21	23	26	31	13
17 x 28	22	25	28	32	15
17 x 32	24	27	30	35	17
17 x 35	26	29	32	37	18
18 x 22	20	22	25	29	12
18 x 25	21	24	27	31	14
18 x 28	23	26	28	33	16

This chart assumes use of the standard roll of wallpaper, eight yards long and 18" wide. Deduct one roll of side wallpaper for every two doors or windows of ordinary dimensions, or for each 50 square feet of opening.

Single-roll wallpaper requirements
Figure 13-19

tions, material used and supervision will all affect the output. On difficult jobs, the painting time may be up to three times higher than the estimates in Figure 13-16.

You might want your CEF to show drying times, as in Figure 13-17. Figure 13-18 shows labor for installing wall coverings and Figure 13-19 covers wallpaper requirements for most room sizes.

Chapter 14

Scheduling Work Flow

Have you ever had the electricians show up to rough in the wiring before the ceiling joists were installed? Have you ever tried to keep the plumber waiting while you relocated a partition wall? Whose fault is it that the inspector shows up for final inspection before the electricians are finished?

Every construction project has to be scheduled — every step is a link in the chain that must follow some prior link and precede some later link. Scheduling is just making sure that each link falls neatly in the correct place so work goes from start to finish by the most direct, most profitable route possible.

The easy way to schedule is with some type of control board. That makes it easy for everyone to grasp immediately what work is finished, what work still has to be done, and when it should be completed. Your control board should be simple, easy to revise, and take as little effort as possible.

To create the schedule, we'll just divide the project into its logical parts and then link those parts together like a chain. We'll use a control chart to show progress to date and the type of work currently being done. With our chart you'll be able to see the current status of a project at a glance, and will know what's supposed to happen next, and when. If something's slipping off schedule, you'll see it immediately and can make adjustments — *before* it costs you a bundle.

Scheduling work is the key to coordinating any construction project. Scheduling includes getting enough contracts or planning enough spec projects to keep crews and equipment busy. It includes having the necessary labor, materials, and subcontractors lined up and on the right job at the right time. Good scheduling makes prompt completion more likely and reduces idle or wasted time. For the spec builder, good scheduling helps sell your houses promptly, so money isn't tied up in unsold property.

Coordination and scheduling are important responsibilities for every manager. In the smallest construction company, one person can keep track of nearly everything that happens. As a company grows, scheduling usually becomes more of a problem. That's the time to begin using a calendar or checklist to remind you of important dates. As your work becomes more complex and timing becomes critical, someone has to begin laying out work schedules and charting progress.

This chapter will cover the basics of scheduling construction projects: How schedules are compiled, the common types of schedules, and how work is controlled and schedules are changed once work begins. Specifically, we'll discuss the Critical Path Method (CPM) and show how to construct a CPM diagram. I'll include brief definitions of terms used with CPM, such as *restraint, event* and *float.*

You probably know already that CPM is a valuable management tool used for planning, scheduling, and controlling construction operations. It's used on nearly every major project and has saved millions of construction dollars for builders all over the world. But before we get to CPM and more complex schedules, let's talk about a few basic concepts.

Introducing the Schedule

A *schedule* is simply a plan for carrying out a project, indicating when each operation should begin and when it should end. Scheduling (also called echeloning) is used to plan both the sequence of trades on the job and the sequence of work done by each trade. Work schedules show expected start date, the time required for each operation, and the expected completion date. They're the basis for determining when and how much manpower and equipment you'll assign for each portion of the work. Once a schedule is drawn up, you'll use it to set delivery times for materials as well as the arrival of each trade and subcontractor. Finally, you'll use the job work schedule to prepare progress and performance reports.

Elements of Scheduling

The elements used in scheduling work include:

- Work item number

- Item description

- Unit of measurement (CY, SY, ton, each, etc.)

- Quantity of work to be performed

- The relation of each item to the whole in terms of work to be performed (such as percentage of the total work required for each item)

- Units of time to be used in the schedule

- Starting date

- Time required for each item

- Completion date

You'll use similar elements for scheduling equipment and manpower, adding the number of pieces of equipment and number of men.

Techniques of Scheduling

When you begin a project schedule, the first step is to list the work elements. Next, you'll determine the construction sequence. Excavation must come before foundation, wall construction before roof framing, subbase and base preparation before paving, and so on. The starting date for the project is, of course, the date the first work element begins.

To find the time required for each work element, first estimate the number of man-days it'll take to complete that element, then divide by the number of men you've scheduled to work on it.

Schedule each work element in its proper construction sequence, showing starting and completion dates. Frequently, it isn't necessary to wait until one element is finished before starting the next. For example, plumbing and wiring rough-in can begin before the last roofing shingle is laid. By doubling up like this whenever possible, you can save many days and more than a few dollars.

Progress Control

There are three steps in progress control:

1) Measure actual production against planned production.

2) Determine causes of discrepancies, if there are any.

3) Take action to correct any deficiencies in production and to balance activities to avoid slippage in the completion date.

Daily Reports— Many builders collect daily progress reports. The foreman or superintendent submits a labor report showing the work done each day. Some contractors, however, find it's more convenient to report work only as it's completed. For example, if 2,000 square feet of wall forms are required for a section of concrete wall, it may be hard to estimate the portion of the work done after the first day. Some forms have been set, others are still being fabricated and some plywood is still being cut to size. It's probably better to delay the report until all formwork is set and ready for concrete.

Work elements which usually show a fairly steady production rate per manhour are suitable for daily reporting. These include laying concrete block, placing concrete or asphalt paving, or the excavating or hauling of large quantities of cut and fill. Framing components may be broken down into production components such as floor, wall, joists, partition, sheathing, and rafters, for example. When possible, use daily reports to provide a continuous, running check on progress.

A daily report should show the manhours expended on each work element. Prepare weekly or monthly reports by recording daily reports in ledger form and totaling for a week or month.

The daily report doesn't have to be elaborate. Figure 14-1 shows one way to do it. In addition to detailed reports, I like using a pocket calendar to chart work flow and hours on a daily basis. At the end of the month I have a bird's-eye view of what happened on any given day, and how the work progressed. Such a record might look like Figure 14-2.

There's another common way to chart progress on a job. When you make up the work schedule, leave blank spaces next to the estimated completion percentages for each work element. As the job progresses, fill in the actual work completed in these spaces. This gives you a direct comparison of estimated and actual job progress. You can spot any problems immediately.

The Critical Path Method

The Critical Path Method of scheduling was one of the outgrowths of the *Program Evaluation Review Technique* (PERT) developed by the Navy. CPM deals with project planning, scheduling, and control. It has come into widespread use in the construction industry, particularly by engineers and programmers.

The object of CPM is to combine all the information relevant to the planning and scheduling of project functions into a single master plan. The plan is intended to:

1) Coordinate all of the work needed to finish the job

2) Show how all the work will be coordinated

3) Point out which efforts are critical to timely completion

4) Promote the most efficient use of equipment and manpower

There's a lot to know about CPM. We can't cover all the details in this chapter. We'll go deeply enough, however, to help you interpret critical path schedules drawn up for jobs under your supervision, and develop critical path schedules for future construction projects. If you're not familiar with CPM, look at Figure 14-3 before reading any further. It shows a typical Critical Path Network for building a house. It isn't as complicated as it looks at first glance.

Arrow Diagramming

Bar charts were once popular with builders. But many charts were needed to handle a larger project. CPM used arrows linking events rather than bars to depict the work schedule. The tail of the arrow represents the start of the element; the head represents the finish.

The length of the longest path through the diagram equals the total time required to complete the project. Shorter paths include arrows which indicate elements that may be performed simultaneously with other elements. But the length of the arrow has no relation to the time it takes to do the job. All arrows may be about the same length, even though they represent jobs that take varying lengths of time.

Every element on the longest path is *critical:* any delay in one of these elements delays the whole project. An element on a shorter path is noncritical: a delay in one of these (within limits, of course) will not delay the whole project.

CPM Planning

Here are two of the basic ground rules of CPM: First, planning and scheduling are separate operations. Second, planning always comes first. The project is initially planned without any consideration of time or the availability of resources. This planning consists of analyzing the project, breaking it into work elements, and arranging these elements in the *arrow diagram* which becomes the working model of the project. As each work element is defined, three questions are considered:

1) What element must immediately precede this element?

2) What element must immediately follow this element?

3) What other elements, if any, can be done simultaneously with this element?

An arrow is drawn for each work element.

FOREMAN'S DAILY REPORT JOB NO. A-33

DATE	CLASS OF WORK	CODE NO. LABOR	CODE NO. MATERIAL	QUANTITY WORK DONE	DESCRIPTION OF WORK	NO. OF EMPLOYEES	REG	HOURS O.T.	TOTAL	RATE	TOTAL
1/20/XX	MASONRY	042111	04211	12000	BUILDING EXT. WALL (CB)	12	96	48	144	2200	316800
	MASONRY	042111	04212	4000	BUILDING ENTRYWAY (FB)	4	32	8	40	2200	968 00
	CARPENTRY	061001	06100	3200 SF	NAILED ON ROOF DECKING	4	64	0	64	2100	134400
	CARPENTRY	061002	06100	3200 SF	APPLIED FELT	2	16	0	16	2000	32000
	CARPENTRY	073001	07300	33 SQ.	INSTALLED SHINGLES	3	51	0	51	2000	102000

MATERIALS USED

12600	COMMON BRICK	04211
4200	FACE BRICK	04212
75 SACKS	CEMENT-LIME MORTAR	04100
7 TONS	SAND	04100R
20 LBS	WALL TIES	04450
36 PCS	4x8x½ PLYWOOD	06100
35 SQ.	3 TAB SHINGLES	07300
9 ROLLS	#15 FELT	06100
50 LBS	7/16 ROOFING NAILS	06050
25 LBS	6d BOX NAILS	06050

Daily report
Figure 14-1

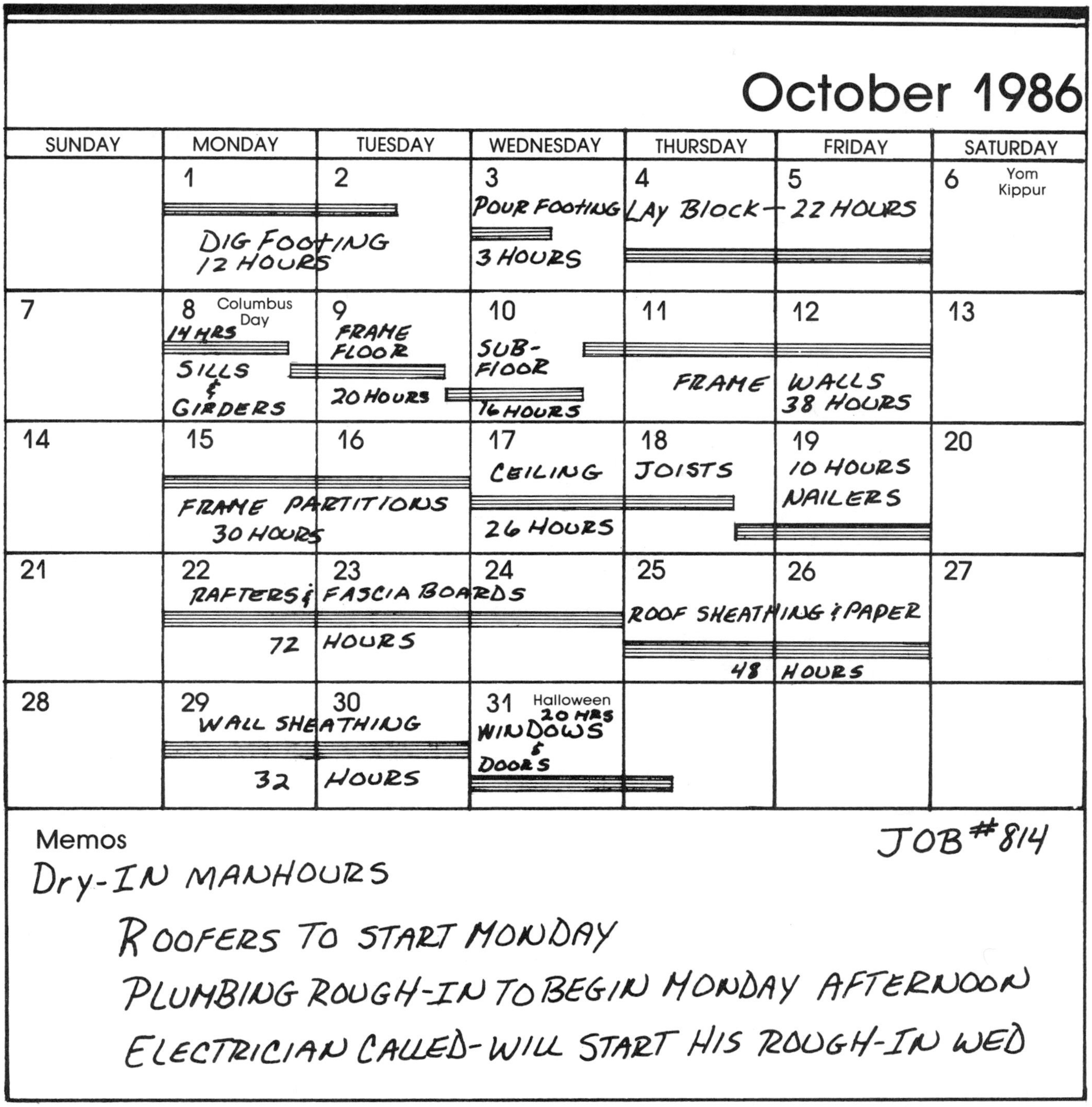

October 1986

SUNDAY	MONDAY	TUESDAY	WEDNESDAY	THURSDAY	FRIDAY	SATURDAY
	1 DIG FOOTING 12 HOURS	2	3 POUR FOOTING 3 HOURS	4 LAY BLOCK — 22 HOURS	5	6 Yom Kippur
7	8 Columbus Day 14 HRS SILLS & GIRDERS	9 FRAME FLOOR 20 HOURS	10 SUB-FLOOR 16 HOURS	11 FRAME	12 WALLS 38 HOURS	13
14	15 FRAME PARTITIONS 30 HOURS	16	17 CEILING 26 HOURS	18 JOISTS	19 10 HOURS NAILERS	20
21	22 RAFTERS & 72 HOURS	23 FASCIA BOARDS	24	25 ROOF SHEATHING & PAPER 48 HOURS	26	27
28	29 WALL SHEATHING 32 HOURS	30	31 Halloween 20 HRS WINDOWS & DOORS			

Memos

JOB #814

Dry-IN MANHOURS

ROOFERS TO START MONDAY

PLUMBING ROUGH-IN TO BEGIN MONDAY AFTERNOON

ELECTRICIAN CALLED — WILL START HIS ROUGH-IN WED

Calendar daily report
Figure 14-2

Remember that the head of the arrow represents the completion of the element, while the tail represents the beginning. The tail is connected to the heads of all arrows representing work elements which must be completed immediately before the element under consideration can begin. The head is connected to the tails of all arrows representing elements which cannot begin until the element under consideration is completed.

Look at Figure 14-4. In this diagram, job A must

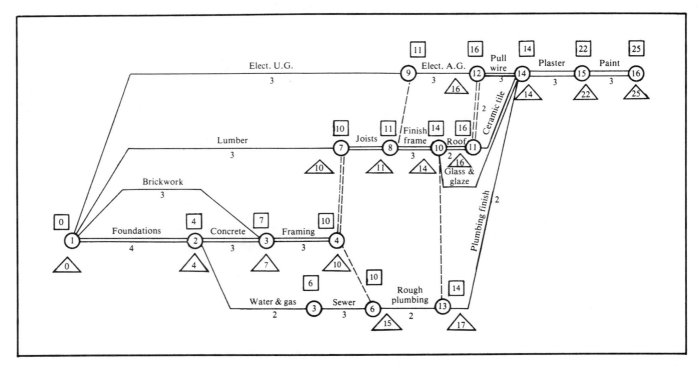

Critical path network of a house
Figure 14-3

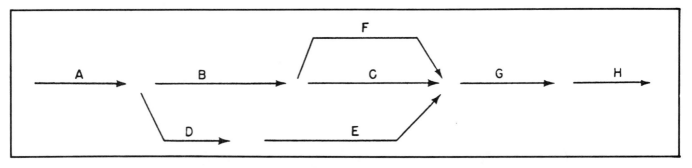

Arrow diagram for Critical Path Method
Figure 14-4

be completed before jobs B and D can start. When job B has been completed, jobs F and C can start. Jobs E, C, and F must be completed before G can begin. Finally, job H can begin when job G is finished.

Figure 14-5 shows an arrow diagram for a very simple military project consisting of the construction of an arch-type high explosives magazine. The project includes the following work elements:

- Excavate foundations

- Reinforce, pour and cure footings

- Pour and cure floor slab

- Form, reinforce, pour and cure front and rear walls

- Waterproof top side of arch

- Install ventilator

- Place and compact magazine earth cover

- Perform final grading and cleanup

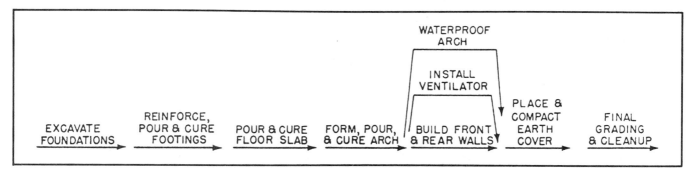

Work elements in a project
Figure 14-5

Obviously, the foundations must be excavated before anything else can be done, so you draw an arrow at the left and label it *excavate foundations*. When the foundations are excavated, the footings can be poured; so the tail of an arrow marked *reinforce, pour and cure footings* is connected to the head of the previous arrow. When the footings have set and cured, the floor slab can be poured. The tail of an arrow marked *pour and cure floor slab* is connected to the head of the footings arrow.

When the floor slab has set and cured, the arch can be formed, poured, and cured, and the tail of an arrow marked *form, pour and cure arch* is connected to the head of the floor slab arrow.

With the arch cured, the front and rear walls can be built. At the same time, two other operations can be carried on: installing the ventilator and waterproofing the arch. Since all three simultaneous elements must be completed before anything else can be done, the heads of the three arrows all converge at the same point. When all three jobs are done, the earth cover can be placed and compacted, and then the final grading and cleanup.

Parallel Work Elements

Suppose a project consists of laying a two-mile weld-joined pipeline from A to B. Work elements might be:

1) Trenching
2) Pipe laying
3) Welding
4) Backfilling

Of course, it isn't necessary to complete one of these before starting the next. As soon as a few hundred feet of trenching are done, pipe laying can begin; as soon as a section of pipe is laid, welding can begin; and so on until all four elements are being performed simultaneously. However, no one element can be *completed* until the previous element has been completed. Welding, for example, cannot be completed until all the pipe has been laid.

In a job like this, you break each work element arrow into two parts, one showing the start and the other the finish, as shown in Figure 14-6.

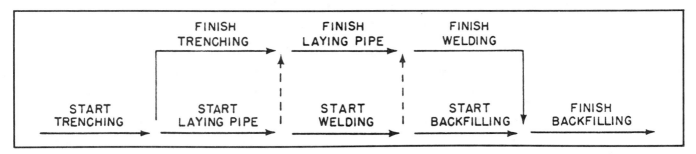

Arrow diagram showing parallel work elements
Figure 14-6

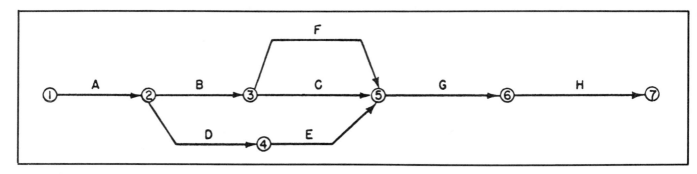

Events and event numbering
Figure 14-7

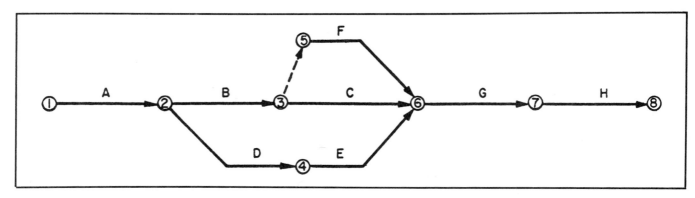

Dummy arrow
Figure 14-8

Events and Event Numbering

The points where arrows connect are called *events.* These events are points in time, marking the time of completion of some work and the time of beginning of other work. Events are numbered from left to right through the diagram. Obviously, the event number at the head of an arrow must be larger than the event number at the tail, since it's later in time.

Figure 14-7 shows an arrow diagram with events numbered correctly. It would be incorrect to label the event at the end of job C with the number 4, and that at the end of job D with 5. If you did this, the job E would begin with event 5 and end with event 4. Remember, you can't have the number at the tail of the arrow larger than that at the head.

Each work element is identified by the event numbers on its arrow. For example, job B in Figure 14-7 would be identified as job (2, 3).

Use of Dummy Arrows

Since there may be two or more jobs going on simultaneously, the arrows for these jobs would have the same event numbers. CPM uses a special device to avoid this. *Dummy arrows* and *artificial event numbers* are inserted, as shown in Figure 14-8. The dotted dummy arrow extending from 3 to 5 and the artificial event number 5 have been introduced so that job F will not have the same event number as job C.

Dependence

If job B can't start until job A is completed, job B is said to be *dependent* on job A. Dummy arrows may be used to indicate this relationship. The dummy arrow from 3 to 4 in Figure 14-9 indicates that job E, or job (4, 6), can't begin until job B, or job (2, 3), is completed. The fact that the tail of the arrow for job E lies at the head of the arrow for job D already indicates that E cannot start until D is completed. The addition of the dummy arrow indicates that before E can start, B must be completed as well.

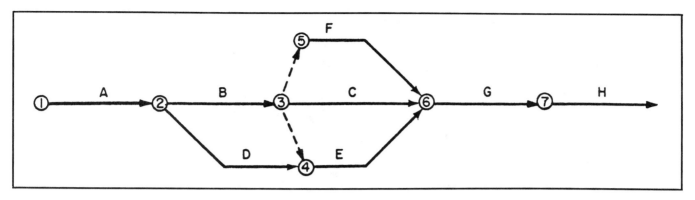

Dummy arrow (3, 4) indicates dependency
Figure 14-9

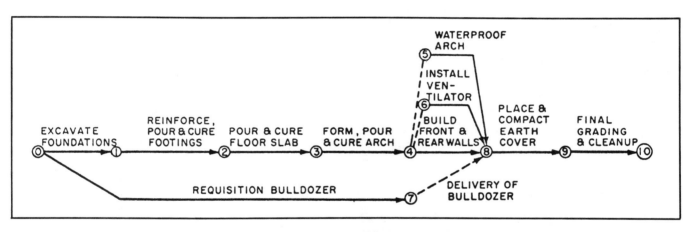

Dummy arrow indicating restraint
Figure 14-10

Restraints

A condition which isn't a work element, but which must be fulfilled before a work element can begin, is called a *restraint*. Suppose, for example, that a bulldozer is required before a certain element can begin. Getting the use of the bulldozer is a project work element. Delivery, however, is not a project work element, but a restraint, or condition which must be fulfilled before the element involving the bulldozer can begin. Figure 14-10 shows how to use a dummy arrow to indicate a restraint.

Final Check of the Diagram

Completion of the arrow diagram marks the end of the planning stage of CPM planning and scheduling. If you're uncertain about the sequence of work elements, find out the proper sequence before you try to create the diagram. Before you assign event numbers, go over it carefully and make sure that the sequence is correct and that all necessary dummies for artificial events and restraints are shown. Don't use any unnecessary dummies, however, as these only confuse the diagram.

CPM Scheduling

When a project has been planned on an arrow diagram, the next step is to schedule it — that is, to place it on a working timetable. When this has been done, you can determine when each of the various jobs must be performed, when deliveries must take place, how much (if any) spare time there is for each job, and when to expect completion of the whole operation. You can also determine which jobs are critical and to what extent a delay in one job will affect succeeding jobs.

You begin scheduling by marking below each job

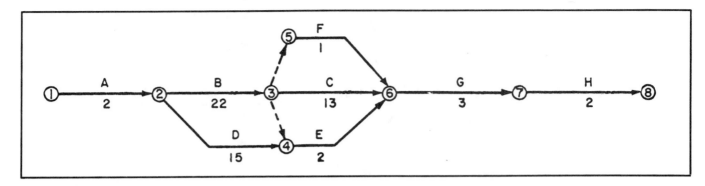

Job durations on arrow diagram
Figure 14-11

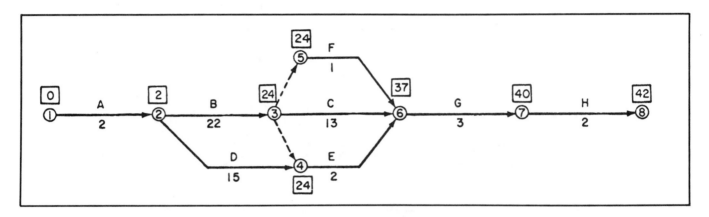

Earliest event times on arrow diagram
Figure 14-12

arrow the expected length of time that job takes, as shown in Figure 14-11. The dummy arrows (3, 5) and (3, 4) have no duration times, since they're not really jobs. A dummy indicating a restraint, however, would be marked with the time required for performance of the restraint.

Earliest Event Times
The earliest time at which an event can occur is the sum of the durations of the work elements on the *longest path* leading up to the event. This time is entered in a rectangular box next to the event on the arrow diagram, as shown in Figure 14-12.

The times shown are, of course, project times. That is, successive *working* days, not successive calendar days, established from a zero at the tail end of the first work element arrow.

The duration of the first job in Figure 14-12 is two days; therefore, event 2 occurs at project time

2. The time for event 3 is the sum of the duration times of (1, 2) and (2, 3), or 24. However, there are two paths leading to event 4: one from 1 through 2 for a total of 17, the other from 1 through 2 and 3, for a total of 24.

Following the rule of selecting the longest path, the event time for event 4 is 24. Similarly, three paths lead to event 6. We select the longest (from 1 through 2 and 3), giving an event time for 6 of 37. (Add 2, 22 and 13 to get 37).

Latest Event Times
You also need to know the latest time at which an event can occur. To determine this, you begin at the end of the project and work backward. To calculate the latest time at which an event can occur, subtract the duration of the last job from the previous latest event time. Enter the latest event time in a small triangle adjacent to the rectangle

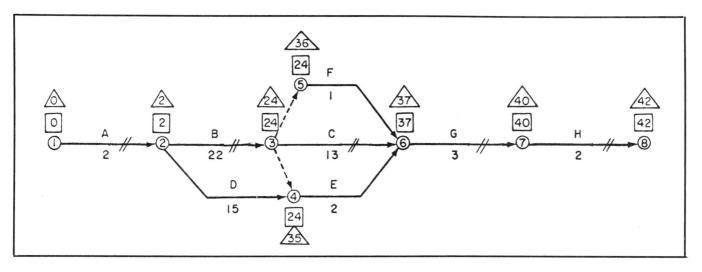

Latest event times on arrow diagram
Figure 14-13

containing the earliest event time, as shown in Figure 14-13.

In Figure 14-13, the *latest* event times for events 6, 7, and 8 are the same as the *earliest* event times. The latest event time for event 7, for example, equals the latest event time for event 8, which is 42, minus the duration of (7, 8), which is 2. The remainder is 40.

But the latest event time for event 4 equals the latest event time for 6, which is 37, minus the duration of (4, 6), which is 2, or 35. The latest event time for event 5 equals the latest event time for 6, which is 37, minus the duration of (5, 6), which is 1, or 36. The latest event time for 3 equals the latest event time for 6, which is 37, minus the duration of (3, 6), which is 13, or 24.

Note that for any event on the critical path, the earliest event time and the latest event time are the same. It's only for events not on the critical path that these event times differ. It follows that identical earliest and latest event times are another means of identifying job arrows on the critical path. For a job to be critical, both of the following conditions must exist:

• At each end of the arrow, the number in the rectangle (earliest event time) and that in the triangle (latest event time) must be the same.

• The job duration must equal the difference between a number at the head of the arrow and a number at the tail of the arrow.

When the job durations have been placed on the arrow diagram, you can determine the critical path. The critical path is the *longest* path through the diagram, in terms of time. In Figure 14-13, there are three possible paths through the diagram: the bottom path from 1 through 2, 4, 6, 7, and 8 totals 24 days, and the top path from 1 through 2, 3, 5, 6, 7, and 8 totals 30 days. The middle path is the longest. Therefore, this is the critical path. It's indicated by marking the arrows with small double slants.

This path represents the normal duration of the project. Every work element on the path is critical to the completion of the project in 42 days. If any one of these elements is delayed, the completion of the whole project will be delayed.

Earliest and Latest Job Start and Finish Times
Figure 14-14 shows a fully developed arrow diagram for the project of building the arch-type magazine discussed earlier. It includes all work elements, with the earliest and latest event times established. Now you can determine the earliest and latest starts and finishes.

In Figure 14-15, for example, what are the latest and the earliest days on which waterproofing of the top side of the arch can be started? What are the earliest and latest days on which the installation of the ventilator can be started?

Before either of these jobs can begin, the stripping of the arch forms, which is job (9, 10), must be completed. This job is on the critical path, and it

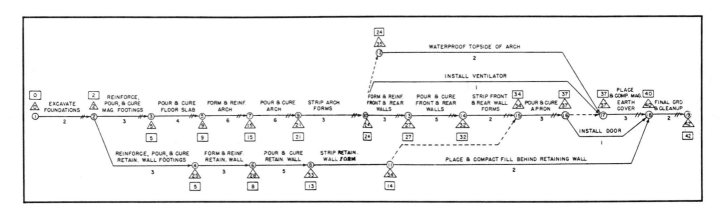

Diagram of high-explosives magazine
Figure 14-14

| JOB | PROJECT DAY | | | | REMARKS |
| | START | | FINISH | | |
	EARLIEST	LATEST	EARLIEST	LATEST	
(1, 2)	1	1	2	2	critical
(2, 3)	3	3	5	5	critical
(2, 4)	3	23	5	25	
(3, 5)	6	6	9	9	critical
(4, 6)	6	26	8	28	
(5, 7)	10	10	15	15	critical
(6, 8)	9	29	13	33	
(7, 9)	16	16	21	21	critical
(8, 11)	14	34	14	34	
(9, 10)	22	22	24	24	critical
(10, 12)	0	0	0	0	dummy
(10, 13)	25	25	27	27	critical
(10, 17)	25	37	25	37	
(11, 15)	0	0	0	0	dummy
(11, 18)	15	39	16	40	
(12, 17)	25	36	26	37	
(13, 14)	28	28	32	32	critical
(14, 15)	33	33	34	34	critical
(15, 16)	35	35	37	37	critical
(16, 17)	0	0	0	0	dummy
(16, 18)	38	40	38	40	
(17, 18)	38	38	40	40	critical
(18, 19)	41	41	42	42	critical

Project schedule from arrow diagram in Figure 14-14
Figure 14-15

will be completed at event time 24. The water-proofing of the arch and the installation of the ventilator must be completed by event time 37, if the project is not to be delayed.

The waterproofing is a two-day job. It can begin as early as day 25 (day of completion of stripping of arch forms plus 1), or as late as day 36 (final deadline for completion minus 2 plus 1). It can be completed as early as day 26 or as late as day 37. Similarly, the installation of the ventilator can begin as early as day 25 or as late as day 37, and can end as early as day 25 or as late as day 37.

Here are the rules for calculating start and finish days for a work element:

- *Earliest start day:* earliest event time at the tail of the arrow plus 1.

- *Earliest finish day:* earliest start time plus job duration.

- *Latest start day:* latest event time at the head of the arrow minus job duration plus 1.

- *Latest finish day:* latest event time at the head of the arrow.

To calculate earliest finish days, work from left to right on the diagram, adding job durations to earliest event times. To calculate latest start times, work from right to left, subtracting job duration from the preceding latest event time.

Enter the results in a schedule as shown in Figure 14-15. This schedule assumes that all jobs will be started as early as possible.

The Concept of Float

The spare time available to perform a task such as the installation of the ventilator in Figure 14-14 is called *float*. Properly controlled, you can use float to determine the most efficient use of manpower, equipment, and materials. The existence of float allows latitude in the timing of the jobs with which it is associated. On the other hand, a job having no float is inflexible; it must start and end precisely at specific times, or the completion of the project will be affected.

In Figure 14-14, the installation of the ventilator has 12 days of float because it's a one-day job and there are 13 days available in which it may be done. Similarly, the waterproofing job in the same figure has 11 days of float. To calculate float, subtract both the duration and the earlier event time at the tail of the arrow from the later event time at the head of the arrow. For job (6, 8) for example, the float comes to 33 minus 8 minus 5, or 20.

Each of the noncritical jobs along the path from event 2 to event 11 has 20 days of float when considered independently. However, there are only 20 days of float available for the whole chain. Here's how to figure it: 35 minus 2 minus 12 (that's 3 plus 3 plus 5 plus 1) equals 20.

Why are there only 20 days of float available? When you calculate float independently for each separate job, you assume that all the preceding jobs were started as early as possible. But as soon as any float is used, the float available to subsequent jobs is correspondingly reduced.

Suppose, for example, that job (4, 6) was delayed for three days. The succeeding job (6, 8) would have three days added to its earliest event time and subtracted from its float. The float for job (6, 8) would then be 33 minus 11 minus 5, or 17.

Using Float to Allocate Manpower and Equipment

In building the structure diagrammed in Figure 14-14, there are three jobs of form stripping to be done. The stripping of the arch is critical, and must

be performed between event times 21 and 24. Similarly, the stripping of the front and rear wall forms must be done between day 32 and 34. However, the stripping of the retaining wall forms is a one-day job which may be done at any time between event time 13 and event time 34. Obviously, the crew should be scheduled to strip the retaining wall at a time when they are not busy with the arch or front and rear wall forms. Similarly, the pouring and curing should be scheduled so as to take advantage of the float in job (6, 8). By starting job (6, 8) one day after its earliest start time, it can be performed concurrently with job (5, 7). Thus, by using up a day of float, more efficient use is made of crew and equipment.

Adjustment of Float

Earliest and latest event times shown in the arrow diagram (Figure 14-14) were changed to reflect the days dropped out as a result of weekend curing. A check was then made to ensure that the critical path was still the same. Shortening the original critical path might cause another path to take its place.

Next, the float was recalculated, and the new float values were entered in the timetable (Figure 14-16). Notice on the timetable that job (11, 18) and job (17, 18), both of which consist of placing and compacting fill (see Figure 14-14), are scheduled about a month apart. Since job (11, 18) shows 20 days of float, however, and since the same equipment will be used for both jobs, the float for job (11, 18) will probably be used to schedule this job to end the day before job (17, 18) begins.

Preparing a Timetable

After the arrow diagram has been completed and the float has been calculated, a timetable like the one shown in Figure 14-16 can be prepared.

This is a timetable derived from the arrow diagram shown in Figure 14-14. Project day 1 falls on March 1, a Thursday. No work is done on Saturdays or Sundays. Therefore, though project days 1 and 2 fall on Thursday and Friday, project day 3 falls on Monday, March 5. As you can see, Saturdays and Sundays are included in the calendar when they can be utilized as curing time for concrete jobs. In this case, Saturday and Sunday become project days. If the days relate to a job on the critical path, the effect is to gain time by cutting a day from the schedule.

In Figure 14-14, five days were cut from the

Job	Project days		Calendar days		Float
	Start	Duration	Start	Finish	
(1, 2)	1	2	March 1, Thurs.	March 2, Fri.	0
(2, 3)	3	3	March 5, Mon.	March 7, Wed.	0
(2, 4)	3	3	March 5, Mon.	March 7, Wed.	16[2]
(3, 5)[1]	6	4	March 8, Thurs.	March 11, Sun.	0
(4, 6)	6	3	March 8, Thurs.	March 12, Mon.	16[2]
(5, 7)	10	6	March 12, Mon.	March 19, Mon.	0
(6, 8)	9	5	March 13, Tues.	March 17, Sat.	16[2]
(7, 9)	16	6	March 20, Tues.	March 25, Sun.	0
(8, 11)	14	1	March 19, Mon.	March 19, Mon.	16[2]
(9, 10)	22	3	March 26, Mon.	March 28, Wed.	0
(10, 12)	—	0	—	—	0
(10, 13)	25	3	March 29, Thurs.	April 2, Mon.	0
(10, 17)	25	1	March 29, Thurs.	March 29, Thurs.	11[2]
(11, 15)	—	0	—	—	0
(11, 18)	15	2	March 20, Tues.	March 21, Wed.	20[2]
(12, 17)	25	2	March 29, Thurs.	March 30, Fri.	10[2]
(13, 14)	28	5	April 3, Tues.	April 7, Sat.	0
(14, 15)	33	2	April 9, Mon.	April 10, Tues.	0
(15, 16)	35	3	April 11, Wed.	April 13, Fri.	0
(16, 17)	—	0	—	—	0
(16, 18)	38	1	April 16, Mon.	April 16, Mon.	2
(17, 18)	38	3	April 16, Mon.	April 18, Wed.	0
(18, 19)	41	2	April 19, Thurs.	April 20, Fri.	0

[1]Curing scheduled for weekend.
[2]Adjusted to reflect weekend curing.

Timetable from arrow diagram in Figure 14-14
Figure 14-16

critical path by scheduling concrete work so that curing could take place on weekends.

For example, job (3, 5) consisted of placing and curing the magazine footings. It was started on Thursday so that curing could be scheduled for the weekend. Since this job was on the critical path, the use of Saturday and Sunday for curing cut two days from the schedule.

Any experienced builder knows that it would be a waste of time and energy to work out a CPM diagram for a small remodeling project. You shouldn't have any trouble keeping the few deadlines in your head. If you do, jotting a few dates down on a notepad may be enough. You know that there's no way to apply roofing shingles until the sheathing is down. But even on a simple project like this, some sort of a plan is needed. Otherwise, the crews will spend more time playing cards than driving nails.

On larger construction projects with many critical dates, scheduling is critical. Consider using CPM. It may save you both time and money, with a bonus of fewer headaches. Try it.

Chapter 15

Successful Management

The subtitle of this chapter might be "The Art of Making a Profit While Your Competitors Go Broke."

Good organization and adequate paperwork are essential parts of your construction company. Unfortunately, most contractors are better at building than they are at running a business. That's usually a prescription for disaster. It takes both construction skill and business ability to survive as a builder. Neither by itself is enough. This chapter explains some of the key points that will keep you out of bankruptcy court, even while your competitors are going belly up or leaving the business in droves.

Every building business is a complex enterprise, using many highly skilled tradesmen and professionals. If you're like most builders, you do practically all the management yourself. That puts the whole load on your shoulders. Fortunately, you don't need a Harvard M.B.A. to run a construction company. A little knowledge here and there will prevent most of the common mistakes.

To run a construction business, you need both administrative and technical skills. Every project requires planning and scheduling of resources: equipment, men and materials. Your field work force has to be reorganized for each job. Someone's got to make arrangements for payrolls, accounting, insurance and tax records.

What does an owner bargain for when he hires you to complete a project? The materials you use are about the same as every other builder in town. It isn't your equipment. Most will probably come from the same rental yard every other contractor uses. Neither is it your tradesmen. Skilled carpenters, plumbers, electricians, etc. are available in every community. What the owner is really buying is your skill as a manager, and your business know-how. This chapter is intended to show how to make your management skills adequate for the challenges you face.

The Function of Management

As a contractor, you have to do much more than supervise your jobs as they progress. You'll have to control several key areas. Some of these management functions would be your responsibility no matter what kind of business you were in. Others are unique to the construction industry. They require specific technical knowledge and experience.

If you don't have this experience and knowledge, there are two choices. The first is to get out of the business while you still have your shirt. The second is to learn enough to defend yourself as soon as possible — tomorrow may be too late.

One way to get the skills you need is to bring a knowledgeable partner into the firm. If you're lucky, you may be able to hire someone who has the special abilities you lack. But most contractors won't be able to find or afford a full time business "pro" to organize and run the construction office.

The business functions you have to perform, or hire someone to perform, fall into several broad areas. They are:

1) General administration
2) Legal matters
3) Personnel
4) Design and engineering
5) Construction
6) Purchasing and subcontracting
7) Financial control
8) Planning
9) Sales

Even one- and two-man construction companies will have regular problems with law, personnel, design, purchasing, financial controls, planning and sales. No one is immune. If, through your skill and hard work, your business grows, these problems will probably get worse, not better. Hopefully, your growing business will be able to hire clerical and supervisory personnel and skilled specialists who can tackle some of the more serious and technical business problems.

But until your company grows, you've got to be able to do it all yourself. That's the reason for this chapter. Let's look at some of the functions you need to master.

Making Decisions
Basically, your job as a business manager is to make decisions on allocation of resources. What types of work do we want? How are we going to get it? What profit margin do we require? What crew do we put on what job today? Who gets paid today and who has to wait?

Whether you succeed or fail will depend primarily on the quality of your decisions. And the validity of your decisions will depend to a great extent on the accuracy and completeness of the information you have as the decision is made.

Information will come to you from many sources, both inside and outside the company. Sources within the company include:

• Your own experience and observations
• Employee input
• Cost records
• Financial reports
• Progress reports
• Personnel records

Every construction company needs some simple and easy reporting system. A system too complex or detailed takes too much time. A system too sketchy misses too much. The object is to keep you informed on all the important facts and to produce enough records to satisfy the requirements of state, federal and local government. Information that takes too long to prepare may not be ready when you need it. Later in this chapter we'll talk about the kind of record keeping you need for a successful construction company.

Sources of information from outside the company include:

• Plans and specifications
• Suppliers' quotes and product information
• General economic data
• Construction reports
• Newspapers, newsletters and magazines
• Your observation of local conditions
• Consultations with professionals
• Contacts with customers and suppliers

On the basis of information from these and other sources inside and outside your business, you'll make both short-range and long-range decisions: choosing designs, estimating costs, purchasing, subcontracting, scheduling construction, hiring and guiding personnel, and planning sales programs, to name a few.

Thinking Ahead
You should look ahead and decide where you want your business to go over the long pull, and how you're going to get it there. You may have to change your plans, but it's better to start with a well-defined goal than to drift in whatever direction circumstances carry you.

Short-range planning is important, naturally. But it shouldn't take too much of your time. Save some time for doing your long-range planning. We'll talk about the importance of good planning a little later in this chapter. But first, let's spend a little time on how you can improve your own management skills.

Building Your Management Ability
First, understand that there are no management supermen. No one is a magic manager. There's no secret formula for success as a construction manager. Most of what you need is just ordinary skills developed by everyday experience and, of course, effort.

Here are some of the characteristics that successful builders have in common.

Integrity

On major building projects, the owner employs an architect whose duty is to watch the job as it progresses, and to see that the builder lives up to specifications. But few small jobs have complete specifications. The only protection the smaller owner has is his confidence in the integrity of the builder.

Be open and aboveboard in your transactions. This doesn't mean you have to tell everyone about all your business, your profits, and so on. Simply treat all with frankness and fairness. Selfishness, unfairness and downright dishonesty may sometimes produce an apparent advantage, but they never bring lasting success. Remember that a list of satisfied clients is about the best advertisement you can have. The fact that you are able to refer to these clients helps convince prospects that you can be trusted. This reputation for square dealing is one of the most valuable business assets a builder can have.

Going the Extra Mile

You'll also find it a good investment to build confidence by giving your client a little bonus in some of your work. By adding a little extra not called for in the contract, you're giving more than the owner expected, and he naturally responds to your thoughtfulness. True, you may not benefit immediately. But someday, the goodwill you've created will work in your behalf. You can charge off these expenses to advertising or to a goodwill fund. It's simply "going the extra mile" and it will *always* work. At least, that's been my experience — and I've been at it for more than a quarter of a century.

Offering Suggestions

Still another method of creating confidence and goodwill is to offer good suggestions. Point out any features of the plans that you don't think will work out the way your client expects. Offer suggestions to improve the plan. If there's an added cost involved, make this clear to the owner, and explain the reason for it.

If you're working from the plans of an architect and come across a defect, don't ignore it because the architect is the one responsible. The owners probably won't blame the architect — they'll blame you, because you're right there, and you're the one who built it.

You owe this service to your customer. He may not see that the limited hall dimensions are going to make it impossible to get large furniture through the front door, or that doors are going to interfere with one another if hung according to plans. It's your job to spot these problems. Point out discrepancies and correct them whenever possible, but not until the owner knows and understands the details of such changes and approves them. This willingness to serve goes a long way in building success.

Fairness with Employees

One hallmark of success in any business is your ability to handle personnel effectively. Start with proper discipline. Insist on professionalism. Assign workers to tasks they are trained to carry out. Provide fair and just reward for work done, unbiased settlement of all disputes between employees, prompt and friendly adjustment of mistakes in time and wages, and sympathetic understanding of their work. Make sure no person loses his job without a fair hearing.

These are methods which develop a well-organized, well-disciplined work crew. Handling personnel matters fairly and consistently results in a satisfied group of workers who produce high-quality work and represent your business well. Resentful employees cost you plenty.

Of course, no employee will have the interests of the company at heart quite as much as you do. But they represent you. Their work and their conduct on the job are important to the success or failure of your business. And if they can see that the more successful your business, the more secure and better paid their jobs are, you've gone a long way toward building a loyal and capable team.

Resourcefulness

The ability to see ahead, to plan delivery of materials as needed, to judge the best way to do the work, can save you plenty on every job. Time, labor, and materials can be saved by good planning and creativity.

A Michigan builder won a contract for the erection of a concrete block factory building. When the excavators reached a foot or so in depth, they discovered extremely sandy soil. The quick-witted superintendent found that a simple rig could wash the soil, leaving clean sand that could be used for pipe bedding and fill. Instead of paying to haul the sand away, it was used to reduce the cost of sand needed for the job. A resourceful person is a money-maker on any building job.

You'll probably never find specification grade sand on any building project. That isn't my point. I'm saying that opportunities can be anywhere. Taking advantage of every opportunity — thinking, planning, and deciding — makes for good management. The more resourceful a builder is, the more likely he is to succeed.

Energy and Vitality
There is probably no business or profession that requires as much energy and vitality as the building business. Agility is important in any business where workers and materials must be kept on the move. The employees must be constantly spurred and inspired to further effort; materials must be kept coming to the job as fast as needed, yet not so fast that they get in the way because of lack of storage space. It's this ceaseless energy and vitality that constantly inspires workers to give their best effort and moves the work on to completion.

Initiative
The builder who has initiative can generally meet unusual situations successfully and wisely. Initiative is simply the ability to start things. As problems present themselves, a person with initiative will devise new ways to overcome obstacles that would stop others. And as new ways of accomplishing things present themselves, his mind is quick to grasp the essential features and put them into effect. Initiative not only *starts* things, but *keeps them going* and *finishes* them.

Awareness of Change
The small builder who operates his own business will generally supervise the work, assuming the duties of the superintendent and of the foreman. The larger organization will have its general superintendent, assistant superintendent, and numerous foremen. Regardless of the type of organization, there must be a continuous flow of new ideas in building, management, materials, supervising employees and all the other factors essential to the operation of the business.

Whether it's the builder himself, the superintendent, or the foreman, the person responsible for the building operations must be cost-conscious, searching for ways to reduce costs and increase productivity. No one ever stops learning in the construction industry — if they're smart. Read the trade literature. Clip relevant articles for future reference. Whether you've got 20 years or 20 days in the building business, keep looking for better, more efficient ways to do what has to be done.

Here are a few magazines or periodicals you can read to keep tabs on what's going on in the building business:

Better Homes & Gardens, Meredith Corp., 1716 Locust Street, Des Moines, IA 50336

Builder, P.O. Box 2029, Marion, OH 43306

Fine Homebuilding, The Taunton Press, 52 Church Hill Road, Box 355, Newtown, CT 06470

Remodeling Contractor, 300 W. Adams Street, Chicago, IL 60606

New Shelter, 33 East Minor Street, Emmaus, PA 18049

There's another good way to keep up with changes. Visit other building sites from time to time. Note any improvements in process, methods, machines, capacities and practices. Staying current on new materials and methods is just good business.

Perseverance
All of the factors we've just discussed are important, but they're of little value unless you're persistent. You'll find that the most successful builders and building executives are highly motivated. They're the type that keep hammering away on a problem early and late, wasting no time in idle talk or useless motions.

If you can carry on with a singleness of purpose; if you can keep your nerve and keep on long after others have given in; if you can believe in yourself and your ideas; if you can follow the proven rules and methods — then you're on the road to success. And if you keep on climbing and use your head, you'll surely arrive at your goal.

Building a Strong Company
When you start your business, you'll be thinking and planning 60 to 90 or 120 days ahead. Later, as your company grows, you may be planning ahead 6 or 9 or 12 months. You'll be busy with those plans,

of course, but don't let them keep you from looking way ahead. It may be helpful, even at the beginning, to try to answer the questions: What will my company be like in five years? What do I want it to be like?

What your company actually becomes 5, 10 or 15 years from now depends on many things. Among them are:

1) Adjustment to technical and economic changes

2) Building a management team

3) Planning and achieving short-term goals

4) Building for long-term growth

The Changing Industry

Your company's future will be affected by changes in conditions. The home building industry is constantly changing: techniques, economics and organization. You'll need to study the changes. What kind of company you'll be running 5 or 10 years from now depends on how well you can adapt to the changes that are inevitable.

One of the changes you'll have to adapt to is the trend toward doing more labor in the factory and less on the building site. Factory labor is cheaper than on-site labor. And quality control is usually better in a factory than on site. In a factory, conditions can be controlled and waste motions eliminated. The result is cheaper and better prehung doors, roof trusses and roofing systems. Prefabricated components also give you, the builder, better control of work schedules. And this usually means cost savings and faster completion.

Being a good salesman becomes more important as prefabricated components make building skills less relevant. Manufactured components may one day make builders little more than salesmen selling homes that can be assembled quickly from structural sections.

But components or no components, there always will be people who want houses designed and built to their own tastes. Owners who can pay for custom styling usually make good clients. But even here, you may be able to blend components and custom workmanship to give the customer what he wants, plus save yourself a bundle.

Building a Management Team

Everyone who's tried to run a one-man construction company has been discouraged more than once about the heavy workload and the variety of skills needed to get the job done. Most have wondered if owning and operating a business is worth all the time and effort. Drawing a weekly paycheck on someone else's payroll is much easier.

But before you throw in the towel, remember your business doesn't have to be a personal marathon. You can relieve yourself of many details by selective hiring. A job superintendent is probably the highest priority.

The Job Superintendent

Let your lead carpenter work on salary and train him as your job superintendent. He should be a good craftsman who knows (or can learn) the technical details of building. But he (or she) needs to be more than a competent craftsman.

First, he should be able and willing to learn. You'll need to teach him. But the more he learns, the better for both of you.

Second, he should be a self-starter. Pick someone who shows initiative — who finds a way to get the work done without troubling you with all the details and complaints. When you aren't there to make the decision, he should show good judgment — about what you would do under the same circumstances.

Third, he should have the ability (or the potential) to become a manager. If he's to relieve you of details now, and later be responsible for large parts of your business, he'll have to learn to manage his own time and efforts as well as that of other people. Rate your candidate on these points. Then ask yourself, "Could this person, with training and coaching, run the company if I were in the hospital in the middle of July?"

Equally important, how does he get along with people? Can he supervise workers without rubbing them the wrong way? Can he negotiate with subcontractors? Is he considerate and patient with demanding customers who don't want to understand a builder's problems?

You need someone who complements your personality and skills, not a carbon copy of yourself. Try to select someone whose strong points balance your weak points. For instance, if you're good at paper work, pick someone who's good at running a job, but would rather stay out of the office.

Finally, when you find that person, don't expect him to be perfect. Everyone has flaws. If your superintendent has trouble getting to work on time or has a hot temper, make allowances. Don't judge someone on his defects. Get the whole picture. If he's a good superintendent and a loyal worker, overlook the minor faults.

Train your assistant by delegating work to him. While he's learning, coach him and check his work. However, as soon as he's shown that he can handle certain tasks well, back away and let him handle them. Give him the authority, material, equipment and labor he needs to do the job and turn him loose. If you've trained him properly, he'll know when to ask for help.

If your company continues to grow, you'll need additional management personnel. Add them as necessary, measuring them against the same yardstick you used when picking your first superintendent. Even better, let that superintendent help you make the selection.

A good manager knows that building employee skills is as important to the success of the company as building with mortar, cement and lumber. Employees can either save or cost you money. Trained, competent employees reduce your management workload. They're the only real inventory most construction companies need — the skill inventory you can call on. The success of your company depends on your ability to find and train good employees and to delegate the right tasks to them. That frees up your time to take care of more important tasks.

The Crew

As a builder, you do a lot of hiring. Most builders use employment agencies, follow the suggestions of friends and employees, and use classified ads in local newspapers to find good applicants. But the best source of new manpower will always be the pool of employees that have worked with you in the past. You probably already know the best workers in each trade in your community.

Unless you can be on the site every minute, which is unlikely, you'll need to have someone in charge when you're away. This key employee is usually the lead carpenter or superintendent-in-training, the first salaried employee you hired. He'll be the key to the quality of your work and to your costs on the job. As your operation grows and you're away from the site more and more, his ability can make or break you.

It isn't easy to assemble an efficient work crew. But if you've found a crew that works well together, try to keep work lined up for them. Laying them off for more than a few days will probably make getting the crew together again difficult or impossible.

How to Plan a Job

Ask any experienced and successful contractor to name the two most important steps in bringing in a profit on a construction job. Odds are, the answer will be something like:

1) The work must be carefully planned and scheduled.

2) The work must be kept on schedule.

In all my years in construction, I've never seen a job that was poorly planned and behind schedule that yielded the contractor a respectable profit. But I've seen lots of well-planned, on-time jobs that made money. Believe me, it's no coincidence. Do your planning and stick to the plan if you want to make money in construction.

Let's start with the planning. As soon as the construction contract is signed, it's time to begin setting goals: When can we start? When will the permit be issued? When can I start delivering materials to the site? An experienced contractor doesn't need an elaborate CPM chart to know what's supposed to happen and when on a small job. He already knows what to do and when to do it. The plan would fit on the back of a napkin. But the contractor knows full well when something's going wrong and when the schedule isn't being kept. On a small job, keep the schedule in your head. A larger job's another matter.

A large job takes more systematic and careful planning. Records must be kept in detail. No one can rely on memory alone to keep all the facts straight.

The first step in any large job is to prepare a rough or tentative plan. This is no more than a list of the work elements, the approximate time (in days) needed for each one, and the approximate dates at which each work element should be started and finished. In the beginning, you probably don't know the exact date when work can start. No problem. Call that "Day One" in your plan and number every subsequent day sequentially from that day.

After you've listed the time needed for each task, list them in the order that they will be done — overlapping tasks that can be done at the same time (such as rough plumbing and wiring) and laying end to end tasks that must be done one after another (such as plastering that has to follow lathing). Next, determine the types and sizes of construction equipment required. Decide what equipment is available and what equipment you may need to rent.

List the classes of labor, both skilled and unskilled. Determine the approximate date or times they'll be needed and the number of workers in each class. Plan in a general way how many regular employees you'll use and how many employees will be hired for the job. Also, select your leadership team, the superintendent and general foremen.

Establish when and where the orders for materials should be placed, when deliveries should be made, and in what quantities. Give consideration to the subcontractors and their work, noting the approximate starting and completion dates. Think about overhead for the job and your cash flow. Sketch out payment dates for materials and labor, and be sure the money required will be available.

After you've completed the tentative or rough plan, start preparing the detailed plans.

Planning Office Work

As part of the planning for a large job, think about support needed from the office staff. The superintendent needs help with purchasing, payroll and records if the work's to go on as scheduled.

If there are two or more owner-contractors, agree on who will direct this particular job. It's best if one person accepts responsibility for dealing with the architect, owner, city departments, material suppliers, etc. on each job. All orders, decisions and correspondence are handled by this individual. Several "managers" or bosses can't run one job successfully. There will be confusion in ordering material, conflicting orders given to the foreman and subs, and conflicting statements made to the architect or owners. The net result will be confusion in the work and loss of confidence by everyone connected with the project. Put one responsible person in charge.

If a superintendent has been appointed to run a particular job, all orders and correspondence relating to that job should be directed to or through the superintendent. This is good management. Your superintendent has to be informed about everything that happens on his job.

In your office, the support work may be divided among several people: payroll, purchasing, cost recording, etc. But it should all be under your

general supervision. Office staff will prepare drawings, handle material orders, permits, payroll, and prepare a reminder or suspense file for controlling routine matters on the job. But that doesn't alter the superintendent's role. He has to know about everything that moves on his job.

Planning Labor Requirements

Once you have a day-by-day schedule for the job, it's easy to plan for labor. Order of appearance will depend on the particular job. It's usually not necessary for any one task to be wholly complete before another task or trade starts. But, the trades and crews should be scheduled so they'll have the least interference with each other and cause the fewest delays. On almost any job there will be some conflict between crews, and some delays. But a little planning can reduce interference to a minimum.

How closely one crew can follow another depends on the job. For example, on a concrete-paving job, it may work best to keep the grading crew at least two days ahead of the concrete workers and one day ahead of the roller. On another job, the grading crew may be one day ahead of the concrete workers. When erecting a reinforced concrete building, the form and steel crews should be several days ahead of the concrete workers.

In building work, the rough plumbing, wiring, and ventilating should be installed as soon as the rough framework of the building is completed. The finished plumbing, heating, and ventilating work should be installed after the completion of the finished walls, floor, and marble work. Interior finish such as wood trim, flooring, and glazing isn't installed until last. After the finish walls and ceilings are completed, the building should be cleaned and kept clean thereafter.

Planning Subcontracted Work

When you've signed the building contract, it's time to let the subcontracts, especially those for work needed almost immediately. Among these would be stone, steel and iron work. In these cases, work must be done on the materials before they can be delivered for installation in the building. The 1/8" or 1/4" scale drawings, with larger scale plans and sections, are adequate for taking off quantities. Sign these contracts as soon as possible so the subcontractors can buy materials and schedule their own labor.

Prepare a time and work schedule for all the

work to be done by subs. This may be a separate schedule or it may be included with the general job schedule. The time and work schedule for the subs should show the kinds of work to be done, the names of the subs, the dates for starting and finishing the work, and notes regarding the rate of progress and other details.

Planning Materials

Once you've laid out what *must* be done to complete the work on time, get in your orders for materials. Request deliveries on specific days or ask if delivery can be made on 24 hour notice after some set date.

Price isn't the only consideration when buying materials. Think about payment terms, returns policy, delivery cost, promises of cooperation, selection, and hundreds of other factors that can make your job either easier or a headache.

Good material deliveries arrive at the job in the right order, in the correct amounts, and at the right time. Foundation forms should be at the top of the lumber pile, not at the bottom. Rafters should be at the bottom. And have the delivery made as close to the point of installation as possible. Your crew will make a hundred trips to the lumber pile. Make those trips as short as possible.

Of course, materials must meet specifications or they're not the right materials at all. A good buyer can save you a mint. Have someone who's familiar with prices, dealers, delivery costs, lead times needed, how to get cooperation from dealers, tracing of shipments, and the reliability of local producers.

The cost of delivery will average about 3% of the cost of materials on many jobs. Usually that cost will be at least partially hidden because your suppliers will absorb delivery expenses on orders that exceed a certain dollar volume. Be sure to get "delivered prices" whenever possible.

Be aware that an F.O.B. price can be deceiving. F.O.B. means only "freight on board." It doesn't say on board what and where. It may be F.O.B. the manufacturing plant, F.O.B. at some railway siding, or on a truck at a job site. And is that F.O.B. price for a carload, less than a carload, crated or uncrated? Someone in your company has to know about minimum quantities, extra prices for small quantities, odd sizes, special deliveries, discounts from list prices, returns policies, restocking charges, and discounts for prompt payment. Like I said, a good buyer can save you a mint.

Give thought to the overhead work required in checking purchases, testing materials, insurance, checking shipments, tracing shipments during transportation, unloading, storage, and guarding stockpiled supplies. Those are all real costs and you have to pay them.

On larger jobs, it's a good idea to contract for all materials before the work starts. When practical, make some allowance for variations in quality, quantity and delivery. It's common for job conditions and requirements to change after work starts.

Watching for Price Changes

When prices are flat, don't stockpile materials beyond the immediate need. Contract for materials a reasonable time in advance. Then take the discount for prompt payments if your finances permit it. A 1% discount for paying in 10 days is like earning 36% interest on your money.

When market prices are rising, buy whatever you can store and pay for. Try to get price commitments from suppliers. If you can't, your contract should provide escalation protection — any rise in material prices can be recovered as a contract extra cost. Don't get caught in the middle, guaranteeing to do the work at a set price, but unable to get suppliers to quote firm prices for future deliveries. No building contractor has to do that. It's pure gambling. You're a builder, not a commodities speculator.

Prepare a material time schedule for all medium and large-sized jobs. It can be a simple tabular or graphic form, but it should show the kind of materials, total quantities of each kind, delivery dates for each kind, and amounts of each kind to be delivered on each date. It may also show dates of first delivery, and minimum amounts of each kind of material to be kept on hand at the job after the first delivery date, and until all the required material is used.

Consulting with the Architect

If an architect is involved in the job, he or she furnishes the initial drawings and specifications, along with any additional drawings, details, and directions required as the work proceeds. Shop drawings will be done by others. The architect usually has the authority to make decisions concerning all the contracting parties and on all controversies arising under the contract.

Many disputes arise because of inadequate specifications. It's the duty of the architect to leave no uncertainty about the work you have to do, the quality of that work, and how much money the owner must pay for it. If your architect isn't doing that, he or she isn't doing an adequate job.

If, after studying a drawing or specification, you see anything that's wrong or not clear, or anything that can't possibly be, or a conflict between the plans and specs, resolve the problem with the architect. Failure to do so can lead to an expensive lawsuit or arbitration.

Superintending the Job

All the planning and ordering should be done before work starts. Plan the work well in advance. Sketch storage and equipment areas at the site. Complete material orders and decide on delivery dates. Get all the necessary permits and let (or be ready to let) subcontracts. Be sure all discrepancies or errors in the plans or specs have been resolved with the architect.

Every time you visit the job, look, think, and ask questions. Check what's being done by your employees and subcontractors. Look at the materials on site to be sure that what's needed is on hand and in good condition. The last thing you want is a crew waiting for materials or a foreman chasing down parts at the local hardware store.

But looking isn't enough. Get a feel for the construction pace. Are workers busy or are they waiting on someone to complete a task? Are they idle because of a lack of lumber, block, nails or mortar? Is there duplication of work or material handling? Nonproductive hands are lost profits. A construction site is a production line. If the momentum in the line is broken, dollars are being wasted.

If work is falling behind schedule, your first step is to find out why. Then start correcting the problem.

Planning to Reduce Costs

Plan from the beginning to minimize costs. Don't wait until construction begins. That's too late.

As your business grows, you'll find that management, financing, and getting new work draw you away from practical construction problems. If you don't have the time to watch for wasted labor, material and effort, delegate the responsibility. Someone has to plan the job so equipment is used efficiently, site layout is appropriate for the work, and wasted effort is kept to a minimum.

The next time you're on another contractor's job, notice the crew sizes that contractor uses. Usually, the larger the crew, the faster the work

gets done, but the more manhours are wasted.

As a rule, the most efficient crew is the smallest crew needed to do the job. Often that's one man and seldom more than two. A slab pouring and finishing crew and a framing crew are obvious exceptions. But many larger crews are really composed of two or more crews working separately most of the time and together less often. Don't let your crews add an extra man or two. That's sure to be manhours wasted.

On one job I visited recently, two laborers were keeping several bricklayers supplied from material piles fully 150 feet from the hoist. I was on the job two days later and found one laborer working where two had been before. The difference: the material piles had been moved to an area immediately beside the hoists. Someone had the sense to make a small change — and save at least $100 a day in wasted labor. Your projects offer the same type of savings, if someone is alert to the opportunities.

True, every job is different. And there's no single correct way to lay out your material and equipment storage areas. But there's always a better way to get the job done. Find it! Experiment. Keep trying new concepts. If it works, do more of the same. If it doesn't work, try something else. Let experience be your teacher.

Personally, I'm always looking for new labor-saving equipment: lifts, hoists, concrete pumps and cranes. Let machines reduce your labor cost when possible. Power equipment is expensive, of course, but your rental yard probably has a good selection of tools and equipment that can cut labor costs. If you know what equipment is available, the kind of work to which they are best suited, and the cost, you can judge where savings are possible. As an efficient builder, your costs will be lower. You can bid lower and get more profitable work.

The time to do this analysis and choose the best building methods is when you're figuring the costs. Without some thinking about organization and equipment, any cost estimate is guesswork. What equipment will be used? How will it be placed? How many operators are needed? How many days will the equipment be needed and where will it be stored? These are the practical questions that you should answer in advance. Ignore questions like these and you'll watch costs soar and profits dwindle. On many building sites you can simplify supervision and scheduling and reduce construction time and costs by using prefabricated components. An obvious step is buying precut lumber. Some builders have found it pays to have all lumber precut at the yard. And many yards offer this service at a reasonable charge. Larger yards offer a wide range of built-up trusses, prefabricated sections, and other components that require less fabricating and erecting on the site.

Many spec builders have found they can build more houses with a work force of a given size by using entire prefabricated sections. Construction then is mainly site preparation, erection of the prefabricated sections, and interior finishing. Two basic scheduling requirements will have to be met:

1)   Sites must be ready for the new sections as they are delivered from the factory.

2)   Your construction crew must be kept busy at a steady pace as it moves from one site to another. Using fabricated sections, you'll need a higher volume of work per year to keep your crews fully occupied.

As an independent builder, you can't afford to gamble thousands of dollars and hundreds of manhours on unproven methods or materials. You'll have to decide whether the time is right for use of a new product or new method. My advice is to avoid being the first to use new materials or equipment. But that doesn't require that you ignore new developments. Stay informed. If something is working for others, be ready to adapt it to your business.

Planning for Profits, Design Ideas for Professional Remodelers and Home Builders, is a service offered by the Georgia-Pacific Corporation. Get more information, including the current subscription price, by writing to:

Director, Marketing Communications
Building Products
Georgia-Pacific Corporation
133 Peachtree Street, N.E.
Atlanta, GA 30348

Record Keeping

Hand-in-hand with good planning goes accurate record keeping. Good records help build a profitable business. Sloppy or non-existent records will ruin you.

Every new contractor starts with an office in his home or car. Today you may need only a file box in the trunk of your car. As your company grows, you'll need more file space and equipment: a phone and answering machine, typewriter, forms, file cabinet, etc. Eventually, you'll want a small office of your own. How much rent you can afford will determine the office size and location. Consider also whether customers will need to visit your office.

Keep office expenses as low as possible. The contractor with the lowest overhead can bid the lowest and get the most work. Used but functional furniture and equipment will be better than new. Remember, the important thing about your office, whether it's in your home or elsewhere, is that it be functional. Very few contractors need an office that looks like a Beverly Hills showroom.

Office Personnel

Who'll look after your office? Many owner-managers of a small building business will have no office personnel other than family members. They answer the phone, take care of routine correspondence, keep records up-to-date, and handle the bookkeeping and accounting.

If you don't know bookkeeping, have a public bookkeeping service set up your books. They'll install whatever system you need and train you or your spouse to keep the necessary daily records. The accountant will spend only one or two days a month bringing your books up to date.

Regardless of how you do it, keep adequate records. Both state and federal tax laws require it, and good sense demands it. Records are the footprints left by money as it wanders in and out of your business. Keep track of those footprints and you're on the trail to better profits.

What Kind of Records?

Back in Chapter 2 we talked about cost records and how they can help you keep costs in line and make accurate bids. In this chapter, we'll concentrate on the financial records that show you how healthy your company is.

Here's the key to record keeping: Spend just the minimum time needed to keep records that are just adequate for your needs. Don't spend one minute more than needed on bookkeeping — but don't spend one minute less either. As a minimum, have the records needed to make up annual operating and financial statements (profit and loss statements) and balance sheets. You'll need these to file tax returns.

You'll also need these statements when you want to borrow money. No bank or S&L is going to lend a cent without seeing your business records. Banks want to see profits so they know you can pay them back. Without a profit and loss statement, no business looks creditworthy.

Suppose next week you plan to bid on a certain job — or need a bond to qualify to bid. You, or your accountant, should be able to prepare a current financial statement. Among other things, your buyer, lender and bond underwriter will want to know the ratio of your company's outstanding debts to its net worth. We'll talk about that a little later. Without a current financial statement, that ratio is only speculation.

Your Cash Schedule

One of your chief responsibilities as manager is to see that your company has the cash it needs when bills are due. After you've estimated how much you'll need, say for the next month, use your estimate sheets to keep control of your cash. You may want an accountant to set up a system that will track your cash flow — to anticipate problems before they become a crisis.

All cash received and all cash paid out in your business must be recorded. You do this by posting the items to the appropriate journal. It might be possible for you to combine several books into a single combined journal or cash book. Ask your accountant about simplified record-keeping systems.

A Word of Caution

Regardless of how you set up your books, remember that record keeping is like score keeping in a baseball or basketball game. The performance of your company is being written in numbers every day. If you don't collect those numbers and take the time to understand what they mean, you don't understand who's winning and who's losing.

Remember also that records become more important as your company grows. Management (that's you) has to rely more and more on what the numbers say and less and less on firsthand observation. A small company that has a good record-keeping system is more likely to grow into a larger company. A small company that has no records or only a poor record-keeping system isn't likely to survive much beyond the next tax audit.

Don't wait until the lumber yard calls to report

that your check bounced. Have an accountant set up a simple but efficient bookkeeping system — and then follow the system like your business depends on it. It probably does.

Don't Fail to Act

One more thought: Records must be used, not just kept! Read and understand what those numbers mean. Compare this month's results with last month's results and the results from this month a year ago. If the trend is positive, understand why. Do more of what you've been doing. If everything's going to pot, figure out why while there's still time to make changes. When the reports show unexpected gains, figure some way to make these improvements a permanent part of your operation.

What Your Records Can Tell You

Most builders know construction pretty well. They wouldn't be in business if they didn't. But few have the accounting, legal and financial background needed to administer a large and growing business. Unfortunately, every construction company has to face some pretty serious problems:

1) What markup percentage do we require? How do we allocate overhead expense among jobs in progress so we can tell where we're making and where we're losing money?

2) Who do we grant credit to and how much? What do we do when they don't pay?

3) How much work in progress can we carry without running out of working capital? How much equipment can we carry?

4) There's only so much money available. Where should we invest it (land, advertising, executive salaries, equipment) to get the best return?

Your basic guide in making decisions like this will be the current balance sheet and profit and loss statement. Any successful builder needs to know how to interpret these documents — and the best way is with ratio analysis.

A balance sheet tells how a business stands at one given moment. It's like a snapshot. A profit-and-loss statement sums up the results of operations over a period of time.

In themselves, these two types of financial documents are a collection of lifeless figures. But when you can interpret and evaluate them, they begin to talk.

A single balance sheet is like the opening chapter of a book — it gives the initial setting. Thus, one balance sheet will show how the capital is distributed, how much is in the various accounts, and the balance of assets (what you own) and liabilities (what you owe).

A single profit-and-loss statement shows sales volume for a given period, the money spent to earn those sales, and the amount left after paying all the bills. That's your profit.

But try this. Take 12 month-end balance sheets for 12 consecutive months. Arrange the figures in 12 vertical columns across a page so you can compare asset and liability account totals for each month in the year. You'll begin to see trends in the figures. Then compare the month of July for this year and last year. Suddenly you begin to see how things are changing — and the direction of the change. That's important information for anyone running a business.

Comparing balance sheets isn't like looking at a snapshot or X-rays. It's like seeing a movie. Things are happening and you can guess why. Similarly, comparing profit-and-loss statements will show changes in expense categories and the profit margin. Did you cut prices to meet competition? Then look for a lower gross profit — unless construction costs were reduced proportionately. Did sales go up? If so, what about expenses? Did they remain proportionate? Was more money spent on office help? Where did the money come from? How about fixed overhead? Was it controlled? By comparing operating income and cost account items from one period to another, you'll find the answers.

Comparing Operating Ratios

Operating ratios (your balance sheet and expense ratios) have many uses. You can use them to analyze collections, check the condition of finances, and pinpoint potential or actual problems in expense categories. Even better is to compare ratios for your company with ratios for other companies, or the construction industry taken as a whole.

The figures that follow let you do exactly that: compare your company with industry average. Be aware that these are *averages.*

The first question, then, is this: Do you want to be just average? Your objective should be to adjust

operations so that you are at least as good as, but preferably better than, the typical construction company.

When dealing with balance sheet ratios, you want to be better than average, of course. But you'll be reassured to know that the averages may represent a margin of safety. If your organization's key ratios are close to the proportions reflected by standard ratios, you're not likely to get into trouble.

A construction company may, of course, go below the average in certain ratios. Its investment in fixed assets (I'll define these terms shortly) may be above average, for instance, but be offset by a high degree of liquidity of current assets. Or the fixed assets may be comfortably financed on a long-term basis. Perhaps a low rate of turnover of tangible net worth and working capital is due to an existing surplus of capital. A builder doing $500,000 annual sales volume on a capital of $100,000 would show higher turnover, and somewhat greater financing problems, than a builder doing the same volume with $200,000 in capital. The latter construction company might even show an above-average ratio of fixed assets to tangible net worth, simply because it hadn't replaced its equipment with newer, more costly, but more efficient machinery.

So take the averages I'll give you here as guidelines, but not as a map of how your business should operate. Your company is unique. Only you can say if your ratios make sense. But do take the time to read this material and work out the ratios for your business. Compare your ratios at set intervals to spot the trends. There's no better way to evaluate your company's financial health.

Interpreting Expense Ratios

The expense ratios in Figure 15-1 are based on figures compiled and published by the Internal Revenue Service, U.S. Treasury Department, and major lending institutions. They apply to small contracting operations with sales less than $1,000,000 per year. Expense ratios are always taken as a percentage of total revenue (sales). Use these average ratios to evaluate the condition of your company at nearly any stage of its development. List the averages on a worksheet, and opposite each figure compute the ratio for your company. After making the comparison, note the items that are seriously out of line.

But don't try to take drastic steps to bring your

ratios in line with the averages. Instead, think about why your ratios are different. Your expense ratios might be out of line with Figure 15-1, for instance, because of special services to customers, compensated for by higher prices. Or a good sales program might bring a better-than-average margin of gross profit. So consider each item in the context of your business.

	General contractors	Sub-contractors
Sales (contract revenues)	100%	100%
Materials and subcontracted work	44	44
On site wages (excluding owner)	19	23
Gross profit	37	33
Controllable expenses		
Off site labor	2	1
Operating supplies	3	2
Repairs and maintenance	1	1
Advertising	1	1
Auto and truck	3	2
Bad debts	Less than .5	Less than .5
Administrative and legal	.5	.5
Miscellaneous expense	2	1
Total controllable expense	13	9
Fixed expenses		
Rent	1	1
Utilities	1	1.5
Insurance	1	1
Taxes and licenses	.5	1
Interest	Less than .5	Less than .5
Depreciation	2	2
Total fixed expenses	6	7
Total expenses	19	16
Net profit	18	17

**Typical expense ratios for small contractors
Figure 15-1**

Once you've analyzed these comparisons, work out corrective steps to bring expenses into balance with sales. These don't necessarily have to be negative steps. You may see a way to improve your showing by *adding* expenses. For example, maybe your advertising budget was too low. An increase might bring in more income.

Gross profit will rise or fall because of cost variables in materials, labor and subcontracted work. If your gross profit is seriously out of line with construction industry averages, it's a good idea to examine your labor and material costs and bidding practices.

Interpreting Balance Sheet Ratios

If you're like most builders, you probably focus on your income account (sales) when you're trying to increase profits or reduce losses. But a careful review of your balance sheet may show other areas that need your attention. Compare the distribution of assets and liabilities in your company with the way assets and liabilities are distributed in other companies. That should open your eyes to what you're doing right and wrong. It may show you ways to change the financial structure of your company and improve operating performance. Figure 15-2 shows a typical balance sheet and income statement.

Even if your balance sheet shows signs of progress, don't rationalize a bad situation by saying, "We just don't have enough capital," or even "Let our suppliers carry us for a while; look at the business we give them." All too often, a cash shortage isn't the result of poor sales. It's a sign of an unhealthy situation that threatens the business. No matter how little cash and working capital you have available, there's some level of business you can maintain with a reasonable cushion of safety.

A careful study of your balance sheet may show slow collections, an ill-thought-out expansion program, or too much money tied up in equipment or vehicles. These are all conditions you can change to improve your company's performance.

All construction companies experience peaks and valleys in business activity. Your business should be able to adapt to changes in prices and sales. There's always a chance of trouble developing from some unforeseen event such as a material shortage, labor dispute, or a string of bad weather that delays completion of a key job. If your liabilities are too heavy, you may not be able to weather the storm.

Here are brief descriptions of the key balance sheet ratios, including average ratios for general and subcontractors.

Current assets to current liabilities: Widely known as the *current ratio,* this is one test of solvency. It measures the liquid assets available to meet all debts falling due within a year's time.

For each ratio, I'll give the formula for working it out, and an example based on a typical small construction firm. Of course, you need to substitute your own figures from your records.

Middleville Construction Company

December 31, 19—

BALANCE SHEET

Cash	$ 948	Notes payable, bank	$ 7,000
Notes receivable	2,438	Notes receivable, discounted	2,421
Accounts receivable	48,728	Accounts payable	76,120
Work in progress and inventory	78,411	Accruals	2,720
Total current assets	$130,525	Total current liabilities	$88,261
Depreciation, land and buildings	23,129	Mortgage	5,000
Equipment and fixtures	5,729	Total liabilities	$93,261
Prepaid expenses	639	Net worth	66,761
Total assets	$160,022	Total liabilities and net worth	$160,022

INCOME STATEMENT

		Dollars	Percent
Net sales		$363,558	100.00
Cost of production		308,334	84.81
Gross profit on sales		$ 55,224	15.19
Operating expenses:			
Wages	$30,689		8.44
Delivery	4,698		1.29
Bad debts allowance	1,807		0.49
Depreciation allowance	3,600		0.99
Total expenses		$ 40,794	11.21
Net profit before other charges		$ 14,430	3.98
Other charges:			
Interest	$ 350		0.09
Nonoperating charges	500		0.13
Drawings	7,490		2.06
Provisions for taxes	4,328		1.19
Total other charges		$ 12,668	3.47
Net profit for period		$ 1,762	0.51

Typical balance sheet and income statement
Figure 15-2

Here's the first one:

$$\frac{\text{Current assets}}{\text{Current liabilities}} = \frac{\$37,867}{\$19,242} = 1.97 \text{ times}$$

Current assets are those normally expected to flow into cash in the course of one year. They usually include cash, notes and accounts receivable, and the value of work in progress. While some firms may consider items such as the cash surrender value of life insurance as current, it's probably wiser to post them as noncurrent.

Current liabilities are short-term obligations due for payment within a year: notes and accounts payable for materials, loans payable, short-term installment loans, taxes, and contract advances received but not earned.

Most small general contractors have a current ratio ranging from 1.1 to 1 through 1.5 to 1. More solvent firms have about twice as much in current assets as current debts. Current assets are normally 50% to 60% of total assets. Cash should be about 6% of assets.

Subcontractors should have a current ratio of 1.5 (assets) to 1 (liabilities). Even 2.5 to 1 wouldn't be excessive. Current assets should be 60% to 80% of total assets, and cash should be about 10% of total assets for subcontractors.

Current liabilities to tangible net worth: Like

the current ratio, this is another means of evaluating financial condition by comparing what is owned to what is owed.

The formula looks like this:

$$\frac{\text{Current liabilities}}{\text{Tangible net worth}} = \frac{\$19,242}{\$33,970} = 56.6 \text{ percent}$$

Tangible net worth is the total asset value of a business, minus any *intangible* items such as goodwill, trademarks, patents, copyrights, and leaseholds. In a corporation, the tangible net worth would consist of the sum of all outstanding capital stock, preferred and common, and surplus, minus intangibles. In a partnership or proprietorship, it could include the capital account, or accounts, less the intangibles.

A word about intangibles: In a going business, these items frequently have a substantial but undeterminable value. Until these intangibles are actually liquidated by sale, it's difficult to evaluate what they might bring. In a profitable business up for sale, the goodwill conceivably could represent the potential earning power over a period of years, and actually bring more than the assets themselves. But many small businesses have no way to cash in on their goodwill.

Many general contractors have current liabilities which are 140% of tangible net worth. A more reasonable figure would be about 80%. General contractors who subcontract less of their work, and most subcontractors, usually have current liabilities of about 70% of net worth.

Turnover of tangible net worth: Sometimes called "net revenues to tangible net worth," this ratio shows how actively invested capital is being put to work by indicating its turnover during a certain period. It helps measure the profitability of the investment. Both overwork and underwork of tangible net worth are considered unhealthy.

The formula uses net sales for a year and the tangible net worth:

$$\frac{\text{Net sales (year)}}{\text{Tangible net worth}} = \frac{\$189,754}{\$33,970} = 5.6 \text{ times}$$

Find the turnover of tangible net worth by dividing the average tangible net worth into net revenues for the same periods. The ratio is expressed as the number of turnovers in the given period. General contractors should "turn" their net worth about 5 to 5½ times a year, and subcontractors should average 6 to 6½ times.

Turnover of working capital: Also known as the ratio of "net sales to net working capital," this ratio measures how actively the working cash in a business is being put to work in terms of sales. Working capital or cash are assets that can readily be converted into operating funds within a year. It doesn't include invested capital. A low ratio shows unprofitable use of working capital; a high one, vulnerability to creditors.

Here's the formula:

$$\frac{\text{Net sales (year)}}{\text{Working capital}} = \frac{\$189,754}{\$37,867 - \$19,242} = 10.1 \text{ times}$$

To find the *working capital,* deduct the sum of the current liabilities from the total current assets. Working capital includes the business assets which can readily be converted into operating funds. A builder with $100,000 cash, receivables, and work in progress, and no unpaid obligations, would have $100,000 in working capital. A business with $200,000 in current assets and $100,000 in current liabilities also would have $100,000 working capital. Obviously, however, items like receivables can't be liquidated overnight.

Most businesses require a margin of current assets over and above current liabilities to provide for stock and work-in-progress, and also to carry receivables after the work is finished until the receivables are collected. General contractors should "turn" their working capital about 14 times a year, while subcontractors average 6 to 8 times a year.

Net profits to tangible net worth: As the measure of return on investment, this is one of the best measures of profitability, often the key measure of management ability. Profits after taxes are the final source of growth. If this return on capital is too low, the capital involved could be better used elsewhere.

You divide net profits for a given period by tangible net worth for that period:

$$\frac{\text{Net profit (after taxes)}}{\text{Tangible net worth}} = \frac{\$5,942}{\$33,970} = 17.5 \text{ percent}$$

The ratio is expressed as a percentage. It relates profits actually earned in a given length of time to the average net worth during that time. *Net profit* means the revenue left over from sales income after all costs have been paid. These include costs of goods sold, writedowns and chargeoffs, federal and other taxes accruing over the period covered, and whatever miscellaneous adjustments may be necessary to reduce assets to current, going values.

General contractors should show a profit after taxes of at least 12% of net worth. A 20% profit is a reasonable expectation. The top 10% of the industry shows about a 40% profit on net worth. Subcontractors should show a profit of at least 15% of net worth, although some reach 35% or more. Small concrete and electrical contractors have averaged above 20%, but contractors in the mechanical trades have averaged less than a 15% profit on net worth in many years.

Fixed assets to tangible net worth: This ratio, which shows the relationship between investment in land, office and equipment, and the owner's capital, indicates how liquid the net worth is. The higher this ratio, the less the owner's capital is available for use as working capital, or to meet debts.

To find the ratio:

$$\frac{\text{Fixed assets}}{\text{Tangible net worth}} = \frac{\$15,345}{\$33,970} = 45.2 \text{ percent}$$

Fixed assets are the sum of assets such as land, buildings, leasehold improvements, fixtures, furniture, machinery, tools and equipment, less depreciation. General contractors should not have more than 25% of their tangible net worth invested in fixed assets. However, heavy construction contractors needing heavy equipment on a daily basis may often have 50% of their net worth invested in fixed assets.

Subcontractors who don't need expensive equipment should have no more than 25% of their tangible net worth invested in fixed assets.

Net profit on net sales: This ratio measures the rate of return on net sales. The ratio is expressed as a percentage which indicates the number of cents of each sales dollar remaining after deducting all income statement items and income taxes. There's a variation in which you divide net profit by net sales. This ratio reveals the profitability of sales — that is, the profitability of the regular operations of a business.

Many people think a high rate of return on net sales is necessary for a business to be successful. But this isn't always true. To evaluate the significance of the ratio, consider such factors as the value of sales and the total capital employed. A low rate of return compared with rapid turnover and large sales volume, for example, may result in satisfactory earnings.

Here's an example:

$$\frac{\text{Net profit}}{\text{Net sales}} = \frac{\$5,942}{\$189,754} = 3.1 \text{ percent}$$

Most general contractors and subcontractors show a net profit on sales of between 1% and 2%. An exceptionally profitable operation would show about a 4% profit. These figures may be misleading because "profit" in a small company often disappears into the owner's pocket before the final figures are prepared for each year.

Knowing Where You Stand

Understanding and using the management techniques in this chapter will help you control your construction business — instead of being controlled by it. Skilled workers and the finest materials don't necessarily make a success. You have to manage these resources to keep them working for you.

Know where your business stands in comparison with others in the same field. Search for ways to improve your profits. Is your expense ratio too high? Find out why and take steps to bring it into line. If your balance sheet shows high labor costs, correct the problem or suffer loss of profits.

As stated at the beginning of this chapter, organizing, planning, and careful record keeping are absolutely essential if you want to keep your company thriving. Without giving these adequate attention, no matter how good your building skills, your chances of operating a successful construction company for long are slim. Whatever you do, don't let the act of building, which you probably get great satisfaction from, lure you away from these management responsibilities, which will take many tedious hours. Successful management is a continuing process. You'll never complete the job. But treat it as though your business depended on it.

It probably does.

If you don't feel you have the necessary management skills, find someone who can help. Remember, running a construction company is a complex, highly competitive business. Your management ability probably counts for more than your trade skills.

Chapter **16**

Finding the Work to Stay Busy

It's not enough to be a good builder and a good manager. You also have to be good at selling your services. Business seldom just walks in the door — at least not good business, the type every contractor likes to have in abundance. You have to look for it, promote it, keep your company name before likely prospects, and then convince the decision-maker that you can do the work at a fair price.

Don't think that you can build a successful construction company with little or no selling effort. Very few builders survive in the competitive construction industry by being the lowest bidder on every project. And no one gets all the high-margin work they can handle without looking for it.

Most contractors have learned ways to identify likely prospects and keep their name and phone number in front of those prospects. They've discovered what it takes to persuade a prospective client to accept their proposal, even if their bid was not the lowest — and sometimes without any competitive bidding at all. That's usually good work, if you can get it, and that's what this chapter is all about — getting all the good, profitable work you can handle.

Let me make a point before we begin. First, there's an economic cycle in this country that rewards those who can anticipate it and punishes those who can't. If you've been in the construction industry for more than about 10 years, you can recall times when every builder in town was hip

deep in work. There just weren't enough builders and tradesmen to do all the construction that owners wanted done. At times like that it doesn't take much selling to stay busy. Basically, you're just taking orders and can pick and choose what you handle. But don't count on the "up" times to last very long. They're great while they last — if material and labor shortages don't drive you up the wall, and if big increases in labor and material costs don't force your back to the wall, and if you price your work correctly. But selling construction work in the good times is like shooting fish in a barrel — no special skills required.

Then there are the times when nearly all construction dries up. It's a buyer's market. Nobody's buying, remodeling or building. There are about 100 bidders even on small jobs that would have gone begging in better times. Builders who were overextended or who thought the good times would last forever are forced out of the industry or go belly up. But if you've cultivated a few loyal customers, found some good, long-term, steady work (even at low profit), have paid down your debts and trimmed your overhead, you'll sail right through to better times.

Selling is pretty easy at the top of the economic cycle. And at the bottom, even the best sales team in the business won't set any records. But these times are the exceptions. In between the top and the bottom, selling will make a big difference in your company.

Let me make one additional point before I get into the meat of this chapter. You'll hear claims by some contractors that they never do much selling. They say their reputation for craftsmanship and fair dealing are all they need to stay busy. I won't claim that isn't true. In time, of course, every builder can build a good reputation. It's your most valuable sales asset. Certainly it's the most valuable asset I've built in my years in the industry. But no one starts out with a great reputation. Even after your business is well established, a good reputation alone won't be enough to keep your firm growing in a weak construction market. Moreover, maintaining and spreading your reputation is the essence of good salesmanship. The contractor who claims he relies on his reputation to bring in business is probably doing everything possible (and doing it very well) to enhance and spread his reputation. What's that but an excellent form of advertising.

Start with the proposition that you have to keep advertising and selling to keep growing in the construction business. That's probably as true for a one-man construction company as it is for the nation's largest engineering and construction firms. Whether you do it yourself or have a marketing department that handles nothing but promoting your services, selling is an important part of every construction business.

For the speculative builder, anticipating what buyers want and will buy is the key to success. That's a form of marketing usually known as market research. If you specialize in modernization and repairs, you'll need an aggressive advertising campaign and a continuous sales effort. Custom builders usually spend less on promotion. Getting to know some local architects and building your reputation with them will be a primary focus of your sales effort. But even in the custom building field, you'll need to do some advertising and active selling. An "institutional" campaign, aimed at creating goodwill and prestige for your company rather than making a specific sale, will usually pay dividends.

The Basics of Advertising

It pays to advertise — you've heard that often enough. But the question is, how *much* does it pay? Or, looking at the other side of the coin, how much should you spend on advertising?

A figure of 2% of gross sales is frequently quoted as a guide. For a spec builder putting up ten $100,000 houses a year, that's $20,000 a year for advertising. In practice, though, few builders grossing $1,000,000 a year spend that much. There's a strong temptation to skimp on the advertising and promotion budget.

That's particularly true for a new company with limited capital. But as a new firm, you need a well-planned campaign to sell your products or services, even more than an established firm does. Dollars spent in advertising may save more dollars later in interest on the investment in unsold houses or in idle crews.

Successful selling requires knowledge and training. By this time, you've come to realize that you need to consult a lawyer on legal matters, an architect for plans, and an accountant to keep your records in order. If you feel like selling isn't your strong point, maybe it isn't. Some builders find that hiring a specialist in sales promotion produces better sales results for each dollar spent than designing and carrying out their own sales campaigns.

But even if you hire specialists to design brochures, or a public relations firm to promote your company name, the information in this chapter will be valuable. You need some way to judge the value of the services you're buying. This chapter should provide all the basic education you need in advertising and promotion.

Planning Your Advertising

Advertising is simply a form of selling by means of publicity. Its goal is to make the public familiar with your name and the service you offer, and to associate that service with your name. Advertising can also make personal sales efforts more productive and create goodwill. It's an economical means of keeping in touch with clients and prospective builders when it would be impossible to do so by personal calls. It brings in inquiries, allowing you to build a prospect list of potential customers.

One of the first decisions you have to make is about the type of client you want. What type of work do you want? Will you try to build a reputation for high class work, or high speed at low cost? Or will you go after any class of work that may come your way? Will you accept repair work, or limit yourself to new construction? In what way does your service differ from that of your competitors?

Considering the class of work you prefer, what

are the possibilities in your vicinity? Are your opportunities purely local, or do they extend over a large surrounding area? How will the seasons affect your work?

Has the market for building been limited in your territory for any reason in the past? Are there people who can use your services who don't know it yet? Are there undeveloped markets where opportunities can be created for your class of work? What is your competition? How strongly entrenched is it? Does the competition offer what you have to offer — the experience, the training, the knowledge of plans and specifications, and familiarity with materials and newer construction ideas? How does the quality of their service compare? Consider all of these questions. They'll suggest areas where you can effectively advertise.

There's a definite advantage in selecting one or two points about your service and emphasizing these regularly in your advertising. What do you want prospective clients to think about when they hear your company's name? What features of your service are most important? Perhaps it's your ability to handle a contract from the preparation of plans to the final erection of the building. Maybe you have at your command the services of some specialist in interior design. Use these or any other "bonus points" to identify advantages to potential clients.

Effective Forms of Advertising

Of course you can't use every form of advertising that's available. Select the form of advertising that best suits your community, your advertising budget and your type of business. Try to visualize your prospects. What type of advertising will reach them most economically? Here are some of the possibilities:

- The Yellow Pages
- Newspapers
- Local radio and TV stations
- Billboards and posters
- Bus stop benches
- Direct mail letters
- Advertising novelties (calendars, cards, matchbooks, pencils, rulers)

Custom builders and builders specializing in modernization do most of their advertising through the Yellow Pages, newspapers, local TV and radio, and personal contacts. Find out what other successful builders in your area are doing and adapt their methods to fit your business. The one that proved most successful for me was a Yellow Pages display ad. But that was in *my* town. The picture in yours is probably different. We'll cover some advertising ideas for spec builders in the next chapter.

Building a Prospect List

Every contact generated by your advertising is a prospect, even if there's no chance of an immediate sale. And prospects are gold in the building business. Keep a record of every name you find; they are all potential sources of business. Build up as large and as complete a mailing list of prospects as possible. Carefully keep the name of every man and woman you have any hope of ever doing business with at any time. Keep a list of your past customers: they're your best source of prospects.

Business won't come to you without invitation. If you want more business, you must go after it, work for it, earn it. Use your imagination in developing possible outlets for your services.

The Basics of Selling

Effective advertising brings in the prospects — but that's all it can do. It takes a personal selling effort to change a prospect into a client.

Here's something you should know up front: the old idea that selling is a mysterious gift just isn't true. While there may be "natural-born salesmen," anyone can sell, even you, if you're willing to study the principles and apply yourself.

The chief problem in selling is to persuade the prospect to feel the way you do about your proposition. The most practical way to do this is with good common sense, plain honest talk, and a sincere desire to solve your prospect's problems. It's a matter of rendering personal service, and not necessarily a question of a highly persuasive personality.

Routine and Creative Salesmanship

Selling can be divided into two categories: routine salesmanship and creative salesmanship. The former is simply a matter of filling an existing want. When a homeowner telephones you to ask the cost of adding a room and your price quote gets the job — that's routine salesmanship. You'll get a few jobs that way. But will just a few jobs feed your family, pay your bills, and make your business grow?

Here's what the creative (and effective) salesman would do in the above situation: Make an appointment to see the homeowner, bringing a plan for a room *and bathroom* addition ideally adapted to his property. He'd point out how it would add to the appearance of the house and increase the property value. He'd suggest the extra room could be rented out if necessary and would soon repay the investment in any event. Closing *this* sale would be creative salesmanship.

Routine salesmanship is handing out what a customer asks for. Creative salesmanship is educating him to want what you think best for him. The true salesman *creates* wants so compelling that they overcome any other want at the moment. You do this by talking the prospect's language.

People don't buy *things,* they buy *benefits.* We don't buy an automobile — we buy travel, convenience, the open road, ease of operation, low upkeep, ego gratification. We don't buy a vacant lot, we buy a potentially profitable investment or a prospective home site. And we don't buy a house. We buy a home, plus the opportunities for gardening, decorating or whatever means home and comfort to us.

You Sell an Essential Service
Before you can sell a prospect, you have to sell yourself — that is, convince yourself that you're selling something you can be proud of. Have a high regard for your occupation. Remember, without the builder, people would have to go back to living in caves. And there aren't that many caves left.

After you've satisfied yourself that you have something worthwhile to sell, go on to the next question: "Why can I give this service better than the other builder?" If you realize that another firm can build a better building and do it cheaper, or give better service to the prospect than you can, you have two choices. Equip yourself so you can provide a similar service, or get into another line of building where you could give reasonable or better service.

The Process of Selling
Before you can make the sale, there are five steps you have to take to bring the prospect around to your way of thinking:

1) Secure his attention.

2) Develop this into interest in your proposition.

3) Bring the prospect to believe in the value of what you're offering.

4) Arouse a desire to possess it.

5) Convert this desire into a decision to buy.

Before any of these can happen, you must approach the prospect and bring your proposition to his attention. With this in mind we can actually divide the complete selling process into the preparation, the approach, the interview (or presentation), and the close. Throughout these four steps, you'll make various arguments mingled with proofs, objections, reassurances, and human appeal.

The Preparation
Before the approach, take time to prepare for the interview. That means getting as much information about the prospect as you can, in advance of the call. Try to avoid going to see a prospect "cold turkey" — that is, without knowing anything about him. If you're making a call to sell remodeling, find out about the homeowner's circumstances, the type of work he wants done, and how many people will be involved in making the decision. You can get a lot of this kind of information by careful screening during the initial phone call.

The preparation includes making yourself fully informed on your proposition, ready to modify your arguments, meet objections, and maneuver around a cold turn-down.

Approaching the Prospect
Even your approach is a sale in one sense — you must "sell" your prospects on the idea of giving you a certain amount of their time. Many people automatically resist any sales attempt. It's probably fatal to say frankly: "My name is Joe Bloe, I'm a builder, and I want to interest you in remodeling your kitchen." In fact, what answer could a person give such an approach other than "Sorry, I'm not interested."?

Here's the approach that gets you somewhere in the shortest possible time: Give the prospects some idea of what you want to accomplish for them. Arouse their desire to know more. A little careful thinking and planning will help you to do this. You need some knowledge of the prospects, and of what your proposition will do for them.

Take the salesman who wants to sell you siding for a house you're building. Does he come at you with, "I'd like to sell you the siding you're going to need on the job?" Not at all. He eases around with a suggestion: "Suppose I could show you a way to save 25% on the cost of that siding job, with a better piece of work and a more lasting job — would you be interested?" That's his approach, and he certainly gets your attention. He'd get mine.

A successful approach builds a sense of confidence and understanding — the willingness to listen. And that's the first step in making a sale. The approach is based on self-interest. You have to make them think, "What will this do for me? What will it do for those close to me?" So in order to approach your prospects most effectively, ask yourself this question: "Why should these people give me fifteen minutes of their time?" Your honest answer will give you the key to your opening approach.

Apply this question to the hypothetical kitchen remodeling project. What would be your answer? Wouldn't it be, "Because I know such a kitchen would make the house more attractive and efficient, plus it would increase the resale value of the property far more than the cost of construction."

That's your opening approach. You're going to sell them because what you suggest will *benefit* them. You're going to make *their* property more useful, attractive and valuable. Get them interested by explaining it in a way they can believe, by believing it yourself.

Presenting or Demonstrating Your Service
Having sold your prospects on the idea of listening to your story, you're ready for the presentation or demonstration. Bear in mind during your presentation that there are two avenues of appeal you can use. One is an appeal to the prospect's reason and intellect through logic and sound argument; the other is an appeal to the prospect's emotions and imagination by positive suggestions and word pictures.

Few of us decide a matter by logic alone. In fact, there's usually a mixture of both reasoning and emotion, although it's surprising how many sales are made on the emotional appeal alone. Try to introduce a suggestion into the prospect's mind around which he can construct a mental picture in which he's the central figure.

Another point to bear in mind is that a human being can't be influenced except through one of the senses: sight, hearing, taste, smell, and touch. And every idea or feeling that enters the mind creates a reaction. These reactions are expressed in the face, hands, eyes, shoulders — in the whole body. These expressions tell the experienced salesman what is going on in the prospect's mind and indicate whether the right line of attack is being followed. So note carefully all reactions to your arguments or suggestions as you go over them with the prospect.

People understand pictures more quickly than words or verbal description. Pictures are direct, simple, uninvolved and concrete. Use words and phrases which bring up pictures of the idea to be conveyed. They work on the imagination more quickly and effectively than high-sounding terms. Your goal is to establish pictures in the mind of the prospect.

If you were selling pots and pans, you'd demonstrate how well-made they are, and how best to use them. Since you can't do that with a building or a remodeling job, use the plan to help create the pictures. The plan can claim the attention and hold it, allowing the mind to focus on something definite. It keeps the mind from wandering. And there's another advantage. If you present a plan for a two-story frame building, for example, he immediately registers "yes" or "no" to the idea of frame construction, two stories, and so on. He begins to make decisions, and each decision is one step closer to saying yes to the project you're proposing.

Make use of words and terms familiar to the prospect; speak his language. It may be perfectly clear to you to speak of the "lintel supports," but it would not mean half so much to the prospect as the phrase "the beam or support across the opening." And while you could instantly picture a "steel casement window" it would be far more clear to your listener were you to say "to make sure you get plenty of light, we plan to use three windows set in a metal frame."

Most people don't know the language of the construction business. They'll understand your ideas better when you use the language of seeing and hearing — phrases they can picture. For example, the homeowner will immediately be able to picture a garage "just big enough to hold two cars comfortably." That's probably not true if you describe a garage 20'0" deep by 18'4" wide. And again, use a plan to focus attention on the work you're describing.

If you're doing custom building, we'll assume

your prospect is relatively well-educated, has enough money to build, and is likely to take keen pride in owning a home that's individual and striking. The design or plan you submit may be original. You may wish to work with a local architect or designer to have a house designed to fit the customer's own particular needs. Or, you may take advantage of the many ready-made or stock plans that are reasonable and attractive.

Consider taking several plans along with you to use in your demonstration. This will mean considerable work or expense, but bear in mind that it's an investment on your part. You're waging an aggressive campaign to secure customers, instead of waiting for them to come to you. Choose a plan or plans that have some special feature that should appeal to this particular prospect. And make yourself familiar with them. Go over the specifications carefully, making a written or mental note of the points you'll bring out in the right order. You don't want to dither when the customer asks you a question about the house you're saying you'll build for him.

Explaining Your Plans

Remember, it's up to you to do the visualizing. Plans are for the workman. Don't expect the prospect to be able to look at the plans and visualize a house. You're using the plans to focus his attention while you explain and dramatize the features of the home.

Gather your facts and arrange your talking points to lead to a definite end. Point out the special features you think will appeal to this particular person. Your prospect may have a fancy for French windows opening out on a screened-in porch. He may have long dreamed of a room set aside for a library, with a fireplace flanked on either side with built-in bookcases. These are the little touches, the "selling points," that serve to center the attention, and that put the job in your pocket.

Naturally, the house has all the standard refinements that one should expect to find. But the other features, the special points such as rustic fireplace, sunroom, formal dining room, attractive kitchen arrangement — these give you an opportunity for special sales appeal.

For example, explain about the special features you plan for the kitchen, including undercabinet lighting of counters, ornamental valances to conceal lighting over the sink or the range, and special wall shelving for condiments, etc. Pull out a copy of Figure 16-1. Use it as the focal point for a discussion about the kitchen design and decor that the prospects want. The perfect kitchen is always a strong selling point, so make the most of the opportunity. You might also sell a bigger job and more appliances than if you tried to stick to a standard kitchen plan.

Of course, your plans won't be accepted without some changes. The prospects will have some definite ideas about their dream house. But if you've done your research into their budget and tastes, you should be able to lay before them some basic recommendations that can be accepted.

Don't go to the expense and the trouble of making these plans final working drawings. Certain features can be roughly suggested. This makes changes easier. And listen carefully to the suggestions your prospects make, instead of opposing them. Of course, if the ideas are entirely impractical, better make a counter-suggestion rather than an abrupt criticism. You might say, "That would be an idea worth considering, but if we do add it to the plan it may interfere with your ideas on . . . ." If the idea was bad, they can withdraw it gracefully and not lose face. Your tact helps the prospect develop a confidence in you that's the foundation for a successful sales effort.

Answering Questions and Objections

Throughout your presentation, you'll be interrupted with questions. Answer them briefly but completely. Then go over the questioned area carefully again. It doesn't matter if you think the question is foolish. Your prospects must be able to understand the answer. Remember this. If you don't overcome every objection they make, there's no sale.

These "objections" are often merely honest doubts on points not clear to them. Supply the facts to clear up these doubts in order to make the sale. The best way to overcome these objections is to anticipate them whenever possible.

The prospects may object that they can't finance the type of building you recommend. You can anticipate this objection by pointing out during the interview that the cash outlay would be approximately so many dollars, while the mortgage at the bank would take care of the remainder, and can be repaid in low monthly payments.

Kitchen Preference Checklist

Name_____ Phone_____

Address_____ Date _____

1. My personal kitchen should have:
☐A quiet corner where I can relax.
☐Soft lighting for a "meal for two" or a quiet moment.
☐A radio.
☐A television set.
☐An intercom system to the rest of the house.
☐A view of my garden or the outside.
☐A telephone extension.
☐Access to a convenience bath.
☐Open view to the family room.
☐Other _____
_____

2. My favorite colors are:
☐Bright tones   ☐Soft tones   ☐Medium tones
☐A mixture of colors for soft contrast
☐A mixture of colors for bright contrast
☐White____ Green____ Yellow____ Orange____
Red____ Blue____ Purple____

☐Comments _____
_____

3. I want my kitchen to have the following:
☐Refrigerator with freezer compartment in the:
Top____ Bottom____ Side by side____
☐Refrigerator with Ice Maker____, Ice and water available on the outside____
☐A double sink ☐A single sink
☐Trash compactor
☐Microwave oven
☐Dishwasher
☐Garbage disposal unit
☐Ceiling fan
☐Lazy Susan shelves
☐Breakfast bar ☐ Breakfast nook
☐Molded counter top ☐Tile counter top
☐Cabinets
Type _____
Style _____

☐Lowered ceilings
☐Separate freezer
☐Chopping block
☐Built-in ironing boad
☐Broom closet
☐Adjusable lighting
☐Pantry

4. Of the four basic kitchen plans I like the following best:
☐U-shaped   ☐L-shaped   ☐Corridor   ☐One-wall

5. My favorite color for appliances is:_____

6. My favorite theme is:
☐Contemporary (clean lines, bold colors, natural wood)
☐Early American (pewter, copper utensils, traditional woods)
☐Traditional (flexible style, simple, informal)
☐Spanish/Mediterranean (wrought iron, massive wood)

7. My choice of floor covering is:
☐Vinyl   ☐Carpet   ☐Inlaid style   ☐Tile
☐Resilient tile ☐Other_____

8. How many in your family? Adults_____?
Children_____? Ages of children_____

9. The height of the person who will use the kitchen the most:____

10. I entertain adult guests ☐seldom ☐often (about____ times a month.)

11. I would like to have the following in the kitchen:
☐Laundry area ☐Eating area ☐Desk/art area ☐Bar
☐Children's play area

Kitchen preference checklist
Figure 16-1

Have your selling points marshalled and under control; be prepared to answer any questions that arise. If the price question comes up before you're ready for it, try to defer it until later. Here's a statement that usually works: "Why not see first whether this is what you want. When we have the actual requirements before us, we can figure out how the costs will run. And I'm sure I can bring it in at a price you'll find satisfactory." Then go back to the presentation you have planned.

Objections offered may embrace price, or service, or countless other factors. Frequently, they won't have anything to do with the merit of your service. For example, when a prospect says he was misled in the past when a builder promised a completion time that he couldn't meet, this momentarily becomes the leading consideration. Clear this up by explaining when you could finish this job, and then lead back into the main considerations.

When the resistance begins to ease up, when your prospects pretty well agree with your ideas and begin asking supplemental questions, it's time to stop your demonstration and get down to closing.

Let it be understood that we want to sell a prospect something he wants; not something he doesn't want. You want to build him something he can be proud of; something that will directly benefit him. Oversell is sometimes possible since there are people who can be talked into almost anything. I never want to build for someone like that unless it was his idea in the first place. I prize my talents too highly to push them on folks, even for their own benefit. Besides, I despise high-pressure selling techniques. With that said, let's move on to closing.

Closing the Sale

You don't want to formally ask the prospect to sign a contract at this point. It's more effective to take the sale for granted and test it with questions like this: "I'm just finishing up a job on the west side of town and could get my crew started on this work by the first of the month. Would that be soon enough?"

If he balks, try to find the reason, then focus on this point until it's eliminated. You may wish to show why *now* is the best time for a decision, presenting sound, logical reasons for your advice.

Your prospect may want to "think it over." Find out if there are any lingering doubts, any information you've failed to supply. If the demonstration has been complete, there's no reason why he shouldn't agree now, assuming finances permit and you're not trying to sell him something that he doesn't want in the first place.

Often a person will say "no" to a proposition when he really doesn't mean it. He's actually waiting for additional facts, further reasons to justify a definite order. Many salesmen accept that no, only to learn that later the person said yes to an identical proposition from someone else. Supply facts, test for the reason behind the refusal. It may simply be a stall. If so, it's a logical starting point for a new, well-directed sales effort. Briefly cover again all the points that the prospect's already agreed to.

You may have a prospect who has listened carefully. He may have agreed that your plan is just what he's looking for — but still no sale. "You certainly have an attractive proposition, but I can't decide right now. Drop in Tuesday, will you?" Well, that doesn't mean a thing except the person is putting off a decision. Try to find out why he's delaying it, and center your arguments on that reason.

When you get the final agreement you've worked so hard for, be prepared with a contract. Figure 16-2 is a typical proposal and contract for a building job. Notice the bold type near the bottom of the first page. It's the notice to the buyers of their right to cancel the contract within three days. Figure 16-3 is the Notice to Customers Required by Federal Law. Every buyer must receive a copy of this when he signs the contract. But if you've done a good selling job, meeting every objection and laying to rest every doubt in the buyers' minds, there's no reason for them to cancel the contract. They're getting the work they want, done at a price that's reasonable.

When to Excuse Yourself

When your prospect has agreed to your proposition and signed your contract, *get out!* Make a few courteous comments, then leave when you've put the sale across. Don't talk yourself out of it. Don't stay to discuss points that might lead to doubts and excuses. The last thing you want is for the buyer to put off the decision that's already been made. Your prospect is now your client. Resist the urge to tell him what cute things your child or grandchild did — until you've finished the job and received payment. By then the client will have told you all the cute things *his* child or grandchild said and did and you can get back at him!

Proposal and Contract

Date_____19_____

To _____

Dear Sir:

We propose to furnish all materials and perform all labor necessary to complete the following:

_____

_____

Job Location:

All of the above work to be completed in a substantial and workmanlike manner according to the floor plan, job specifications, and terms and conditions on the back of this form for the sum of

Dollars ($_____)

Payments to be made as the work progresses as follows:_____

_____

the entire amount of the contract to be paid within_____days after substantial completion and acceptance by the owner. The price quoted is for immediate acceptance only. Delay in acceptance will require a verification of prevailing labor and material costs. This offer becomes a contract upon acceptance by contractor but shall be null and void if not executed within 5 days from the date above.

By_____

"YOU, THE BUYER, MAY CANCEL THIS TRANSACTION AT ANY TIME PRIOR TO MIDNIGHT OF THE THIRD BUSINESS DAY AFTER THE DATE OF THIS TRANSACTION. SEE THE ATTACHED NOTICE OF CANCELLATION FORM FOR AN EXPLANATION OF THIS RIGHT."

You are hereby authorized to furnish all materials and labor required to complete the work according to the plans, job specifications, and terms and conditions on the back of this proposal, for which we agree to pay the amounts itemized above.

Owner _____

Owner_____Date_____

Accepted by Contractor_____Date_____

Proposal and Contract
Figure 16-2

1. The Contractor agrees to commence work within (10) days after the last to occur of the following, (1) receipt of written notice from the Lien Holder, if any, to the effect that all documents required to be recorded prior to the commencement of construction have been properly recorded; (2) the materials required are available and on hand, and (3) a building permit has been issued. Contractor agrees to prosecute work thereafter to completion, and to complete the work within a reasonable time, subject to such delays as are permissible under this contract. If no first Lien Holder exists, all references to Lien Holder are to be disregarded.

2. Contractor shall pay all valid bills and charge for material and labor arising out of the construction of the structure and will hold Owner of the property free and harmless against all liens and claims of lien for labor and material filed against the property.

3. No payment under this contract shall be construed as an acceptance of any work done up to the time of such payment, except as to such items as are plainly evident to anyone not experienced in construction work, but the entire work is to be subject to the inspection and approval of the inspector for the Public Authority at the time when it shall be claimed by the Contractor that the work has been completed. At the completion of the work, acceptance by the Public Authority shall entitle Contractor to receive all progress payments according to the schedule set forth.

4. The plan and job specification are intended to supplement each other, so that any works exhibited in either and not mentioned in the other are to be executed the same as if they were mentioned and set forth in both. In the event that any conflict exists between any estimate of costs of construction and the terms of this Contract, this Contract shall be controlling. The Contractor may substitute materials that are equal in quality to those specified if the Contractor deems it advisable to do so. All dimensions and designations on the plan or job specification are subject to adjustment as required by job conditions.

5. Owner agrees to pay Contractor its normal selling price for all additions, alterations or deviations. No additional work shall be done without the prior written authorization of Owner. Any such authorization shall be on a change-order form, approved by both parties, which shall become a part of this Contract. Where such additional work is added to this Contract, it is agreed that all terms and conditions of this Contract shall apply equally to such additional work. Any change in specifications or construction necessary to conform to existing or future building codes, zoning laws, or regulations of inspecting Public Authorities shall be considered additional work to be paid for by Owner as additional work.

6. The Contractor shall not be responsible for any damage occasioned by the Owner or Owner's agent, Acts of God, earthquake, or other causes beyond the control of Contractor, unless otherwise provided or unless he is obligated to provide insurance against such hazards, Contractor shall not be liable for damages or defects resulting from work done by subcontractors. In the event Owner authorizes access through adjacent properties for Contractor's use during construction. Owner is required to obtain permission from the owner(s) of the adjacent properties for such. Owner agrees to be responsible and to hold Contractor harmless and accept any risks resulting from access through adjacent properties.

7. The time during which the Contractor is delayed in this work by (a) the acts of Owner or his agents or employees or those claiming under agreement with or grant from Owner, including any notice to the Lien Holder to withhold progress payments, or by (b) any acts or delays occasioned by the Lien Holder, or by (c) the Acts of God which Contractor could not have reasonably foreseen and provided against, or by (d) stormy or inclement weather which necessarily delays the work, or by (e) any strikes, boycotts or like obstructive actions by employees or labor organizations and which are beyond the control of Contractor and which he cannot reasonably overcome, or by (f) extra work requested by the Owner, or by (g) failure of Owner to promptly pay for any extra work as authorized, shall be added to the time for completion by a fair and reasonable allowance. Should work be stopped for more than 30 days by any or all of (a) through (g) above, the Contractor may terminate this Contract and collect for all work completed plus a reasonable profit.

8. Contractor shall at his own expense carry all workers' compensation insurance and public liability insurance necessary for the full protection of Contractor and Owner during the progress of the work. Certificates of insurance shall be filed with Owner and Lien Holder if Owner and Lien Holder require. Owner agrees to procure at his own expense, prior to the commencement of any work, fire insurance with Course of Construction. All Physical Loss and Vandalism and Malicious Mischief clauses attached in a sum equal to the total cost of the improvements. Such insurance shall be written to protect the Owner and Contractor, and Lien Holder, as their interests may appear. Should Owner fail so to do, Contractor may procure such insurance, as agent for Owner, but is not required to do so, and Owner agrees in demand to reimburse Contractor in cash for the cost thereof.

9. Where materials are to be matched, Contractor shall make every reasonable effort to do so using standard materials, but does not guarantee a perfect match.

10. Owner agrees to sign and file for record within five days after substantial completion and acceptance of work a notice of completion. Contractor agrees upon receipt of final payment to release the property from any and all claims that may have accrued by reason of the construction.

11. Any controversy or claim arising out of or relating to this contract shall be settled by arbitration in accordance with the Rules of the American Arbitration Association, and judgment upon the award rendered by the Arbitrator(s) may be entered in any Court having jurisdiction.

12. Should either party bring suit in court to enforce the terms of this agreement, any judgment awarded shall include court costs and reasonable attorney's fees to the successful party plus interest at the legal rate.

13. Unless otherwise specified, the contract price is based upon Owner's representation that there are no conditions preventing Contractor from proceeding with usual construction procedures and that all existing electrical and plumbing facilities are capable of carrying the extra load caused by the work to be performed by Contractor. Any electrical meter charges required by Public Authorities or utility companies are not included in the price of this Contract, unless included in the job specifications. If existing conditions are not as represented, thereby necessitating additional plumbing, electrical, or other work, these shall be paid for by Owner as additional work.

14. The Owner is solely responsible for providing Contractor prior to the commencing of construction with any water, electricity and refuse removal service at the job site as may be required by Contractor to effect the improvement covered by this contract. Owner shall provide a toilet during the course of construction when required by law.

15. The Contractor shall not be responsible for damage to existing walks, curbs, driveways, cesspools, septic tanks, sewer lines, water or gas lines, arches, shrubs, lawn, trees, clotheslines, telephone and electric lines, etc., by the Contractor, subcontractor, or supplier incurred in the performance of work or in the delivery of materials for the job. Owner hereby warrants and represents that he shall be solely responsible for the condition of the building with respect to moisture, drainage, slippage and sinking or any other condition that may exist over which the Contractor has no control and subsequently results in damage to the building.

16. The Owner is solely responsible for the location of all lot lines and shall if requested, identify all corner posts of his lot for the Contractor. If any doubt exists as to the location of lot lines, the Owner shall at his own cost, order and pay for a survey. If the Owner wrongly identifies the location of the lot lines of the property, any changes required by the Contractor shall be at Owner's expense. This cost shall be paid by Owner to Contractor in cash prior to continuation of work.

17. Contractor has the right to subcontract any part, or all, of the work agreed to be performed.

18. Owner agrees to install and connect at Owner's expense, such utilities and make such improvements in addition to work covered by this Contract as may be required by Lien Holder or Public Authority prior to completion of work of Contractor. Correction of existing building code violations, damaged pipes, inadequate wiring, deteriorated structural parts, and the relocation or alteration of concealed obstructions will be an addition to this agreement and will be billed to Owner at Contractor's usual selling price.

19. Contractor shall not be responsible for any damages occasioned by plumbing leaks unless water service is connected to the plumbing facilities prior to the time of rough inspection.

20. Title to equipment and materials purchased shall pass to the Owner upon delivery to the job. The risk of loss of the said materials and equipment shall be borne by the Owner.

21. Owner hereby grants to Contractor the right to display signs and advertise at the job site.

22. Contractor shall have the right to stop work and keep the job idle if payments are not made to him when due. If any payments are not made to Contractor when due, Owner shall pay to Contractor an additional charge of 10% of the amount of such payment. If the work shall be stopped by the Owner for a period of sixty days, then the Contractor may, at Contractor's option, upon five days written notice, demand and receive payment for all work executed and materials ordered or supplied and any other loss sustained, including a profit of 10% of the contract price. In the event of work stoppage for any reason, Owner shall provide for protection of, and be responsible for any damage, warpage, racking, or loss of material on the premises.

23. Within ten days after execution of this Contract, Contractor shall have the right to cancel this Contract should it be determined that there is any uncertainty that all payments due under this Contract will be made when due or that any error has been made in computing the cost of completing the work.

24. This agreement constitutes the entire Contract and the parties are not bound by oral expression or representation by any party or agent of either party.

25. The price quoted for completion of the structure is subject to change to the extent of any difference in the cost of labor and material as of the date and the actual cost to Contractor at the time materials are purchased and work is done.

26. The Contractor is not responsible for labor or materials furnished by Owner or anyone working under the direction of the Owner and any loss or additional work that results therefrom shall be the responsibility of the Owner. Removal or use of equipment or materials not furnished by Contractor is at Owner's risk, and Contractor will not be responsible for the condition and operation of these items, or service for them.

27. No action arising from or related to the contract, or the performance thereof, shall be commenced by either party against the other more than two years after the completion or cessation of work under this contract. This limitation applies to all actions of any character, whether at law or in equity, and whether sounding in contract, tort, or otherwise. This limitation shall not be extended by any negligent misrepresentation or unintentional concealment, but shall be extended as provided by law for willful fraud, concealment, or misrepresentation.

28. All taxes and special assessments levied against the property shall be paid by the Owner.

29. Contractor agrees to complete the work in a substantial and workmanlike manner but is not responsible for failures or defects that result from work done by others prior, at the time of or subsequent to work done under this agreement.

30. Contractor makes no warranty, express or implied (including warranty of fitness for purpose and merchantability). Any warranty or limited warranty shall be as provided by the manufacturer of the products and materials used in construction.

31. Contractor agrees to perform this Contract in conformity with accepted industry practices and commercially accepted tolerances. Any claim for adjustment shall not be construed as reason to delay payment of the purchase price as shown on the payment schedule. The manufacturers' specifications are the final authority on questions about any factory produced item. Exposed interior surfaces, except factory finished items, will not be covered or finished unless otherwise specified herein. Any specially designed, custom built or special ordered item may not be changed or cancelled after five days from the acceptance of this Contract by Contractor.

Proposal and Contract
Figure 16-2 (continued)

Notice To Customer Required By Federal Law

You have entered into a transaction on_____which may result in a lien, mortgage, or other security interest on your home. You have a legal right under federal law to cancel this transaction, if you desire to do so, without any penalty or obligation within three business days from the above date or any later date on which all material disclosures required under the Truth in Lending Act have been given to you. If you so cancel the transaction, any lien, mortgage, or other security interest on your home arising from this transaction is automatically void. You are also entitled to receive a refund of any down payment or other consideration if you cancel. If you decide to cancel this transaction, you may do so by notifying:

_____
(Name of Creditor)

at   _____
(Address of Creditor's Place of Business)

by mail or telegram sent not later than midnight of_____. You may also use any other form of

written notice identifying the transaction if it is delivered to the above address not later than that time. This notice may be used for the purpose by dating and signing below.

I hereby cancel this transaction.

_____
(Date)   **(Customer's Signature)**

Effect of rescission. When a customer exercises his right to rescind under paragraph (a) of this section, he is not liable for any finance or other charge, and any security interest becomes void upon such a rescission. Within 10 days after receipt of a notice of rescission, the creditor shall return to the customer any money or property given as earnest money, downpayment, or otherwise, and shall take any action necessary or appropriate to reflect the termination of any security interest created under the transaction. If the creditor has delivered any property to the customer, the customer may retain possession of it. Upon the performance of the creditor's obligations under this section, the customer shall tender the property to the creditor, except that if return of the property in kind would be impracticable or inequitable, the customer shall tender its reasonable value. Tender shall be made at the location of the property or at the residence of the customer, at the option of the customer. If the creditor does not take possession of the property within 10 days after tender by the customer, ownership of the property vests in the customer without obligation on his part to pay for it.

Notice to Customer Required by Federal Law
Figure 16-3

These are the selling principles and methods that are being followed by successful builders. They apply whether you're developing a sales plan to sell houses or remodel kitchens. Just adapt the principles to your own operation, and go out there and sell!

Chapter **17**

Spec Building and Land Development

Most construction contractors have considered speculative building at one time or another. Far fewer have ever tried it. There's risk in finding land, building, and then betting that you can sell what you've built for more than it cost you. And it takes money to get started. But the rewards can be as big as the risks.

Understand first that there are times to be a spec builder and there are times to stick with remodeling or custom building. The years 1985 and 1986 were great years for home builders in most parts of the country. A spec builder with an inventory of houses to sell could count on selling nearly everything before construction was even finished. Just the opposite was true in 1981 and 1982 — as you may remember. Unfortunately, building almost anything usually takes at least a year, from the inception of the idea to the sale. Sometimes it's hard to anticipate what the market will be like a full year from now.

Recognize that there's a construction cycle. Don't get stuck with your life savings in a house that has to be sacrificed to satisfy an anxious lender.

The best time to build your spec house is when construction activity is just picking up after a recession. A small builder should be able to build and sell several homes during the two or three years of good times that builders enjoy after a housing bust.

Here's how to find out where we are in the con-

struction cycle at any given time. The U.S. Department of Commerce publishes two excellent reports that make trends in construction activity very clear to anyone who takes the trouble to glance at them:

New One-Family Houses Sold and For Sale (Construction Reports publication C25) costs $21 per year. It's shown in Figure 17-1.

Housing Starts (Construction Reports publication C20) is also $21 per year. Look at Figure 17-2. Subscribe by sending a check to:

Superintendent of Documents
U.S. Government Printing Office
Washington, D.C. 20402

Housing Starts shows how many new, privately-owned housing units are started each month in the U.S. You'll note from the chart published in each issue that starts rise and fall fairly regularly. You wouldn't launch a new ship into the teeth of a gale. So don't start new projects when starts are falling. Instead, close out the work you're doing and reduce the risk that you'll be stuck with a house that won't sell.

Housing Starts will help you visualize where we are in the construction cycle. But that's just part of the story. You can really stack the deck in your favor if you start building when new home sales are rising and the inventory of homes for sale is falling. That's where the other Department of Commerce report comes in.

CONSTRUCTION REPORTS

U.S. Department of Commerce
BUREAU OF THE CENSUS

**U.S. Department of Housing
and Urban Development**

New One-Family Houses Sold and For Sale

May 1986

C25-86-5
Issued July 1986

**New One-Family Houses Sold and For Sale and
Months' Supply at Current Sales Rate**
(Seasonally Adjusted)

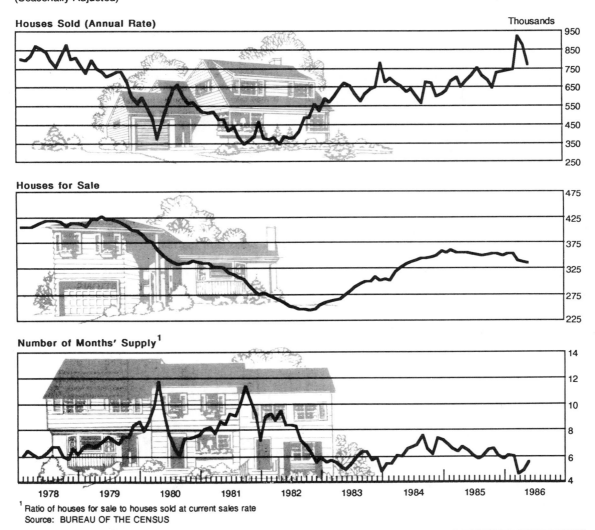

[1] Ratio of houses for sale to houses sold at current sales rate
 Source: BUREAU OF THE CENSUS

Questions regarding these data may be directed to Steve Berman, Construction Starts Branch, Telephone 301-763-7842.

For sale by the Superintendent of Documents, U.S. Government Printing Office, Washington, D.C. 20402. Postage stamps not acceptable; currency submitted at sender's risk. Remittances from foreign countries must be by international money order or by draft on a U.S. bank. Annual subscription price: domestic $21, foreign $26.25. Single copy domestic $1.25, foreign $1.56. Annual report: $7, foreign $8.75.

New One-Family Houses Sold and For Sale
Figure 17-1

CONSTRUCTION REPORTS

Housing Starts

MAY 1986

U.S. Department of Commerce
BUREAU OF THE CENSUS

C20-86-5
Issued June 1986

This issue contains a supplement showing annual statistics on new mobile homes placed by State and selected characteristics of new mobile homes placed by region.

New Privately Owned Housing Units Started
Seasonally Adjusted Annual Rate

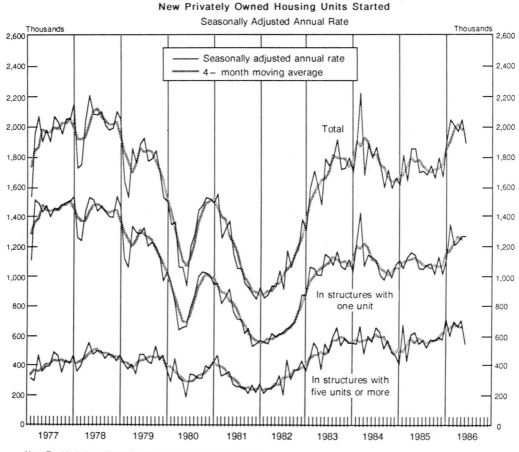

Note: Total includes units started in structures with two to four units.

Questions regarding these data may be directed to David Fondelier, Chief, Construction Starts Branch, Telephone (301) 763-5731.

For sale by the Superintendent of Documents, U.S. Government Printing Office, Washington, D.C. 20402. Postage stamps not acceptable; currency submitted at sender's risk. Remittances from foreign countries must be by international money order or by draft on a U.S. bank. Annual subscription price: domestic $21.00, foreign $26.25. Single copy: domestic $2.00, foreign $2.50.

Housing Starts
Figure 17-2

New One-Family Houses Sold and For Sale shows the annual rate of home sales, how many houses are currently for sale, and the number of months' supply of homes at the current sales rates. Months of supply is the key figure. It's a builder's market when home sales are rising and the inventory of homes for sale is low. A six-month inventory shows that there's a shortage of new homes for sale. When the unsold inventory reaches twelve months, that's panic time for builders with unsold homes.

Plan to have your first home under construction (or expand your operation) when the inventory of new homes for sale is less than seven months. The good times will probably last two or three years, if the previous 20 years is a reliable guide to the future. In these two or three years you should make good money on every home you turn out.

The hole in any building plan is, of course, the interest rate. There seems to be no realistic method of planning for erratic (and often incomprehensible) changes in the cost of money!

Do the Research Before You Build

Three R's are often quoted as the essentials of selling houses:

1) Right market
2) Right plan
3) Right price

For the speculative builder, merchandising starts with the selection of a site for the project and continues through the design of the house, the color scheme of the interior, and the landscaping of the exterior. Even technical details of construction and the utilities and mechanical equipment must be considered from the standpoint of merchandising.

Study Your Market

It's pretty easy to buy a lot and build a house (or find a landowner who will subordinate the lot to you and split the proceeds when the house is sold). It's much harder to sell that home at a profit. You can avoid many mistakes by analyzing your market carefully before you invest your money and time. Ask yourself these questions:

• What's the general economic outlook for my community? Is the community growing?

• What keeps the economy growing? Farming?

Industry? Services? A military base? A combination of all these?

• How are most people of the area employed? What wage levels predominate? What kind of people do I expect to sell to?

• Are there any inducements for new companies to come into the area? Are there any active trade promotion groups?

• How many houses are sold each year in each of the geographical areas I'm interested in? What kind of home sells best? What price range?

• What part of town is best for the kind of house I plan to build? Where are sites available?

• What financing do competitors in my price range offer to purchasers? What kind of financing can I offer?

• What's working and what's not? You'll usually do better following the lead of a successful spec builder, building in the same community, using similar designs, and selling houses at competitive prices.

At first, you'll probably build within a fairly restricted price range. But you have to make a lot of decisions besides price. Remember that after decisions are made and construction has started, there's no turning back. Assuming that you've done the market research, decided the time is right and bought a lot or two — you still have to define exactly what kind of house you're going to build. Make a study of these points to help you narrow it down:

• What size of house is in demand? How many square feet of floor will you build? Single or double garage?

• What appliances will you include: range, refrigerator, washer, dryer, dishwasher, garbage disposal, microwave oven, ceiling fans, trash compactor?

• Will you include central air conditioning? A heat pump?

• What type of architecture is the most popular?

What will provide the most house for the money? Is the house you're considering appropriate for the neighborhood where you'll build?

• Finally, what type of construction: brick, frame, concrete, prefabricated?

The annual report and other publications of the Federal Housing Administration have a lot of information that every spec builder should consider before making a commitment. These reports are available at local FHA insuring offices.

How Many Jobs to Start With?
You've already thought about the demand in deciding what to build and what your price range will be. You now have to decide on the type of work, the price, style, location, and what the market can absorb. If sales haven't been brisk in your proposed price, style and location, you might be wise to wait until conditions seem more favorable. If you decide to go ahead anyway, be cautious.

If your tract includes more than five or six homes, don't plan to build out the entire project at once. Begin work on one or two homes. These are your models. Use them to sell the rest of the project. Don't begin work on the remaining homes until you have signed purchase agreements from buyers who can qualify for the loans they need.

Here's a rule of thumb for planned unit developments: Don't have more than 1/5th of the project completed and unsold at any time. If your tract consists of 10 homes, complete two and begin work on two more. But don't start work on the next two as long as two remain complete and unsold. Keep construction running parallel to sales. If sales lag, stop building.

Sure, it would be cheaper to keep building and keep the crews busy. But don't let the tail wag the dog. There are no profits without sales. If there are no sales, there's no reason to keep building. Unsold inventory will ruin you, just like it's ruined hundreds of over-optimistic builders in the past.

Merchandising for the Spec Builder
A successful merchandising and promotion program for a speculative construction project probably will include the following elements:

1) An exhibit home. This is the product to be sold.

2) Special exhibits of materials. Use advertising brochures furnished by your suppliers to gain acceptance for your product.

3) An advertising campaign. This brings prospects to the house.

4) A sales program. This sells the house. Give prospects an opportunity to examine the work you've done. A builder's best talking points are: "These are the houses I've built. These are my satisfied customers." The previous chapter covered advertising and sales in more detail.

For most spec builders, a model home is almost essential. And, of course, it helps if your model is completely furnished and landscaped. Just the interior decoration, landscaping and furniture can run $25,000 or more. But try getting a landscaping company to donate the landscaping. A local interior decorator may donate an interior plan at little or no cost and a furniture store may donate furniture just for the advertising value. A cooperative landscaper, interior decorator and furniture store can cut your investment substantially — and you're probably happy to display their names to prospective buyers that go through your model.

If you can't get a furniture store to take you up on your offer of free advertising, and if your budget won't stretch that far, just a few articles of furniture in strategic places may serve as well as a completely furnished house. Here's the approach of one successful builder I know: In his houses with a formal dining room, he completely furnishes only the dining room, and sets the table, napkins and all. His philosophy is that once you catch the belly's attention, the mind is sure to follow. Pictures on the walls and a few other homey touches here and there can also help create a "lived in" atmosphere.

You'll be able to arrange with some of your suppliers to provide materials for special exhibits that you can set up in the garage, the basement, or a breezeway. These exhibits focus attention on quality products used in the house.

Consider putting up small signs in various places around the house to call attention to special products and special construction. This helps to sell the idea of a quality house. It's a good idea, too, to display maps showing the location of the house in relation to nearby schools, shopping center, churches, public transportation, and other desirable community facilities.

Architectural and Engineering Services

As an independent builder, the quality of the houses you build is your best sales pitch. And nothing speaks quality better than good design. A poor floor plan or lack of curb appeal turns buyers off quick. To be a success as a spec builder, you have to understand what goes into planning and designing a house that will appeal to buyers.

Most small home building projects require these architectural and engineering services:

1) Basic design of the house, with the architectural details needed to give the appearance you want.

2) Complete plans and specifications, as required for the building permit and for approval by FHA, VA, or whatever lender you're using.

3) Development of the plot plan and siting the house on the lot.

4) Checking with surveyor's transit to see that the house is put in the right place, and that foundation and walls are true and square according to plans.

5) Control and supervision to make sure that construction is completed according to plans and specs.

Almost all building work requires complete plans and specs. You're not going to get a building permit or a loan without plans.

Do you plan to be your own designer? A trade association survey found that 38 percent of the homebuilders polled designed the houses they built. But unless you have engineering or architectural training, you'd probably be wise not to try it. Besides, the time you spend working up the plans is likely to keep you from other important business. Where, then, do you turn for architectural and engineering planning? There are a number of sources:

An architectural firm— Any architect worth his fee will draw an adequate plan that complies with zoning and building codes and other regulations. But this is also the most expensive source. You may feel you can't afford it, at least in the beginning.

Architectural design services— Many architectural firms have published portfolios of home designs. Plans and specs are offered for fees that are thousands less than the cost of an original design. If you can find a plan that suits your needs, why hire an architect to draw one for you?

Designs from suppliers— If you're planning to use prefabricated components, you can build from plans developed by the manufacturer. These designs have been greatly improved in recent years. But you'll still need a local surveyor or engineer to survey the site, and maybe to modify the plan for the site or to meet local building requirements.

House plan services— Various plan services offer collections of designs with plans and specs that satisfy requirements in some communities. If you decide to use one of these, make sure the plans will be acceptable to your lender.

Style of the House

The style of your houses should be in harmony with other houses in the area. An occasional striking, unusual design may attract prospective buyers. But don't be a style innovator. Build what most of your potential customers want to buy. This is the main reason why the designs used by many spec builders are rather unimaginative.

Custom features are one way to overcome monotony in basic design. In fact, the custom feature may sell the house. An additional patio, an outdoor eating area, an attractively lighted kitchen or bathroom, a built-in dressing table, special closets — any one of these may provide the extra touch that will persuade the customer to buy.

How many levels should you build: one-story, two-story, or split-level or three-story houses? Over the years, fashions and tastes change. Study your market before you make this decision. And if you decide on split-level, study the lay of the land. Any slope will have to be reckoned with when you decide which of the three basic split-level designs you'll use — side, front to back, or back to front.

Equipment and fixtures are second only to good design in the selling of your houses. In many areas, air conditioning is essential. Adequate wiring costs very little more than minimum wiring and is a strong sales point. Ovens, ranges, dishwashers, heat circulating fireplaces and garbage disposals are expected by many of your potential buyers. But keep in mind the price range you've chosen. Consider how much equipment you can provide

Handbook of Construction Contracting

without sacrificing basic quality.

Many a sale has been lost because of a combination of colors the prospective buyer didn't like. It pays to get advice from local interior decorators or from department stores that offer decorating counsel.

Make sure your house has "curbside appeal." The impression your prospects have as they get out of their car may be the most important one. The general design of the house, the way it's situated on the lot, and the landscaping all affect this initial impression. Few small builders can afford to wait for a good lawn to grow. Provide a sod lawn, at least in front, and some planting. A bare house standing on a pile of mud isn't attractive. A good rule is to provide your model with just a little more landscaping than other builders provide in theirs.

Should You Use a Broker?

For the spec builder, a basic question is whether you'll handle the actual selling yourself, or sell your houses through a real estate broker. Selling through a broker has important advantages. A good broker:

- Knows the local market thoroughly

- Has experience in sales and promotion campaigns

- Has wider sales contacts than you have

- Has a wide range of houses to attract prospective buyers to his office

- Has a trained sales force ready to go

- Can assist with customer financing

Using a broker doesn't cost you anything until sales are made, and saves you time to concentrate on actual building.

Of course, your profits may be greater if you do your own selling. You should be aware, however, of the nature of the task you're undertaking. Are you willing to spend the additional time and effort needed to sell successfully? Are you certain you can afford to take this time away from construction?

If you're building in volume, you may decide to set up a sales staff of your own. This way you can keep the sales activities under close control and integrate them completely into your long-term operations.

If, however, as a small and beginning builder, you're thinking of doing your own selling, ask yourself these questions:

1) Do I have the ability to sell houses?

2) Can I afford the time to do my own selling?

3) Could someone else do the job better?

Land Developing

Few contractors become land developers. But most land developers are, or at one time were, construction contractors who accumulated enough capital to buy and hold large parcels of land. You may never have the assets needed to get into land development, but some of the principles used by the larger developers can be adopted in small operations. If you're considering land development, even in a small way, the rest of this chapter may give you some good ideas.

Buying Sites from Developers

Begin by buying a site from a larger developer. Many developers will sell undeveloped parcels to contractors with a little more ambition and a few good ideas. Here's why: Generally it's easier to sell a large tract if several builders are operating simultaneously. If that's the case, your competition will be greater. That's the down side. But the greater number and variety of houses available will probably attract more customers to the development. This gives you some of the advantages of a large tract even though your company is small.

A danger in relying on developed sites is that you may be tempted to buy a site just because it's available, rather than because it's the best location for your particular needs. In most communities, real estate developments seem to be moving in certain directions. People want houses in the west end of town, for instance, or the southwest. Houses built in favored areas are easier to sell. Avoid buying lots that no one else is willing to develop.

Don't Be a Pioneer

At the start, you'll be wise to work in areas that are already known for the type of houses you plan. If, for example, you're building three-bedroom ranch houses, buy lots where this type of home is being built. Don't try to pioneer with limited capital. Follow a trend, don't try to start one. As a matter of fact, your best building sites often will be as close to your competitors as you can get. This is especially true if your houses offer better quality than theirs or sell at a lower price.

In established areas, it's easier to get people out to look at your houses. In fact, you'll be benefiting from your competitors' advertising. People who come out to look at their models will look at your houses too.

Developing Your Own Land

If you have adequate capital, you can gain many advantages through developing your own building sites. However, bear in mind that many of the following steps *must* be taken to satisfy local subdivision requirements. Some of them are required by the agencies that lend money (or insure loans) on houses to be built. The others are necessary to make sure that your tract meets all the basic demands of your customer.

Before You Buy the Land

You'll need to do several things before you obligate yourself for land. Among them are:

1) Satisfy yourself that the tract is in the proper general area for the kind of houses you plan to build. Also be sure that each lot is free of natural hazards, such as unstable soils, floods and so on.

2) Check all zoning and building codes which may affect the area. Be sure that the codes permit the type of units you plan to build on the lot sizes you plan to use. Find out about any nearby nonresidential uses which may be permitted that might hinder the sale of your completed houses. Construction beginning on the town's new sewage treatment plant on the adjoining lot isn't going to impress many buyers.

3) Check availability of utilities — water, gas, electricity, and sewage hook-ups.

4) Check the local highway and planning bodies to be sure that the tract isn't in the path of any projects — such as a new highway — which might subject it to condemnation proceedings. A future threat of this kind can make it hard to sell your finished houses.

5) Check to see how close the tract is to public transportation and to community facilities like churches and schools.

6) Discuss your project with a local surveyor, architect, or planner. Get preliminary sketches of potential development of the tract, number of sites to be obtained, cost estimates, preliminary time schedules, and so on.

7) Ask for a preliminary analysis of the tract from the Federal Housing Administration. This service is free. Its purpose is to aid development of site locations and land planning.

The timing of contacts with FHA is important. If you plan to use FHA mortgage insurance, talk with an FHA representative before you go any further. Don't buy the land, or spend money for plans and exhibits, until you know where you stand on this point.

Developing the Plans

Land planning is a specialized field. Moreover, the quality of your tract will have a great effect on your sale of houses. You'll need architects, land surveyors, and engineers, for example, to prepare land plats. These professional people can help with the following:

• Preparing topographic and land use surveys. These show contour lines and location of streams, roads, woods, rock outcroppings, and other natural features which affect planning.

• Compiling a list of existing protective covenants or deed restrictions, zoning ordinances and other present regulations of local agencies which will affect the tract.

• Developing preliminary layout for streets and lots.

• Showing accessibility to all utilities; for example, if individual septic tanks are proposed, percolation tests will be required. (This is the time to schedule with the FHA the work to be done in subsequent stages.)

• Preparing a detailed development plan showing lot lines and street lines; finished grade elevations, and complete plat of the tract ready for filing.

If you didn't contact the FHA at the time suggested above, you should now request a complete analysis of the subdivision from them. Next file the plat and restrictions with the appropriate local agency or land office. Then proceed with plans for construction of individual houses.

The FHA and Land Planning
The Federal Housing Administration has done some good work in this field. Since its beginning in 1934, the FHA has had an interest in improving the quality of development programs and land planning. It does no planning or design work itself, but it offers many services to help local developers and planners.

Your nearest FHA insurance office can give you details on FHA's land development services, which include professional help as well as publications.

FHA has developed standards for subdivisions and street improvements which improve livability of these areas. These standards assure more lasting values for homeowners. FHA wants to maintain a high standard of community planning as a guarantee for the long-term loans it insures in such areas. You'll find these FHA services valuable even if you don't plan to sell your houses under FHA-insured loans.

For instance, if you develop a subdivision according to the FHA neighborhood requirements, you'll find it easier to get approval when one of your customers wants to buy a house on an FHA-insured mortgage. Even if none of your buyers wants an FHA loan, adherence to FHA standards will make it easier to get other financing. And, of course, easier financing will help you to sell the houses you build.

Chapter **18**

Making a Business Plan

There's an old builder's proverb that I like: *The only thing that happens without a plan is an accident.* Let's spend this chapter exploring what a good business plan can do to make your company stronger and more profitable — and less prone to accidents.

A business plan can be your pathway to profit. To build that pathway, ask yourself these questions: What business am I in? What do I sell? Where is my market? Who will buy? Who is my competition? What's my sales strategy? How much money is needed to operate my firm? How will I get the work done? What management controls are needed? How can they be carried out? When should I revise my plan? Where can I go for help?

No one can answer these questions for you. As the owner-manager you have to answer them and draw up *your* business plan. If done right, your plan will show a logical progression from where you are to an achievable goal. All you need to create a plan like this is common sense and a little help. You provide the common sense and I'll be your helper. Let's get started.

A Note on Using This Chapter
This chapter requires your participation. Along with the text come blanks to fill in. The blanks make this chapter your workbook. When filled in, these blanks become *your* business plan. If you

don't want to write in the book, copy the pages and write on the copies. But whatever you do, keep the worksheets when you've finished the plan. No business plan should be written in granite. Review these pages occasionally and update the plan to fit changing conditions.

It takes time and energy and patience to draw up a satisfactory business plan. Use this chapter to get your ideas and the supporting facts down on paper. If you need more information on business planning, the Small Business Administration offers many guides and aids at modest prices. There's an SBA office in many larger cities.

An important part of every business plan is cash planning: Will you have the money available when you need it? As you do this plan, remember that it's easy to omit some important cost items. Anything you leave out of the picture will be an extra cash drain. Too many of these and your business is headed for disaster.

Remember, too, that any plan, by itself, is worthless until you put it into action. Without your action, nothing happens. We'll cover action steps at the end of this chapter.

What's in This for Me?
The hammer, trowel, pliers, and wrench are common tools of the construction industry. You need them to get the work done. Think of management

---

*Why am I in business?*_____

_____

_____

_____

_____

_____

_____

*What business am I in?*_____

_____

_____

_____

*Why did I answer that way?*_____

_____

_____

_____

_____

_____

_____

Business plan worksheet
Figure 18-1

as another tool you can use. Every job must be planned and organized if your company is going to run smoothly and efficiently. That's the goal of management — and the reason for planning.

When complete, your business plan will help guide your daily business activities. When you know where you want to go, it's easier to plan what you must do to get there. Also, the business plan can serve as a communications device which will orient key employees, suppliers, bankers, and whoever else needs to know about your goals and your operations.

Whether you're just thinking about starting your own firm or are already in business, a business plan can help. It's your roadmap from where you are to where you want to be. It will also point out some of the strengths you can draw on and some of the personal weaknesses that may be a handicap. To be a successful contractor you must not only know your business thoroughly, but must also know your limitations. It's O.K. if you don't know much about law or accounting. But it's a mistake if you don't get professional help when legal and accounting issues come up. Knowing the things you do well and the things you need help with will make you a better, more successful contractor.

My product	Types of customers	Location of customers
_____	_____	_____
_____	_____	_____
_____	_____	_____
_____	_____	_____
_____	_____	_____
_____	_____	_____

What about my operation will make them want to buy?

_____

_____

_____

_____

Customer profile worksheet
Figure 18-2

Why Am I in Business?

Most contractors are in business to make money and be their own boss. Important reasons, to be sure. But don't forget this: You're not likely to stay in business unless you also satisfy a consumer need at a competitive price. Profit is the reward for satisfying consumer needs in a competitive economy.

In the first years of business, your profits may seem like a small return for the long hours, hard work, and responsibility of being the boss. But there are other rewards associated with having your own business. For example, while your employees are helping you make your business a success, you in turn are helping them by paying them a fair wage for their work. Everybody comes out ahead. Or maybe your satisfaction will come from building a business you can pass on to your children.

Take time to answer this question right now. Use Figure 18-1. Why are you in business?

What Business Am I In?

At first glance this may seem like a rather silly question. You may say, "If there's one thing I'm sure of, it's what business I'm in." But wait. Let's look further into the question. Suppose you say, "I build houses." Are you a speculative or custom builder? Are you a remodeler? Are you a subcontractor? Can you schedule a complete job and make money? Your planning depends on your definition of the kind of business you want to run.

Consider this example. Bob Kamehameha started a small construction business shortly after returning from World War II. Because of Bob's special skills and talent for design, he specialized in building small restaurants and ice cream parlors. There was enough call for this type of building to keep him and his crew busy until the early '60's. Then sales began to fall off.

By moving his shop to smaller quarters with less overhead, and laying off half his crew, he was able to maintain his business to his satisfaction the rest of his life. After his death in 1985, his son examined the situation and decided that he wasn't really in the business of building small restaurants. He was in the business of custom finishing.

Today his business is prospering. He's building cabinets and small bars for private homes. His company also does other finishing work which requires the craftsmanship his crew is best at.

On Figure 18-1, state what business you're really in. Then use the rest of the space to give your reasons for that answer.

Making the Marketing Decisions

When you've decided what sort of construction business you're really in, you've just made your first marketing decision. Now, to sell your service or product, you face other marketing decisions.

Your marketing objective is to find enough jobs at the right times to provide a *protifable continuity* for your business. Your job starts must be coordinated to eliminate the down time between jobs. In other words, you want to get enough jobs, starting at the right times, to stay full-time in the construction business.

Unless you can come up with enough ideas to keep a crew working 12 months a year, maybe you're not ready for construction contracting.

Where Is Your Market?

Describe your market area in terms of customer profile (age, school needs, income, and so on) and geography. For example, if you're a custom builder, you may decide to build homes in the $80,000 to $130,000 price range. This would mean that your customers will have to have incomes in the middle- to upper-middle ranges. You may also decide that you can profitably build these homes on the owner's lot if it's located within a radius of 30 miles from your office.

The significance of a customer profile is that it will help you narrow your advertising to those media that will reach the customers you've targeted. In Figure 18-2, describe your market in terms of customer profile and geography.

Now that you've described what you want in terms of customers and location, what is it about your operation that will make these people want to buy your service? For instance, do you offer quality work, competitive prices, guaranteed completion dates, unique design, or other services that set you apart? Use the bottom half of Figure 18-2 to describe your operation's selling points.

Planning Effective Advertising

You've determined what it is you're marketing, who's going to buy it, and why they're going to buy it. Now you have to decide on the best way to tell your prospective customers about your product.

What should your advertising tell about your company? Use the top of Figure 18-3 to answer.

What form should your advertising take? Ask the local media (newspapers, radio and television stations, and advertising agencies) for information about their services and the results they offer for your money.

Use the middle portion of Figure 18-3 to help you determine what advertising is needed to sell your construction service.

How you spend advertising money is your decision, but don't fall into the trap that snares many advertisers. Here's how one consultant describes this pitfall: "It's amazing how many business managers consider themselves experts on advertising copy and media selection without any experience in these areas." Get as much help from experts as you can afford. And review Chapter 16 on advertising and sales strategy.

Beating the Competition

The competition in the construction industry often results in low profit margins. But if you're just starting or are a relatively small firm, this can work to your advantage. The smaller firm can often compete with the bigger outfit because overhead is usually lower in a smaller firm. For example, your office may be in your home, saving that expense. You can often work right out of your truck, saving the expense of a field office. The big boys can't do that.

Competition is largely price competition, although a good reputation for quality and efficiency can lift you above some of the price wars. But because the construction industry is highly competitive, there's a high failure rate for poor planners and poor performers. This points out the need for careful planning, particularly in the areas of estimating and bidding.

Now it's time to assess your competition. Answer the following questions on Figure 18-3: Who will be your major competitors? How will you compete against them?

Tailoring Your Sales Strategy

The market for the construction industry is unique in many ways. As a contractor, your market will be dependent on such variables as the state of the economy, local employment stability, the seasonality of the work, labor relations, good sub-

What should my advertising tell prospective customers? _____

_____

_____

_____

_____

_____

Form of advertising	Size of audience	Frequency of use		Cost of a single ad		Estimated cost
_____	_____	_____	x	_____	=	_____
_____	_____	_____	x	_____	=	_____
_____	_____	_____	x	_____	=	_____
_____	_____	_____	x	_____	=	_____
_____	_____	_____	x	_____	=	_____
_____	_____	_____	x	_____	=	_____
				Total		$ _____

Who are my major competitors? _____

_____

_____

_____

_____

_____

How will I compete against them? _____

_____

_____

_____

_____

_____

Advertising worksheet
Figure 18-3

contractors, and interest rates. How hard you work is merely one factor — it doesn't guarantee you success — unfortunately. You'll also find that you're unavoidably dependent on others, such as customers or financing institutions for payment, and other contractors for performance of their work. Always take your cash flow into consideration when you estimate and bid on a job. The money must be there in time to meet your own obligations.

Planning the Work

When your marketing efforts are successful, you have a job to do, at the right time and the right price. Now how will you plan the work so that the job gets done on time?

No matter how you plan the work, your plan should assist you in two specific ways: First, it should help you maintain your production schedule. Second, it should allow you the flexibility to meet changed conditions, such as bad weather.

In planning the work, keep two things in mind; the timing of starts, and the timing of the various steps in the construction process. If you have sufficient help and dependable supervisory personnel, you can take on as many projects as you can *control*. The size and nature of the job must be considered here also.

The work scheduling will show the various operations in sequence, and assign a working day designation to each, with a space for the calendar day designation. Several operations may be in progress simultaneously. Such a work schedule will show at a glance whether the work is progressing at the right time. Many companies offer commercial scheduling boards designed for this purpose.

Figure 18-4 is a partial work schedule to show how yours could be set up. Note that there is a column that can be filled in with either a solid mark or

an X to indicate either partial or completed work. When you look at a particular calendar day, an X next to it would indicate that you're on schedule. A partial X, like the one in item 3, foundation pour, indicates that the work is three-quarters complete. An open square indicates a delay. Here, then, is a convenient way to see trouble spots that are causing delays. It gives you an opportunity to take corrective action. When you move up to bigger jobs, you may want to use the Critical Path Method of work scheduling we covered in Chapter 14.

Always save your work schedules. They'll form the basis for future estimates. Work schedules for completed similar jobs will give you information on the steps of production, an indication of what materials you'll need and when you'll need them, an indication of how long the job will take, and any peculiarities that may affect the completion of the current job. When you consider all these things, you'll be more likely to submit an accurate bid.

By carefully keeping such records, you'll also have an indication of how many workers you'll need. If the work falls behind schedule, you may need to bring more workers to the job to assure scheduled completion and avoid a penalty, if your contract includes a late clause. Also, such records will give you an indication of the organizational structure you may need for your firm.

Organizing to Get the Job Done

Organization is the key to efficient operation, even if you're still doing everything yourself. As your company grows, you won't be able to keep it up. You'll have to delegate work, responsibility, and authority. The organizational chart is the map that can guide you down the road to expansion. It shows quite clearly who is responsible for the major activities of your business.

Working Day				
Activity	**Start**	**Finish**	**Calendar day**	**Complete**
1. Layout	1	1	15	☒
2. Foundation forms	1	2	16	☒
3. Foundation pour	3	3	19	☒

Typical work schedule
Figure 18-4

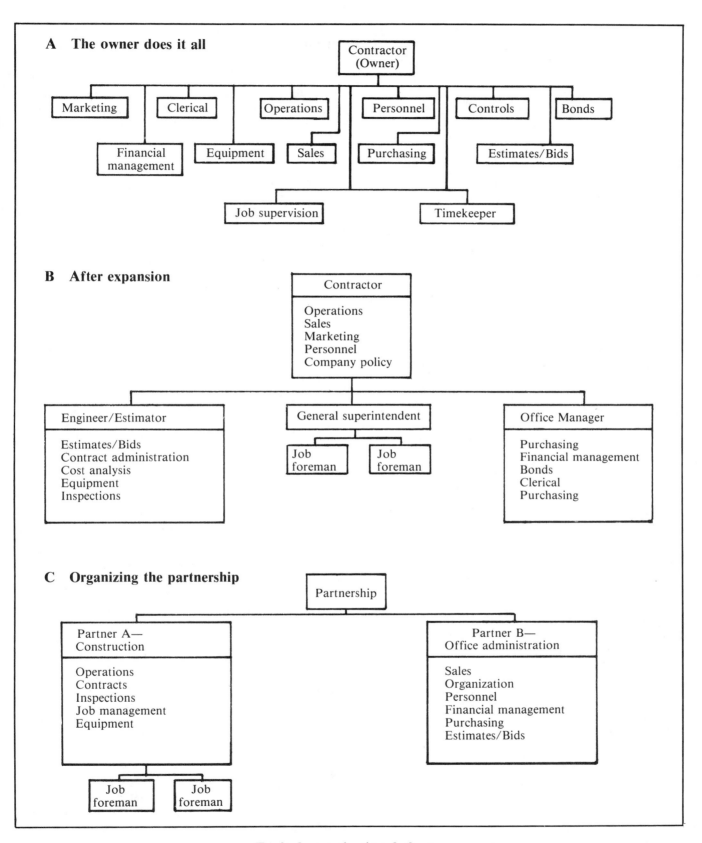

Typical organizational charts
Figure 18-5

When your company is small, it's up to you to do almost everything. In this case the organizational chart might look something like Section A in Figure 18-5. As the company grows, you'll add some specialists to take some of the load off of you. Your organization will change, and your chart will look more like Section B. If you decide to take on partners instead of employees to share the load, you'll need a chart that shows the division of responsibilities among the partners. Look at Section C.

Draw an organizational chart for your company as it exists now, and update it every time you make significant changes. Post it in a prominent place, so you and your employees are constantly aware of the scope of responsibilities.

What Are Your Bonding Requirements?

Getting bonds is a necessity of life for many construction contractors — especially those handling larger jobs. Not being able to get a bond (or liability insurance) keeps many contractors from bidding on the more profitable jobs. Bonding companies provide bonds for a certain percentage of the contract price. There are three main types of bonds:

1) *Bid bonds* assure that the bidder is prepared to perform the work according to the terms of the contract if successful in the bid.

2) *Performance bonds* assure completion of the job according to plans and specifications.

3) *Payment bonds* assure those dealing with the bonded contractor that they will be paid.

Bonding plays an important role in the competition between construction contractors. The owner, by requiring that the contractor is bonded, is more or less assured of adequate completion of the job. Inexperienced contractors with too little assets to meet bonding requirements can't bid. That puts all bidders on about the same footing when the owner is comparing bid prices.

Answer the questions on bonding in Figure 18-6. How often will you need a bond? Where will you get it? On what terms?

Bonding companies won't issue a bond unless the contractor can show experience and the organizational and financial capacity to complete the project. This can be a real stumbling block to the new construction firm. But when you get a bond, you may find that the banks you're trying to borrow from are more willing.

The Small Business Administration has a surety bond program designed to help small and emerging contractors who have been denied bonds. SBA is authorized to guarantee up to 90 percent of losses incurred under bid, payment, or performance bonds on contracts up to $1,000,000. Application for this assistance is available from any SBA field office.

What Are Your Personnel Requirements?

You'll have to decide whether to carry a permanent crew or hire workers as needed. Will you use union or nonunion labor? How many workers will you need? What will they cost in wages and fringe benefits? What about a foreman? Will you need clerical help? Answer these questions on Figure 18-6.

What Equipment Do You Need?

What special equipment will you need (assuming that your work force will supply their own hand tools)? Will you rent or buy? What about office space? Answer on Figure 18-6.

Put Your Plan into Dollars

An important part of management is financial management. The financial part of the job has to be planned as carefully as the actual construction. Your contract price must cover the direct and indirect construction costs, a share of overhead, and if you've planned well, a little profit.

Your accountant will help you set up the accounting system that meets your needs. But you have to do most of the financial planning yourself. You establish the goals and set your own limits.

In your financial planning, the first consideration is where the dollars will come from. In dollars, how much business (sales) will you be able to do in the next 12 months? Fill in the top line of Figure 18-7, the cash forecast worksheet, with the amount of cash you expect to take in for each of the months in the next year. Try to be as realistic as possible, and remember to anticipate the weather in your area.

Expenses

When 12 months of income are projected, you've anticipated annual sales volume. Now you need to think about expenses. For example, if you plan to do $100,000 worth of work, how much will it cost you? And even more important, what will be left over as profit at the end of the year?

Profit is your payoff. Even after you pay

Bonding

*Will I need bonding: often*_____ *occasionally*_____ *seldom*_____

*Where will I get the bond?*_____

*What will the terms be?*_____

Personnel

*Will I carry a permanent crew or hire workers as needed?*_____

*Will I use union or nonunion labor?*_____

*How many workers will I need?*_____

*What is the hourly rate I will pay?*_____

*What will fringe benefits cost?*_____

*Will I supervise the work myself or hire a foreman?*_____

*If I hire a foreman, what will his salary be?*_____

*Will I need clerical help?*_____ *What will it cost?*_____

Equipment

Equipment	Rent	Buy		My cost
_____	_____	_____	$	_____
_____	_____	_____	$	_____
_____	_____	_____	$	_____
_____	_____	_____	$	_____
_____	_____	_____	$	_____

*Will I need an office or use my home?*_____

*If I will need an office, what will the rent and other expenses be?*_____

_____

_____

Bonding, personnel and equipment requirements
Figure 18-6

yourself a salary for living expenses, your business must make a profit if it's to survive and pay back the money and time you invest in it. Profit helps your firm grow strong — and that means having a financial reserve for any lean periods.

Figure 18-8, the expenses worksheet, is designed to help you figure your yearly expenses. To use this worksheet, you need some figures, the typical operating ratios for your kind of contracting. They'll be more accurate than the sample ratios on

Estimated Cash Forecast	Jan.	Feb.	Mar.	Apr.	May	Jun.	Jul.	Aug.	Sep.	Oct.	Nov.	Dec.
Expected available cash												
Cash balance												
Expected receipts:												
Job A												
Job B												
Job C												
Bank Loans												
Total expected cash												
Expected cash requirements:												
Job A												
Job B												
Job C												
Equipment payments												
Taxes												
Insurance and bonds												
Overhead												
Loan repayments												
Total cash required												
Cash balance												
Total loans due to bank												

Cash forecast worksheet
Figure 18-7

the form. To find the figures for your specialty, check with the trade association that serves your area of the construction industry. If you can't find better ratios, look back to Chapter 15. Figure 15-1 has typical expense ratios for small general contractors and subcontractors.

Matching Money and Expenses
After you've planned for your month-to-month expenses, here's the next question: Will there be enough money coming in to meet these expenses and to keep you going if there's any down time between jobs?

The cash forecast is a management tool which can eliminate much of the anxiety that can plague you during lean months. Use the estimated cash worksheet, or ask your accountant to use it, to estimate the cash that will flow through your business during the next 12 months.

Is Additional Money Needed?
In your planning you may find periods when you'll be short of cash. For example, you need materials and supplies to start a new job, even though it will be a month or two before your first payment. What do you do in the interim if material suppliers won't ship without at least partial payment?

Your bank may be able to help with a short-term loan. Any banker you approach will want to know whether your company's financial condition is weak or strong. The bank officer will ask to see a balance sheet.

Expenses Worksheet

	Sample Figures for Specialty Contractors*	% of Your Sales	Your Annual Sales Dollar	Your Dollars Jan	Your Dollars Feb	Your Dollars Mar	Your Dollars Apr	Your Dollars May	Your Dollars Jun	Your Dollars Jul	Your Dollars Aug	Your Dollars Sep	Your Dollars Oct	Your Dollars Nov	Your Dollars Dec
Sales		100.00%													
Cost of sales		44.45													
Gross profit		55.55													
Controllable expenses															
Outside labor		1.15													
Operating supplies		2.34													
Gross wages		22.78													
Repairs and maintenance		.59													
Advertising		1.12													
Car and delivery		2.04													
Bad debts		.03													
Administrative and legal		.48													
Miscellaneous expenses		1.03													
Total controllable expenses		31.56													
Fixed expenses															
Rent		1.00													
Utilities		1.41													
Insurance		1.16													
Taxes and licenses		.85													
Depreciation		.10													
Total fixed expenses		6.18													
Total expenses		37.74													
Net profit (before income tax)		17.81													

*These percentages are taken from *Barometer of Small Business*, Accounting Corporation of America. These figures are presented only as a sample and refer to specialty contractors with an annual gross volume between $50,000 and $200,000. The percentages vary from one business to another.

Expenses worksheet
Figure 18-8

```
          Current Balance Sheet
           for (name of your firm)

          as of _____
                 (date)

                                             Assets

  Current assets                     $ _____
    Cash                             $ _____
    Receivables                      $ _____
    Inventories of supplies & tools  $ _____
      Total current assets           $ _____
  Fixed assets                       $ _____
  Other assets                       $ _____
  Total assets                       $ _____


                                           Liabilities

  Current liabilities                $ _____
    Notes payable                    $ _____
    Accounts payable                 $ _____
    Miscellaneous current liabilities $ _____
      Total current liabilities      $ _____
  Equipment contracts                $ _____
  Owner's equity                     $ _____
  Total liabilities                  $ _____
```

Balance sheet
Figure 18-9

Figure 18-9 is a blank balance sheet. Even if you don't need to borrow, use it. Or have your accountant use it to draw the picture of your firm's financial condition. If you don't need to borrow money, you may want to show your plan to the bank that handles your company's account. It's never too early to build good relations with your banker. Inevitably, the time will come when you'll have to borrow.

Control and Feedback
To make your plan work, you'll need feedback at the various stages of your management process. As the business manager, you need to plan the job, direct the job, and control the job. That always includes financial planning. The management controls you set up should supply you with the information you need to keep your operation "on the money."

During the planning stage, you'll be estimating costs and bidding. Use your job cost analysis to make sure that the job is going to make a profit. And then make sure estimated costs and actual costs are similar. This requires good supervision and a work schedule, competent tradesmen, and your personal follow-up.

Is Your Plan Workable?
Now that you've planned this far, step back and take a look at your plan. Is it realistic? Can you do enough business to make a living?

Now's the time to revise your plan if it isn't workable, not after you've committed time and money. If you feel that some revisions are needed before you start implementing your plan, then make them. Go back to the cash flow and adjust the figures. Better yet, show your plan to someone who hasn't had a hand in making it. You may need fresh ideas. Your banker, contact man at the SBA, or any experienced contractor, will be able to point

Action	Completion date
_____	_____
_____	_____
_____	_____
_____	_____
_____	_____
_____	_____
_____	_____
_____	_____

Action worksheet
Figure 18-10

out strong points and weaknesses in the plan.

If you have any strong doubts about your business or your ability to run it, it might be better to delay going into business until you feel as comfortable with the tools of management as you are with the tools of your trade.

Put Your Plan into Action

When your plan is as near on target as possible, you're ready to put it into action. Keep in mind that action is the difference between a plan and a dream. If a plan isn't acted upon, it's of no more value than a pleasant dream that evaporates when you wake up.

The first action step would be acquiring enough capital to get started. Do you already have the money? Will you borrow it from friends, relatives, or a bank? Where and when will you hire your employees?

What else needs to be done? Look for positive action steps that will get your business rolling. For example, where and how will you get whatever licenses you need to be a contractor? The regulations for businesses may vary for different kinds of business and certainly will vary from state to state. You can find out by contacting the various levels of government, including the IRS for federal tax regulations — but your best bet is to go to your local SBA office. They can give you specific information for all levels of government. Local chambers of commerce can also often help you in this area.

Use Figure 18-10 to list the things that you must do to get your business off the drawing board and into action. Give each item a date so it can be done at the right time.

Keep Your Plan up to Date

How many people in this world can predict the future? Very few indeed! You can expect things to change. You can expect circumstances to be different from what you expected. This is only natural. The difference between successful and unsuccessful planning is often only the ability to keep alert and watch for changes. Stay on top of changing conditions and adjust your plan accordingly.

In order to adjust your plan to account for changes, you must:

1) Be alert to the changes occurring in your industry, your market, and in your customers.

2) Check your plan against these changes.

3) Determine what revisions, if any, are needed in your plan.

Once a month or so, go over your plan. See whether it needs adjusting. If revisions are needed, make them and put them into action. You'll be keeping your business on track *and* in the money.

Chapter 19

Selecting the Legal Structure for Your Firm

When you're thinking of going into business for yourself, one of the first things you have to decide is what legal structure is best for your company. This chapter will help you make that decision. But even if you've been in business for a while, keep reading. There are many reasons why you might want to take a second look at the legal structure you're operating under. Tax laws and your company change as time passes. You may want to consider a change in the organization of your business.

Each form of business organization has its advantages and disadvantages. In this chapter I'll briefly discuss the pros and cons of each structure. To make the change, you'll probably need the assistance of both a lawyer and an accountant. But understanding the information in this chapter should help reduce your legal and accounting bills.

If you were to compare starting a business with playing a card game, you might say, ''The game is just for fun, but business is business. They're not the same at all.'' Well, you'd be right. But there are some important similarities.

Playing a card game well requires skill, strategy, planning, and most important, a thorough knowledge of the rules. Going into business requires skill (the knowledge of your craft or trade), and it also requires strategy and planning. Most important, to be successful in business, you must understand the rules (or the laws) by which you must conduct your business. Your planning and

strategy have to consider all local, state, and federal laws, and business customs that affect every construction company. And then, like the card game, luck comes into play.

Like many card games, business can be played several ways: as a proprietorship, partnership, or corporation. In fact, there are even variations on these three basic forms of business organization. Let's look at each of the three so you can choose the legal structure that fits your business best. Each has certain general advantages and disadvantages and has to be considered in the context of your circumstances, goals, and needs. Let's start with the sole proprietorship.

The Sole Proprietorship

The sole proprietorship is usually a business which is owned and operated by one person or a husband and wife. To establish a sole proprietorship, you need only a business license (if required by your city, county or state). It's the simplest and most common form of business organization.

Advantages of the Sole Proprietorship

There are five general advantages of a sole proprietorship over the other legal forms. We'll take them one at a time:

Ease of formation— There's little paperwork and few legal restrictions associated with running a proprietorship. It needs little or no governmental approval and is usually less expensive to set up than a partnership or corporation.

Sole ownership of profits— You're not required to share profits with anyone.

Control and decision-making rests in one owner— There are no co-owners or partners to consult (except possibly your spouse).

Flexibility— You can respond quickly to business needs. The proprietor makes all the decisions.

Taxes and control— You're relatively free from government control and special taxation. Income and expenses are reported on the proprietor's form 1040 C. The business itself doesn't file a tax return.

Disadvantages of the Sole Proprietorship

Unlimited liability— You, as proprietor, are responsible for the full amount of business debts, which may exceed your total investment. This liability could put all of your assets, including your house and car, in jeopardy. For other kinds of liability, such as physical loss or personal injury, be sure to carry adequate insurance coverage.

Unstable business life— The business may be crippled or terminated when the proprietor can't work or dies.

Less capital— If money is needed, the proprietor has to borrow it or contribute it himself.

Borrowing difficulty— It's usually very hard to borrow the working capital the business needs.

Relatively limited viewpoint and experience— There's only one owner to bring knowledge and experience into the business.

If you're just starting a small business, you'll probably start as a proprietorship. That reduces start-up expenses. Later, if the business grows and you feel the need, you can form a partnership or corporation.

The Partnership

The Uniform Partnership Act, adopted by many states, defines a partnership as "an association of two or more persons to carry on as co-owners of a business for profit." Though not specifically required by the Act, written Articles of Partnership are customarily executed. This is the contract that spells out the rights and duties of each partner: the contribution by the partners into the business (whether financial, material or managerial) and the roles of the partners in the business. Here are the issues usually covered in a partnership agreement:

Name, Purpose, Domicile
Duration of Agreement
Character of Partners (general or limited, active date)
Business Expenses (how they'll be handled)
Authority (each partner's authority in conducting of business)
Separate Debts
Books, Records, and Method of Accounting
Division of Profits and Losses
Draws or Salaries
Rights of Continuing Partner
Death of a Partner (dissolution and winding up)
Employee Management
Release of Debts
Sale of Partnership Interest
Arbitration
Modification of Partnership Agreement
Settlement of Disputes
Required and Prohibited Acts
Absence and Disability

Some of the characteristics that distinguish a partnership from other forms of business organization are the limited life of a partnership, unlimited liability of at least one partner, co-ownership of the assets, mutual agency, share of management, and share in partnership profits.

State law requires at least one general partner who accepts unlimited legal liability for the partnership. If the partnership is sued, the general partner can lose his investment, as well as everything he owns personally. There may also be limited partners who take no part in how the partnership is run. They're purely passive investors. But their losses are limited to whatever they've invested in the partnership. So long as they don't participate in conduct or control of the business, limited partners won't have the same liability exposure as a general partner.

Advantages of the Partnership

Ease of formation— Legal formalities and expenses are few compared to the requirements for creation of a corporation.

Direct rewards— All partners will want to contribute to the success of the business so they can share in the profits. That usually brings together the talents of several people — and may help guarantee success where a person working alone might not be able to succeed.

More capital available— The more partners, the more cash may be available. Some partners may be brought in just for the cash they can contribute.

Flexibility— A partnership may be more flexible in the decision-making process than a corporation (but perhaps less flexible than a sole proprietorship). Of course, the more partners, the more difficult it may be to get an agreement.

Taxes and control— More freedom from government control and special taxation than a corporation, but more restrictions than a proprietorship — especially if partners are brought in solely as investors. The partnership itself pays no taxes. It merely files an "information return" showing a gain or loss. Each partner takes his share of the gain or loss on his personal tax return.

Disadvantages of a Partnership
Unlimited liability of at least one partner— All partners other than limited partners should have liability insurance for their own protection.

Unstable life— The death of any partner usually ends the partnership. A new partnership agreement will be needed if the business is to continue.

Relatively hard to raise large sums of capital— This is particularly true when cash is needed for a long term. Partnership shares can't be sold as easily as shares in a corporation. Partners usually don't want to be locked in forever. After a few years, most partners will usually want out. If the agreement doesn't give them some way to withdraw and take their money, many prospective partners won't want to participate. That can be a problem in a construction business that depends on the funds partners have contributed.

Binding acts— The firm is bound by the acts of just one partner as agent.

Difficulty of disposing of partnership interest— The buying out of a partner may be difficult unless specifically arranged for in the written agreement.

The Corporation
The corporation is by far the most complex of the three business forms. Corporate law is a complex issue. I'm not a lawyer and won't give legal advice. But I'll summarize most of what you need to know about corporations in the next few paragraphs.

As defined by Chief Justice Marshall's famous decision in 1819, a corporation "is an artificial being, invisible, intangible, and existing only in contemplation of the law." In other words, a corporation is a separate legal entity, distinct from the individuals who own it.

Forming a Corporation
A corporation is formed by the authority of a state government. Corporations that do business in more than one state must comply with the Federal laws regarding interstate commerce, and with the laws of the states where they do business. But they need to be incorporated in only one state.

Ordinarily, the first step in forming a corporation is to take subscriptions for capital stock and create a tentative organization. Then approval must be obtained from the Secretary of State in the state in which the corporation is formed. This approval is in the form of a charter for the corporation, stating the powers and limitations of the particular enterprise.

Advantages of the Corporation
Limits the stockholder's liability to a fixed amount of investment— Stockholders generally aren't liable for the debts of the corporation.

Ownership— Stock usually can be sold to any interested investor.

Stability and relative permanence of existence— Since the corporation has a separate legal existence, it continues to exist and do business even after the death of an owner.

Ease of raising capital— Investors get stock in return for their money. A corporation with good prospects can often raise large amounts of cash. Lenders such as banks usually consider well-capitalized corporations to be good risks as borrowers and probably won't require personal guarantees from stockholders.

Delegated authority— The owners often delegate authority to hired professional managers who are responsible for the results of the company. Of course, one or more of the owners may manage the company on behalf of the rest of the owners.

Ability— The corporation has the ability to draw on the expertise and skills of more than one individual.

Disadvantages of the Corporation

Manipulation— Minority stockholders often have little effective voice in how the corporation is run.

Paperwork— It usually costs at least $1,000 in fees and taxes to set up a corporation.

Incentive— There may be less incentive to produce if the manager doesn't share in profits.

Taxes— There's a double tax: corporate profits are taxed, of course. And if any of those profits are distributed as dividends to shareholders, the shareholders also have to pay additional income tax. However, a Subchapter S small business corporation escapes double taxation. Like a partnership, the Subchapter S corporation pays no taxes. Instead, stockholders take gains and losses on their personal income tax return.

To qualify for the Subchapter S election, the corporation must have ten or fewer shareholders, all of whom are individuals or estates. There can be no nonresident alien shareholders. There can be only one class of outstanding stock. All shareholders must consent to the election, and a specific portion of the corporation's receipts must be derived from active business rather than certain passive investments. No limit is placed on the size of the corporation's income and assets.

If you're interested in taking this route, your attorney and accountant can help you handle the paperwork. It's a complicated matter to be handled by experts.

Consider These Questions

In deciding which legal structure is best for you, review these eight questions:

1) What is the size of the risk? That is, what is the amount of the investors' liability for debts and taxes?

2) Would the firm continue if something happened to the principal or principals?

3) What legal structure would insure the greatest adaptability of administration for the firm?

4) What is the influence of applicable laws?

5) What are the possibilities of attracting additional capital?

6) What are the needs for and possibilities of attracting additional expertise?

7) What are the costs and procedures in starting up?

8) What is the ultimate goal and purpose of the enterprise, and which legal structure can best serve its purposes?

As a small business owner, you're required to wear many hats, but you're not a lawyer, certified public accountant, marketing specialist, production engineer, environmental specialist, and so on. So get the facts from the experts before making decisions. Find professional counsel so you don't misunderstand technical or legal issues. Good advice can help you avoid making bad decisions and false starts that require backtracking and added expense. This is especially true when you're deciding what legal form to adopt.

Chapter 20

Your Business and the SBA

The U.S. Small Business Administration is a small, independent federal agency, created by Congress in 1953 to assist, counsel, and guide the millions of American small businesses which form a large proportion of the country's economy.

The function of the SBA is to help people get into business and stay in business. To do this, the SBA carries out programs and policies that will help small business. The SBA provides prospective, new, and established people in the small business community with financial assistance, management counseling, and training. It also helps get and direct government contracts for small firms. The agency makes special efforts to assist women, minorities, the handicapped, and veterans to get into business, and stay in business.

What Is a Small Business?
The SBA generally defines a small business as one which is independently owned and operated, and is not dominant in its field. To be eligible for SBA loans and other assistance, a business must meet a size standard. Any construction company with fewer than 500 employees is likely to qualify as a small business.

Who Is Eligible?
Most small, independent businesses are eligible for SBA assistance. Under the Disaster Loan Recovery Program, owners of both small and large businesses are eligible for SBA Disaster Loan Assistance. So are homeowners, renters, and nonprofit organizations.

Helping Women Get into Business
Women, of course, are eligible for all SBA loan and assistance programs and counseling services. But helping women become successful entrepreneurs is a special goal of SBA. Women make up more than half of America's population, but they own less than a fourth of its businesses.

Kinds of Financial Assistance Available
The SBA is able to offer several different kinds of financial assistance to small businesses. These include regular business loans, small general contractor loans, and surety bonds.

Regular Business Loans
The SBA offers a variety of loan programs to eligible small businesses that can't borrow on reasonable terms from conventional lenders without government help.

Most of SBA's business loans are made by private lenders and then guaranteed by the agency. Guaranteed loans carry a maximum of $500,000 and SBA guarantees as much as 90 percent. Maturity may be up to 25 years. The average size of a guaranteed business loan is $155,000 and the average maturity is about eight years.

A few business loans are made directly by SBA, up to a maximum of $150,000. The SBA, under law, cannot consider making a direct loan unless a private lender (usually a bank) refuses to make a loan itself or take part in an SBA-guaranteed loan. Funds authorized for direct loans are limited, and demand usually exceeds supply.

Small General Contractor Loans

Small general contractor loans are available to assist small construction firms with short-term financing. Loan proceeds can be used to finance residential or commercial construction or rehabilitation of property for sale. Proceeds can't be used for owning and operating real estate for investment purposes.

Surety Bonds

SBA is committed to making the bonding process accessible to small and emerging contractors who, for whatever reasons, can't get the bonds they need. It can guarantee to a qualified surety up to 90 percent of losses incurred under bid, payment, or performance bonds on contracts up to one million dollars. The contracts may be used for construction, supplies, or services provided by either prime or subcontractor for government or nongovernment work.

Loan Administration

After a loan has been made, SBA personnel in district offices service the loan to help assure borrower success. In the participation loans, SBA works with banks in troublesome situations. In the case of direct loans, SBA personnel work directly with borrowers.

An increasing number of banks today take part in what SBA calls its *Certified Lending Program.* Under this program, banks acting under SBA supervision handle much of the necessary paperwork and review client financial status. This speeds up loan processing and frees SBA personnel for other assistance to small businesses.

When loan recipients have difficulty repaying the loan, the SBA attempts to mitigate losses, both to the government and borrowers, by a variety of means. They may provide Service Corps of Retired Executives (SCORE) counseling or remedial loan adjustments. In the event of business failure, they recover funds through disposition of the business assets and other collateral security, or through reliance on the pledge of any guarantors.

Management Assistance Program

Statistics show that most small business failures are due to poor management. That why the SBA is anxious to improve the management ability of small business owners.

The SBA management assistance program includes free individual counseling, courses, conferences, workshops, problem clinics, and a wide range of publications. Counseling is provided through programs established by the SBA's management assistance staff, the Service Corps of Retired Executives and its corollary organization of active business men and women, the *Active Corp of Executives* (ACE), and other professional associations. SBA tries to match the need of a specific business with the expertise available through its counseling programs. It's also constantly trying to involve private-sector organizations and institutions in overall management assistance. Here's a brief summary of what these programs include:

SCORE and ACE: These volunteer groups help small business executives solve their operating problems through one-on-one counseling. But counseling isn't limited to small businesses in trouble. It's available as well to managers of successful firms who want to improve their performance.

Small Business Institutes: SBI's have been organized through the SBA on almost 500 university and college campuses as another way to help small businesses. At each SBI, senior and graduate students at schools of business administration, and their faculty advisors, provide on-site management counseling. Students are guided by the faculty advisors and SBA management assistance experts and receive academic credit for their work.

Small Business Development Centers: SBDC's draw from resources of local, state, and federal government programs, the private sector, and university facilities to provide managerial and technical help, research studies, and other types of specialized assistance of value to small business. These university-based centers provide individual counseling and practical training for small business owners.

Business Management Courses: Courses in planning, organization, and control of a business are co-sponsored by SBA in cooperation with educational institutions, Chambers of Commerce, and trade associations. Courses generally take place in the evening and last from six to eight weeks. In addition, conferences covering such subjects as working capital, business forecasting, and marketing, are held for established businesses on a regular basis. SBA conducts Pre-Business Workshops, dealing with finance, marketing assistance, types of business organization, and business site selection for prospective business owners. Clinics that focus on particular problems of small firms in specific industrial categories are held as they're needed.

Management, Marketing, and Technical Publications: The SBA issues management, marketing, and technical publications on hundreds of topics. The publications are available to established managers of small firms concerned about specific management problems and various aspects of business operations. Several SBA publications are available to builders free of charge. You can get others for a small fee from the U.S. Government Printing Office. Brochures describing the SBA's programs are available at all field offices.

Assistance to Veterans: The SBA makes special efforts to help veterans get into business or to expand existing veteran-owned small firms. Acting on its own, or with the help of veterans' organizations, it sponsors special business training workshops for veterans. Each SBA office has a veterans' affairs specialist to help veterans with loans and training. If you're a veteran and want to enter the building business or expand your existing business, you might want to check this out.

How to Get Help from the SBA

The Small Business Administration is organized into three operational levels:

• The central office in Washington, D.C., which determines policy, works with the White House, other Executive Branch agencies and departments, and Congress, to provide management and direction of SBA programs nationwide.

• Regional offices, located in 10 major cities around the country: Boston, New York, Philadelphia, Atlanta, Kansas City, Dallas, Denver, Chicago, Seattle and San Francisco. Regional offices don't make individual loans or offer specific assistance to individuals or companies.

• District offices, located throughout the country, are staffed by a team of experts in lending, procurement and assistance. They process loan applications, offering individual management assistance, and coordinating other small business services.

District offices are the real contact point for small businesses needing information or assistance. There's a list of district offices, with addresses and telephone numbers, at the end of this chapter. In addition, help is available through several other sources:

Retired business executives are organized under SCORE. These men and women volunteer their services to small businesses seeking managerial assistance. SCORE volunteers work in each district office. Their services are free. If you need this kind of help, contact your nearest district office.

Small Business Institutes on hundreds of university and college campuses offer free guidance and assistance to troubled small firms. Small Business Development Centers have been organized as a pilot program on a limited number of university and college campuses, to bring under one roof a variety of small business managerial and financial information and assistance. Call your SBA district offices for the SBI's or SBDC's in your area.

There's no charge for any SBA service. Help is there for the asking. Don't wait until you're up to your neck in trouble. Miracles aren't part of their service.

Summary

Success in the construction business doesn't come easy. It requires hard work and constant study. You must know how to estimate, how to build, *and* how to manage. There are no magic formulas.

In these two volumes we've covered a lot of ground. You've gained an insight into construction contracting and pretty well have a handle on how to do the job. But don't expect to absorb all the material at one reading. Study it until it becomes clear. There's a lot to digest.

Here's one of the most important concepts you can take away with you. Think *quality*. Sloppy builders don't last long. Quality doesn't only sell homes — it sells *you.* In building, that's the name of the game. It has people standing in line waiting for you while your competitors hunt for a bankruptcy lawyer. I speak from over 25 years' experience. If I can't build quality, I don't take the job.

Here's a strange fact: Quality builders seem to have little trouble being paid all that's due them — and on time! Call it the Jones' rule of anti-sloppiness. Call it anything you want, but I've always tried to build quality and I've been paid every dime due me. I've even remodeled homes on a handshake — and been paid with no problem.

But I don't recommend you try that. I might have just been lucky. Always get a contract.

I hope you love construction sites the way I do, whether it's an add-on or a high-rise. It's the smell of the wood, the cement and the steel. It's the yells, the talk, the laughter of the busy workers. It's the sound of hammers and saws and the rumblings of trucks.

Building is a good way to make a living. And when all the smells fade and all the sounds have died, you stand back and look at what you've created. It's a thing of beauty. It'll be there for a good long time.

SBA Field Offices

Alabama
2121 8th Ave. N.
Suite 200
Birmingham, AL 35203
(205) 254-1344

Alaska
8th and C Streets
Room 1068
Anchorage, AK 99501
(907) 271-4022

Arizona
2005 N. Central Ave.
5th Floor
Phoenix, AZ 85004
(602) 261-3732

300 W. Congress St.
Federal Bldg., Box 33
Tucson, AZ 85701
(602) 629-6715

Arkansas
320 W. Capitol Ave.
Suite 601
Little Rock, AR 72201
(501) 378-5871

California
2202 Monterey St.
Room 108
Fresno, CA 93721
(209) 487-5791

660 J St.
Room 215
Sacramento, CA 95814
(916) 551-1445

880 Front St.
Room 4-S-29
San Diego, CA 92188
(619) 293-7252

*450 Golden Gate Ave.
Room 15307
San Francisco, CA 94102
(415) 556-7487

211 Main St.
4th Floor
San Francisco, CA 94105
(415) 974-0642

350 S. Figueroa St.
6th Floor
Los Angeles, CA 90071
(213) 894-2977

Fidelity Federal Bldg.
2700 N. Main St.
Suite 400
Santa Ana, CA 92701
(714) 836-2829

Colorado
*Executive Tower Building
1405 Curtis St.
22nd Floor
Denver, CO 80202
(303) 844-5441

712 19th St.
Room 407
Denver, CO 80202
(303) 844-2607

Connecticut
One Hartford Square W.
Suite 201
Hartford, CT 06106
(203) 722-3600

Delaware
844 King St.
Room 5207
Wilmington, DE 19801
(302) 573-6294

District of Columbia
1111 18th St., N.W.
6th Floor
Washington, D.C. 20036
(202) 634-4950

Florida
2222 Ponce De Leon Blvd.
5th Floor
Miami, FL 33134
(305) 536-5521

400 W. Bay St.
Room 261
Jacksonville, FL 32202
(904) 791-3782

700 Twiggs St.
Room 607
Tampa, FL 33602
(813) 228-2594

3500 45th St.
Suite 6
West Palm Beach, FL 33407
(305) 689-2223

Georgia
*1375 Peachtree St., N.E.
5th Floor
Atlanta, GA 30367
(404) 881-4999

1720 Peachtree Road, N.W.
6th Floor
Atlanta, GA 30309
(404) 881-4749

52 N. Main St.
Room 225
Statesboro, GA 30458
(912) 489-8719

Guam
Pacific News Bldg.
Room 508
238 O'Hara St.
Agana, GM 96910
(671) 472-7277

Hawaii
300 Ala Moana
Room 2213
Honolulu, HI 96850
(808) 546-8950

Idaho
1020 Main St.
Suite 290
Boise, ID 83702
(208) 334-1696

Illinois
*230 S. Dearborn St.
Room 510
Chicago, IL 60604
(312) 353-0359

219 S. Dearborn St
Room 437
Chicago, IL 60604
(312) 353-4528

Washington Bldg.
4 N. Old State
Capitol Plaza, 1st Floor
Springfield, IL 62701
(217) 492-4416

Indiana
New Federal Bldg.
575 N. Pennsylvania St.
Room 578
Indianapolis, IN 46204
(317) 269-7272

Iowa
210 Walnut St.
Room COM
Des Moines, IA 50309
(515) 284-4422

373 Collins Rd. N.E.
Room 100
Cedar Rapids, IA 52402
(319) 399-2571

Kansas
Main Place Bldg.
110 E. Waterman St.
Wichita, KS 67202
(316) 269-6571

Kentucky
600 Federal Place
Room 188
Louisville, KY 40202
(502) 582-5976

Louisiana
1661 Canal St.
Suite 2000
New Orleans, LA 70112
(504) 589-6685

500 Fannin St.
Room 6B14
Shreveport, LA 71101
(318) 226-5196

Maine
40 Western Ave.
Room 512
Augusta, ME 04330
(207) 622-8378

*Regional Office

294

Maryland
10 N. Calvert St.
3rd Floor
Baltimore, MD 21202
(301) 962-4392

Massachusetts
*60 Batterymarch St.
10th Floor
Boston, MA 02110
(617) 223-3204

150 Causeway St.
10th Floor
Boston, MA 02114
(617) 223-3224

1550 Main St.
Room 212
Springfield, MA 01103
(413) 785-0268

Michigan
477 Michigan Ave.
McNamara Bldg.
Room 515
Detroit, MI 48226
(313) 226-6075

220 W. Washington St.
Room 310
Marquette, MI 49885
(906) 225-1108

Minnesota
610-C Butler Square
100 N. 6th St.
Minneapolis, MN 55403
(612) 349-3550

Mississippi
One Hancock Plaza
Suite 1001
Gulfport, MS 39501
(601) 863-4449

100 W. Capitol St.
New Federal Bldg.
Suite 322
Jackson, MS 39269
(601) 960-4378

Missouri
*911 Walnut St.
13th Floor
Kansas, City, MO 64106
(816) 374-5288

1103 Grand Ave.
6th Floor
Kansas, City, MO 64106
(816) 374-3419

Federal Court House Bldg.
339 Broadway
Room 140
Cape Girardeau, MO 63701
(314) 335-6039

815 Olive St. .
Room 242
St. Louis, MO 63101
(314) 425-6600

309 N. Jefferson St.
Springfield, MO 65805
(417) 864-7670

Montana
2601 First Ave. N.
Room 216
Billings, MT 59101
(406) 657-6047

Nebraska
Empire State Bldg.
19th & Farnam Streets
2nd Floor
Omaha, NB 68102
(402) 221-4691

Nevada
301 E. Stewart St.
Room 301
Las Vegas, NV 89125
(702) 388-6611

P.O. Box 3216
50 S. Virginia St.
Room 238
Reno, NV 89505
(702) 784-5268

New Hampshire
55 Pleasant St.
Room 209
Concord, NH 03301
(603) 224-4041

New Jersey
1800 E. Davis St.
Room 110
Camden, NJ 08104
(609) 757-5183

60 Park Place
4th Floor
Newark, NJ 07102
(201) 645-2434

New Mexico
Patio Plaza Building
5000 Marble Ave. N.E.
Room 320
Albuquerque, NM 87100
(505) 766-3430

New York
*26 Federal Plaza
Room 29-118
New York, NY 10278
(212) 264-7772

445 Broadway
Room 236-B
Albany, NY 12207
(518) 472-6300

111 W. Huron St.
Room 1311
Buffalo, NY 14202
(716) 846-4301

333 E. Water St.
4th Floor
Elmira, NY 14901
(607) 734-8130

35 Pinelawn Rd.
Room 102-E
Melville, NY 11747
(516) 454-0750

26 Federal Plaza
Room 3100
New York, NY 10278
(212) 264-4355

100 State St.
Room 601
Rochester, NY 14614
(716) 263-6700

100 S. Clinton St.
Room 1071
Federal Bldg.
Syracuse, NY 13260
(315) 423-5383

North Carolina
230 S. Tyron St.
Room 700
Charlotte, NC 28202
(704) 371-6563

North Dakota
657 2nd Ave. N.
Fargo, ND 58102
(701) 237-5771

Ohio
1240 E. 9th St.
Room 317
AJC Federal Bldg.
Cleveland, OH 44199
(216) 522-4180

85 Marconi Blvd.
Room 512
Columbus, OH 43215
(614) 469-6860

550 Main St.
Room 5028
Cincinnati, OH 45202
(513) 684-2814

Oklahoma
200 N.W. 5th St.
Suite 670
Oklahoma City, OK 73102
(405) 231-4301

Oregon
1220 S.W. Third Ave.
Room 676
Federal Bldg.
Portland, OR 97204
(503) 423-5221

Pennsylvania
*One Bala Cynwyd Plaza
231 St. Asaphs Rd.
Suite 640-W
Philadelphia, PA 19004
(215) 596-5889

One Bala Cynwyd Plaza
231 St. Asaphs Rd.
Suite 400-E
Philadelphia, PA 19004
(215) 597-5889

100 Chestnut St.
Room 309
Harrisburg, PA 17101
(717) 782-3840

960 Penn Ave.
Convention Tower
5th Floor
Pittsburgh, PA 15222
(412) 644-2780

*Regional Office

20 N. Pennsylvania Ave.
Room 2327
Wilkes-Barre, PA 18701
(717) 826-6497

Puerto Rico
Federal Bldg.
Room 691
Carlos Chardon Ave.
Hato Rey, PR 00918
(809) 753-4002

Rhode Island
380 Westminister Mall
5th Floor
Providence, RI 02903
(401) 528-4586

South Carolina
1835 Assembly St.
3rd Floor
Columbia, SC 29202
(803) 765-5376

South Dakota
101 S. Main Ave.
Suite 101
Sioux Falls, SD 57102
(605) 336-2980

Tennessee
404 James Robertson Pkwy.
Suite 1012
Nashville, TN 37219
(615) 251-5881

Texas
Federal Bldg.
Room 780
300 E. 8th St.
Austin, TX 78701
(512) 482-5288

400 Mann St.
Suite 403
Corpus Christi, TX 78408
(512) 888-3331

221 W. Lancaster Ave.
Room 1007
Ft. Worth, TX 76102
(817) 334-5463

*8625 King George Dr.
Bldg. C
Dallas, TX 75235
(214) 767-7643

1100 Commerce St.
Room 3C36
Dallas, TX 75242
(214) 767-0605

10737 Gateway W.
Suite 320
El Paso, TX 79902
(915) 541-7586

222 E. Van Buren St.
Suite 500
Harlingen, TX 78550
(512) 423-8934

2525 Murworth
Room 112
Houston, TX 77054
(713) 660-4401

1611 10th St.
Suite 200
Lubbock, TX 79401
(806) 762-7466

100 S. Washington St.
Room G-12
Marshall, TX 75670
(214) 935-5257

727 E. Durango St.
Room A-513
Federal Bldg.
San Antonio, TX 78206
(512) 229-6250

Utah
125 S. State St.
Room 2237
Salt Lake City, UT 84138
(801) 524-5800

Vermont
87 State St.
Room 205
Montpelier, VT 05602
(802) 229-0538

Virginia
400 N. 8th St.
Room 3015
Richmond, VA 23240
(804) 771-2617

Virgin Islands
Veterans Dr.
Room 283
St. Thomas. VI 00801
(809) 774-8530

4C & 4D Estate Sion Farm
Room 7
St. Croix, VI 00820
(809) 773-3480

Washington
*2615 4th Ave.
Room 440
Seattle, WA 98121
(206) 442-5676

915 Second Ave.
Room 1792
Seattle, WA 98174
(206) 442-5534

W. 920 Riverside Ave.
Room 651
Spokane, WA 99201
(509) 456-3783

West Virginia
168 W. Main St.
6th Floor
Clarksburg, WV 26301
(304) 623-5631

550 Eagan St.
Suite 309
Charleston, WV 25301
(304) 347-5220

Wisconsin
500 S. Barrow St.
Room 17
Eau Claire, WI 54701
(715) 834-9012

212 E. Washington Ave.
Room 213
Madison, WI 53703
(608) 264-5261

310 W. Wisconsin Ave.
Room 400
Milwaukee, WI 53203
(414) 291-3941

Wyoming
100 E. B St.
Room 4001
Casper, WY 82602
(307) 261-5761

*Regional Office

INDEX

Absolute volume 94
Accurate labor costs 29-33
ACE (Active Corps of Executives) 291
Action worksheet, business plan 285
Adhesive, resilient floor tile 215
Adjustment of float, CPM 233
Advertising
- basics 253-254
- effective 254
- planning 253, 276
- worksheet 277

Aggregate, plaster 178
Aluminum siding 194
American Softwood Lumber Standard 111
Analysis of unit costs 24
APA
- panel sheathing 130, 132
- plywood siding 186, 188
- Sturd-I-Wall 186, 187

Apron trim, estimating 208
Arch roof, area of 218
Arches, brick, labor 109
Architect, consulting 242
Architectural services 269
Area estimate 5
Area/rake conversion table 149
Arrow diagrams, CPM 223, 226, 229, 230-232
- dummy arrows 228

Asbestos shingles, estimating 156, 157
Ashlar lines 39
Ashlar stone, estimating 101-102
Asphalt roofing
- estimating 149-152
- production 152
- rolls 153

Asphalt shingles 150, 151, 152
Assistance to veterans, SBA 292
Attic stairs 199
Average winter low temperatures 161, 162
Awning windows 170

Back band trim, estimating 208
Backfill estimating 49-50
- by hand 55-56
- example 63, 65-66
- labor CEF 49

Backhoe production table 54
Backhoes 53-54
Balance sheet
- and income statement 248
- blank example 284
- description 282, 284

Balance sheet ratios 246-250
Balloon framing, estimating 126, 127
Bank excavation, labor 60
Bank measure 55
Barriers, vapor 160-161
Base cabinets, kitchen 203
Base trim 210
Baseboard, estimating 210
Basement excavation 45, 46
- example estimate 61, 62, 63

Basement plan 9-10
Basement stairs, estimating 199, 200-202
Batt or blanket insulation
- description 159
- estimating material 161-163
- labor CEF 163

Batterboards
- description 37
- using 38-39

Beating the competition 276
Bench marks 37
Bevel siding
- application details 194
- estimating 191
- materials and labor 193

Bid bonds 280
Bid, including all costs 20
Bituminous roofing, built-up 157
Block, concrete, estimating
- labor 90-91
- materials 87-90
- mortar 90
- sample estimate 91

Block, concrete, sizes & shapes 88
Block, glass, estimating 98
- labor and materials CEF 99

Board and batten siding 194, 195
Board feet
- calculating 113-114
- conversion table 115
- per SF of wall studs 125

Board products, square feet 175
Board sheathing
- application detail 130
- coverage table 132
- estimating 132-133

Bond correction factors, mortar 92, 93
Bonding requirements, planning 280-281
Bonds
- surety 280, 281
- types 280

Breaking load, tie wire 77
Brick
- chimneys, estimating 104, 106-108
- general data, CEF 8
- masonry, estimating 91-98
- modular, CEF 8

Brickwork material and labor 8, 97
Bridging, estimating 122-123
Broker, selling through 270
Building layout 37-40
Built-up bituminous roofing, estimating . . . 157, 158
Built-up girders, estimating 119, 121
Bulldozers 50-51
Business
- loans (SBA) 290-291
- management assistance (SBA) 291-292

Business plan
- analyzing 273-276
- finances 280-285
- marketing 276-278
- organizing work 278-280
- updating 285

Buying land
- before you buy 271
- from developers 270

Cabinets, estimating 202-204, 205
Calculating area
- gables 189, 191
- odd shapes 218

Calculating volume
- excavation 41-43
- irregular shapes 43
- pit 45-47

Calendar daily report 222-223, 225
Cap trim, estimating 208, 209
Carpentry, finish 199-211
Carpentry, rough 7, 110-145
Carpet, wall-to-wall 215
Casement windows 170
Cash forecasting 280, 282
Cash schedule 245
Cast-in-place terrazzo, labor 214
CEF (Construction Estimate File)
- backfilling, labor 49
- batt/blanket insulation, labor 163
- cabinets and tops, labor 205
- concrete block 6
- description 6
- excavating by machine 60
- excavating trench/pier by hand 45
- fireplace mantels, labor 208
- folding doors, labor 167
- glass block, materials/labor 99
- how to use 6-9
- insulation values 164
- modular brick 8
- overhead doors, labor 168
- poured insulation, labor/materials 163
- rigid board insulation, labor 163
- roof trusses, labor 144
- rough carpentry 7
- slabs on grade, labor/materials 68, 85
- sliding glass doors, labor 168
- stairs 202
- steel doors & frames, labor 167
- wall paneling, labor 185
- wood block flooring, labor 207

Ceiling joists, estimating 133-134
Ceramic tile, estimating 212-213
Change, keeping up with 238
Chart, organizational 279
Checking estimates 20-21
Checklist, estimate summary 13-16
Chimneys and fireplaces, estimating 104-109
Circle, area of 218
Clearing sites 40-41
Closing the sale 259
Collar beams, estimating 141, 142
Collar joints, mortar for 92, 93
Column footings 80-81
Column forms
- designing 78, 81
- example problem 78-79

Combining layers, insulation 162
Common
- rafter lengths 138
- roof styles 146
- types of wood flooring 204
- window abbreviations 168

Compiling an estimate 17-20
Concept of float, CPM 233
Concrete block
- CEF 6
- courses 90
- estimating 87-91
- sample estimate 91

Concrete, estimating 67-68
Concrete forms, designing 68-84
- columns 78-79
- floors 79-80
- foundations 79
- stairs 80
- walls 73-78

Concrete forms, estimating
- footings 80-82
- panels 80, 83-84

Concrete, reinforced, estimating 85-86
Concrete slabs, estimating 84-85
Condensation zones 160
Cone, area of 218
Construction Estimate File (CEF) 6
Construction Reports 264-266
Construction Specifications Institute (CSI) 25
Consulting with architect 242
Contingency, in estimate 18
Contract and proposal, sample 260-261
Contract cancellation notice 259, 260, 262
Contractor's labor burden 17-18
Control, of business plan 284
Conversion diagram, rafters 137
Corner bead, estimating 184
Corner bracing, estimating 126, 130
Corner squaring, in layout 38, 39
Corner treatment, siding 191, 193
Cornices, estimating 196-197, 198
Corporation
- advantages 288-289
- disadvantages 289
- forming 288

Cost data, from field 33-35
Cost estimate, excavation 60-66
Cost keeping (cost accounting)
- cost data from field 33-35
- description 23-24
- forms for 25-28
- labor costs 29-33
- unit costs, analysis 24-25

Courses, business management (SBA) 291-292
Critical Path Method (CPM)
- arrow diagrams 223
- description 223
- planning 223-229
- scheduling 229-234

Cross-bridging, estimating 123
Crown molding, estimating 210
Customer profile 275, 276
Cut stone, estimating 102-104
Cutting labor cost, brick 95
Cylinder, area of 218

Daily journal 25, 28, 29
Daily reports, from field 222-223, 224, 225
Datum plane 37
Decimal conversion table 57
Designing
- column forms 78-79
- concrete forms 68-73
- floor forms 79
- wall forms 73-78

Designs from suppliers 269
Determining elevations 37
Developing
- land 270-271
- plans 271-272

Dimension lumber grades 111-112
Dimensions
- lumber 113-114
- stairs 200

Direct overhead, in estimate 18
District offices, SBA 294-296
Door trim, estimating 207-209
Doors, estimating 165-168
Double-hung windows 169
Downspouts/gutters, labor 157, 158
Drag loaders 54, 55
Draglines 54
Drop siding, estimating 191, 193
Drying times, paint coatings 220

*E*ffective advertising ... 254
Elevation, in surveying ... 37
Elevations, plan ... 9, 11
  crew, building a ... 240
  fairness with ... 237
  job superintendent ... 239-240
Equipment requirements, planning ... 280, 281
Escalation (cost increases) ... 18-19
Estimate
  area ... 5
  checking ... 20-21
  compiling ... 17-21
  forms ... 120
  piece ... 5
  summary ... 33, 35
  summary checklist ... 13-16
  tips ... 21
  unit cost ... 5
  workbook ... 12
Estimated labor, rough carpentry ... 144, 145
Estimating. *See also* specific items
  brick ... 91-98
  chimneys and fireplaces ... 104-109
  concrete ... 67-86
  concrete block ... 87-91
  doors ... 165-168
  excavation ... 36-66
  exterior finish carpentry ... 186-198
  fireplaces and chimneys ... 104-109
  glass block ... 98, 99
  insulation ... 159-164
  interior finish carpentry ... 199-211
  introduction to ... 5-23
  lath and plaster ... 177-185
  ornamental plaster ... 185
  roof coverings ... 146-158
  rough carpentry ... 110-145
  specialty finishes ... 212-220
  stonework ... 98-104
  wallboard ... 172-177
  windows ... 168-171
Examining the site ... 12
Example bid ... 20
Excavation, estimating
  backfilling ... 49-50, 55-57
  clearing site ... 40-41
  cost estimate ... 60-61
  general ... 41-43
  power equipment ... 50-55
  quantity estimate ... 57-60
  staking out ... 37-40
  surveying ... 36-37
  trench and pit ... 43-49
  unit cost estimate ... 61-66
  volume, calculating ... 41-45, 47
Excavation lines ... 38
Expense ratios ... 246-247
Expenses ... 280-283
Exterior finish carpentry, estimating
  corner boards ... 195
  cornices ... 196-197, 198
  drip cap ... 195
  porch ceilings ... 197-198
  shakes ... 191, 192
  siding ... 186-190, 191-195
  soffits ... 197-198
  water table ... 195
Exterior framing, estimating
  balloon ... 126, 127
  corner bracing ... 126, 130
  modern braced ... 129
  western ... 124-126, 128
Exterior paint ... 217, 218, 219-220

*F*ace lines ... 39
Fasteners, lath & plaster ... 178-179
Federal law, notice to customer ... 262
Felt, saturated, estimating ... 152
FHA services ... 271-272
FICA (Social Security and Medicare) ... 17-18
Fill, grading and tamping ... 85
Financial assistance, types of (SBA) ... 290-291
Finish carpentry, exterior
  corner boards ... 195
  cornices ... 196-197, 198
  drip cap ... 195
  porch ceilings ... 197-198
  shakes ... 191, 192
  siding ... 186-190, 191-195
  soffits ... 197-198
  water table ... 195
Finish carpentry, interior
  cabinets ... 202-204
  fireplace mantels ... 206-207
  labor ... 210, 211
  molding or trim ... 207-210
  stairs ... 199-202
  wood flooring ... 204-206
Fireplace mantels, estimating ... 206-207
Fireplaces and chimneys, estimating ... 104-109
Firestops ... 126
First floor plans ... 10-11
Flashing, sheet metal, labor ... 157
Float, adjusting (CPM) ... 233
Floor form design ... 79-80
Floor framing, estimating
  bridging ... 122-123
  joists ... 121-122
  subflooring ... 123-124
Floor plan ... 9
Floor tile adhesive ... 215
Floor tile, estimating ... 212-214, 215, 216
Flooring
  ceramic tile ... 212-213
  laminated block ... 204
  nail schedule ... 206
  parquet ... 204
  plank ... 204
  resilient ... 213, 215
  strip ... 204
  terrazzo ... 213, 214
  wood, estimating ... 204-206
Flue lining sizes ... 104, 106
Folding doors, labor ... 167
Footing forms ... 80-82
Footing line ... 38-39
Footings
  column and pier ... 81
  stepped ... 82
  walls ... 81-82
Foreman's daily report ... 33, 34, 224
Form, estimate, sample ... 13-16
Forming a corporation ... 288
Forms, cost keeping ... 25, 28, 29-35
Forms, designing
  column ... 78-79, 81
  concrete ... 68-73
  floor ... 79-80
  footing ... 80-83
  foundation ... 79, 83
  panel ... 83-84
  slabs ... 84, 85
  stairs ... 80
  wall ... 73-78, 82-84
Forms, prefabricated ... 68-73
Forms, sectional ... 83-84
Foundation forms ... 79, 83
Framing, hip roof ... 142
Furring ... 176
FUTA (Unemployment Insurance) ... 17-18

*G*ambrel roof, area of ... 218
General excavation ... 41-43
Girders, built-up ... 119, 121
Glass block, estimating ... 98, 99
Glass doors, labor ... 168
Grading fill ... 85
Grading, lumber ... 110-113
Gutters and downspouts, labor ... 157, 158
Gypsum plaster
  cement plaster ... 181, 182
  finish coat ... 179
  lightweight plaster ... 184
  sanded plaster ... 183
Gypsum wallboard, estimating ... 172-177

*H*and excavation, estimating ... 47-48
Handsplit shakes, estimating ... 156-157
Hardboard estimating
  lap siding ... 189, 191
  panel siding ... 189-191
  shakes ... 191, 192
Hauling excavated dirt ... 54-55
Heaped capacity ... 55
Hip rafters, estimating ... 138-140
Hip roof, framing ... 142
Hip/valley conversion table ... 154
Horizontal
  sliding windows ... 170
  wood siding ... 191, 193
Horizontal application
  sheathing ... 131
  siding ... 187
Horizontal projection, roofs ... 147-149
House plan services ... 269
House style ... 269-270
Housing Starts ... 264, 266

*I*ndirect overhead ... 18
Installing
  doors ... 166-168
  walers ... 69
Insulation
  combining layers ... 161, 162
  estimating ... 161-164
  forms of ... 159-160
  placement ... 161
  values, common materials ... 161
Interior finish carpentry, estimating
  cabinets ... 202-204
  fireplace mantels ... 206-207
  molding or trim ... 207-210
  stairs ... 199-202
  wood flooring ... 204-206
Interior paint ... 216, 217, 219-220
Interior wallboard, estimating ... 172-177

*J*ack rafters, estimating ... 140
Job planning ... 240-244
Job superintendent ... 239-240
Jobs, number of ... 268
Joist spacing
  floors ... 121
  forms ... 79
Joist spans, ceiling ... 133
Joists, estimating
  ceiling ... 133-135
  floor ... 121-122

*K*itchen cabinets, estimating ... 202-205
Kitchen preference checklist ... 258

*L*abor burden, contractor's ... 17
Labor cost cutting, brickwork ... 95
Labor costs, keeping ... 29-33
Labor distribution record ... 29, 32
Labor estimates. *See* item name
Labor requirements, planning ... 241
Laminated block flooring ... 204
Land developing
  FHA ... 272
  plans ... 271
  site selection ... 270-271
Lannon stone ... 101-102
Lap siding
  estimating ... 189, 191
  hardboard ... 189
  plywood ... 186, 188-189
Latest event times, CPM ... 230
Lath and plaster, estimating
  corner bead ... 184
  fasteners ... 178-179
  furring ... 181
  gypsum finish coats ... 179
  labor ... 181, 184
  lath ... 178
  ornamental work ... 185
  plaster ... 177-178, 181-184
  surface area, figuring ... 179-180
Lath nailers, estimating ... 134
Legal structure, company
  corporation ... 288-289
  partnership ... 287-288
  sole proprietorship ... 286-287
Length, rafters ... 136-140
Liability insurance ... 17-18
Lightweight partition block ... 87-88
Lime, hydrated, yield ... 183
Linear feet, lumber measurement ... 114
Linear feet to board feet conversion ... 114, 115
Linings, flue ... 106
Listing
  doors ... 165-166
  rafter material ... 141-143
  windows ... 168-189
Loans, SBA
  regular business ... 290
  small general contractor ... 291
Lookouts, estimating ... 142
Loose fill, insulation ... 159-160
Lumber
  characteristics ... 113
  classifications ... 110
  dimensions ... 113-114
  estimates, taking off ... 114
  grades, dimension ... 112
  grading ... 110-112
  standard ... 111-112

*M*achine excavation ... 50-56, 60
Main stairs, estimating ... 199, 200-202
Management
  courses ... 291
  function of ... 235-236
  planning jobs ... 240-244
  publications ... 292
  team ... 239-240
Management assistance program (SBA) ... 291-292
Manhour guides, using ... 7-9
Mantels, estimating ... 206-207, 208

Markers, survey 36-37
Market, studying 267-268, 276
Marketing decisions 275-278
Marketing publications 292
Masonry, estimating
  brick 91-98
  chimneys 104, 105-109
  concrete block 87-91
  fireplaces 104-109
  glass block 98, 99
  stonework 98-104
Masonry, mortar for 90
Masterformat 12, 25-27
Material
  characteristics, soil 56
  estimate checklist, rough carpentry 116-118
  quantities, mortar 93
  transitions, siding 195
Maximum
  concrete pressure 73
  spans, stringers 79
  stud spacing 74
  tie wire spacing 76
  wale spacing 75
Measuring for excavation 40
Merchandising for the spec builder 268-272
Metal windows, prefabricated 170
Milled lumber 110-113
Millwork, labor, painting 219
Modern braced framing 129
Modular brick, CEF 8
Modular brick and mortar, estimating 92
Moisture content, sand 94
Molding, estimating 207-210
Money and expenses, matching 282
Monuments (survey) 36
Mortar, masonry
  estimating 92-94
  types of 90-94
  yield 93, 94
Mortar, plaster 181-184
Multi-form system, wall forms 72
Multiple coat application, plaster 179
Muntins 168

*N*ail requirements
  rough carpentry 145
  strip flooring 206
  wallboard application 174
Neat lines, staking out 39
New One-Family Houses Sold and For Sale 264, 265
New site grades 50
Nominal size, softwood lumber 111
Non-modular brick table 92
Notice to Customer Required by Federal Law 262

*O*ffering suggestions, management 237
Office personnel, record keeping 244
Operating ratios
  balance sheet ratios 247-250
  comparing 246
  expense ratios 246-247
Organizational charts, typical 279
Organizing to get job done 278-280
Ornamental plaster, estimating 185
Overhead doors, labor 168
Overhead, expenses 18

*P*aint
  binders 216, 217
  drying times 220
  estimating 215-220
  requirements, chart 217
  selection chart 218
Painting, labor 219-220
Paints, exterior 218
Panel forms, estimating 83-84
Panel sheathing, APA 130, 132
Panel siding, estimating 186, 188-189, 191
Paneling, labor 185
Parallel work elements, CPM 227
Parquet flooring, estimating 204
Partnerships 287-288
Payment bonds 280
Payroll summary 29, 31
Performance bonds 280
Performance-rated panels, subflooring 124
Perm ratings, vapor barriers 160
Perseverance, importance of 238
Personnel requirements, planning 280
Piece estimate 5
Pier footings 80-81
Pit, calculating volume 45, 47
Pit excavation 43-48
Pitch and rise, finding 136-137
Plank flooring, estimating 204
Planning
  advertising 253, 276
  CPM 223
  labor requirements 241
  materials 242
  office work 241
  reducing costs 243
  subcontracted work 241
  tract 271
  work schedule 278
Plans, reading 9
Plaster
  aggregate 178
  corner bead 184
  coverage 182, 183
  estimating costs 177, 181-185
  fasteners 178-179
  finish coat 183
  gypsum cements 177-178, 181-184
  ornamental work 185
  portland cement 177-178, 181-183
  surface area, figuring 179-180
Plastering, labor 184
Plot plan 11
Plywood sheathing
  allowance spans 132
  APA specifications 130
  application methods 131
  estimating 132-133
Plywood siding
  APA Sturd-I-Wall 186-188
  application methods 187-189
  nails 186
Plywood subflooring 124
Porch ceilings, estimating 197-198
Portland cement plaster 177-178, 181-183
Post, girder and sill estimating 119, 121
Poured insulation, estimating 163
Power equipment, excavating with
  backhoes 53-54
  bulldozers 50-51
  drag loaders 54
  draglines 54
  power shovels 51-53
Precast terrazzo, labor 213, 214
Prehung doors, installing 166-167
Prefabricated metal windows 170
Presenting your service, selling 256-257
Price changes, watching for 242
Profit 19-20, 23
Program Evaluation Review Technique (PERT) . 223
Progress control, scheduling 222-223
Project schedule, CPM 232
Projected horizontal area, roof surface 147-149
Proposal and contract, sample 260-261
Prospect list, building 254
Pumping and shoring, excavation 43

*Q*uantity estimate, excavating 57-60
Quantity estimate sheets 58, 64
Quarry tile, estimating 212-213

*R*after
  conversion diagram 137
  framing materials 141-143
  framing terms 136
  length, finding 136-138
  length table, framing square 140-141
  pitch 136
  ratios 137-138
  rise 136
  supports 142
  types 135
Rafters
  common 135-138
  estimating 135-142
  hip 135, 137-140
  jack 135, 140
  valley 135, 137-140
Rake/area conversion table 149
Ratio analysis 245-250
Ratios, accounting
  balance sheet 247-250
  current assets to current liabilities 247-248
  current liabilities to tangible net worth 249
  expense 246-247
  fixed assets to tangible net worth 250
  operating 246, 247
  small contractor expenses 247
Ratios, rafter 137
Reading floor plans 9-12
Record keeping 244-245
Rectangle, area of 218
Reflective insulators 159
Regional offices, SBA 294-296
Regular business loans, SBA 290-291
Reinforced concrete, estimating 85-86
Remodelers, contingencies 18-19
Reports, construction, government 264-266
Rescission of contract 259, 262
Research for spec building 267-268
Residential estimate breakdown 35
Resilient flooring, estimating 213, 215, 216
Resourcefulness, importance of 237-238
Restraints, CPM 229
Ridge boards, estimating 141
Rigid board insulation, labor 163
Roll roofing 150, 152, 153
Roof
  areas, estimating 143, 147-149
  covering, estimating 146-158
  hip/valley conversion table 154
  horizontal projection 147-149
  sheathing 143
  styles, common 146
Roof framing, estimating
  ceiling joists 133-135
  rafters 135-143, 144
  trusses 144
Roofing
  asphalt, estimating 149-152, 153
  built-up bituminous, labor 157
  roll 150, 153
  slope limitations 150
Room area chart, wallboard 174
Rough carpentry, estimating
  building paper 124
  ceilings 133-135
  floors 121-124
  framing 124-130
  general 110-119
  labor 7, 144, 145
  posts, girders, sills 119, 121
  rafters 135-143
  roof sheathing 143
  sheathing 131-133
Rough open sizes, windows 169
Rubble stone, estimating 100-101

*S*addle boards, chimneys 141
Sales strategy, tailoring 276
Salesmanship, creative 254
Sand
  moisture content 94
  specific gravity 93-94
  weight 94
Sash bars 168
Saturated felt, estimating 152
SBA. *See* Small Business Administration
Schedule, work, typical 278
Scheduling work flow
  CPM 223-234
  progress control 222-223
  techniques 222
SCORE (Service Corps of Retired Executives) 291, 292
Screens, window, estimating 170-171
Second floor plan, reading 11
Sectional drawing 9
Sectional forms, concrete 83, 84
Selling
  basics 254-255
  procedures 255-263
Selling, how to
  answering objections 257
  approaching the prospect 255
  closing the sale 259
  demonstrating your service 256-257
Service Corps of Retired Executives. *See* SCORE
Setting your price 21-22
Shakes
  application details 192
  coverage 156
  estimating 156-157
  hardboard siding 191
Sheathing, roof
  estimating 143
  labor performance 144
  nail requirements 145
Sheathing, wall
  allowable spans 132
  application 130, 131
  coverage 132
  estimating 132-133
  insulating foam board 132
Sheet flooring, labor 216
Sheet metal work, labor 157
Sheetrock
  application 172-173
  area table 175
  estimating 172-177
  finish materials 174-176
  labor 176-177
  nails 174

Shingles
asbestos . . . 156-157
asphalt . . . 150, 151-152
wood . . . 154-155
Shoring, excavation . . . 43
Shutters, estimating . . . 171
Side casing, estimating . . . 208
Siding
aluminum . . . 194
bevel . . . 194
corner treatments . . . 193
hardboard . . . 189
lap . . . 189
plywood . . . 186-189
Stud-I-Wall . . . 186-187
vinyl . . . 194-195
wood . . . 191, 194
Siding, estimating . . . 186, 189, 190, 193-195
Single-hung windows . . . 169-170
Single-roll wallpaper requirements . . . 220
Site
examination . . . 12
grades, establishing . . . 50
survey . . . 36-37
Slabs
concrete for . . . 85
estimating . . . 84, 85
formwork . . . 79, 84
labor, CEF . . . 68, 85
material . . . 68
Slope limitations, asphalt roofing . . . 150
Small Business Administration (SBA)
definition, small business . . . 290
field offices . . . 294-296
financial assistance . . . 290-291
management assistance . . . 291-292
operational levels . . . 292
Small Business Development Centers (SBDC) . . . 291, 292
Small Business Institutes (SBI) . . . 291, 292
Small contractors
expense ratios . . . 247
SBA loans . . . 291
Soffits, estimating . . . 197-198
Soil characteristics . . . 56
Sole proprietorship
advantages of . . . 286-287
disadvantages of . . . 287
Solid bridging . . . 122-123
Spacing, concrete forms
column yoke . . . 78
joist . . . 79
stringers . . . 79
stud . . . 74
tie wire . . . 76
wale . . . 75
Spacing, floor joists . . . 121
Spec building
construction cycle . . . 264-267
merchandising . . . 268-270
research needed . . . 267-268
Specialty finishes, estimating
carpet . . . 215
ceramic tile . . . 212-213
paints . . . 215-220
resilient flooring . . . 213, 215, 216
terrazzo . . . 213-214
wall coverings . . . 220
Specific gravity, sand . . . 93
Split-faced ashlar stone, estimating . . . 101-102
Square, area of . . . 218
Squaring corners . . . 38
Stair
checklist . . . 201
dimensions . . . 200
forms . . . 80
Stairs, estimating . . . 199-202
Staking out the building . . . 37-39
Standard nail requirements, rough carpentry . . . 145
Starters strips, roofing . . . 152
State Unemployment Insurance . . . 18
Steel door buck . . . 70
Steel doors, labor . . . 167
Stepped footing, estimating . . . 82
Stone
anchors . . . 100
ashlar . . . 101-103
cut . . . 101-103
joints . . . 104
rubble . . . 100-101
Stonework, estimating . . . 98-103
Storm windows . . . 170
Stringers, maximum span . . . 79
Strip flooring
definition . . . 204
labor . . . 205-206
material . . . 205-206
nailing schedule . . . 206
Stripping and storing topsoil . . . 41
Struck capacity . . . 55
Stud spacing, maximum . . . 74
Studs, estimating
balloon framing . . . 126-127
board feet per square foot . . . 125
modern braced framing . . . 129
western . . . 124-126, 128
Study your market . . . 267
Sturd-I-Wall, APA . . . 186, 187
Style of the house, spec building . . . 269
Subchapter S corporation . . . 289
Subcontract work, planning . . . 241
Subflooring, estimating . . . 123-124
Successful management
function of management . . . 235-236
job planning . . . 240-241
management ability, building . . . 236-238
management team, building . . . 239-240
ratios . . . 245-250
record keeping . . . 244-245
strong company building . . . 238-239
Summary of estimate . . . 33, 35
Superintending the job . . . 243
Surety bonds, SBA . . . 280, 291
Surface area
roof . . . 147-149
wallboard . . . 179
Survey markers . . . 36
Swell . . . 55, 56

*T*aking off
concrete quantities . . . 67
excavation . . . 57-58
lumber estimates . . . 114
molding . . . 207-208
Tamping fill . . . 85
Taxes . . . 17-18
Technical publications . . . 292
Temperature zones . . . 162
Terrazzo, estimating . . . 213, 214
Thinking ahead, management skills . . . 236
Tie wire
breaking load . . . 77
spacing . . . 76, 77
Tile
adhesive . . . 215
ceramic . . . 212-213
quarry . . . 213
resilient . . . 213, 215
terrazzo . . . 213, 214
Timetable from arrow diagram, CPM . . . 233-234
Topographical spot . . . 36, 37
Topsoil
excavation . . . 63
stripping and storing . . . 41
Tractors, excavation . . . 50-54
Trench and pit excavation . . . 43-48, 63
Trenching by hand, labor . . . 48
Triangle, area of . . . 218
Trim, estimating
base trim . . . 210
labor . . . 210
lengths of trim . . . 208
patterns . . . 209
standard trim . . . 207
trim for openings . . . 209
Truck hauling . . . 54-55
Trusses, estimating . . . 143, 144

*U*ncoursed rubble stone . . . 100
Underlayment
floor . . . 123-124
roof . . . 143
Unemployment insurance . . . 17-18
Uniform Classification of Operations . . . 12
Unit cost estimates
definition . . . 5
excavation . . . 61, 65
plaster . . . 187
Unit costs, cost keeping . . . 23-25, 61

*V*alley/hip rafter conversion table . . . 154
Valley rafters . . . 135, 138-140
Vapor barriers . . . 160-161
Vertical application
sheathing . . . 131
siding . . . 187, 195
Vertical board siding, estimating . . . 195
Vertical wood siding . . . 194
Veterans, assistance to (SBA) . . . 292
Vinyl flooring, labor . . . 213, 216
Vinyl siding . . . 194
Volume, excavation
calculating . . . 41-42
factors . . . 47
irregular shapes . . . 43
material in piles . . . 41
pit . . . 45, 47
sloping ground . . . 42-43
trenches . . . 44

*W*ale spacing, maximum . . . 75
Wall cabinets . . . 203
Wall coverings, labor . . . 220
Wall footings . . . 81
Wall forms
designing . . . 68, 73-78
estimating . . . 80, 82, 83
prefabricated . . . 68-73
Wall paneling, labor . . . 185
Wall plan, foundation . . . 91
Wall sheathing
allowable spans . . . 132
application . . . 131
board . . . 130
coverage . . . 132
estimating . . . 132
Wall studs
balloon framing . . . 126-127
board feet per square foot . . . 125
modern braced framing . . . 129
western framing . . . 124-126, 128
Wallboard
area table . . . 175
ceiling/wall application . . . 172
estimating . . . 173-174
finish materials . . . 174-176
labor . . . 176-177
nails . . . 174
Wallpaper requirements . . . 220
Wall-to-wall carpet . . . 215
Waste factors, floor tile . . . 215
Water table, estimating . . . 195
Water, weight of . . . 94
Weekly time sheet . . . 29, 30
Weight of sand & water . . . 94
Western framing . . . 124-126, 128
Window
abbreviations . . . 168
dimensions . . . 169
labor guidelines . . . 170
RO sizes . . . 169
screens . . . 170
shutters . . . 171
stools . . . 208
trim . . . 207-209
types . . . 169-170
Windows
awning . . . 170
casement . . . 170
double-hung . . . 169
estimating . . . 168-171
horizontal sliding . . . 170
metal . . . 170
single-hung . . . 169
storm . . . 170
Withholding taxes . . . 17-18
Women, help from SBA . . . 290
Wood flooring
block flooring, installing . . . 207
estimating . . . 204-206
labor . . . 205
material . . . 205
nail schedule . . . 206
types of . . . 204
Wood kitchen cabinets . . . 202-204
Wood shakes, estimating . . . 156-157
Wood shingles
application . . . 154
coverage . . . 155
estimating . . . 155-156
Wood siding . . . 191, 193-194
Work elements, CPM . . . 227
Work flow, scheduling . . . 221-234
Work schedule, planning . . . 278
Workbook, estimating . . . 12
Worker's Comp Insurance . . . 17-18
Worksheets, business plan
action . . . 285
advertising . . . 277
balance sheet . . . 284
bonding . . . 281
business plan . . . 274
cash forecast . . . 282
customer profile . . . 275
equipment . . . 281
expenses . . . 283
personnel . . . 281

*Y*ield of hydrated lime . . . 183
Yoke spacing, column . . . 78

Practical References for Builders

CD Estimator

If your computer has *Windows*™ and a CD-ROM drive, CD Estimator puts at your fingertips 85,000 construction costs for new construction, remodeling, renovation & insurance repair, electrical, plumbing, HVAC and painting. Quarterly cost updates are available at no charge on the Internet. You'll also have the *National Estimator* program —a stand-alone estimating program for *Windows*™ that *Remodeling* magazine called a "computer wiz," and Job Cost Wizard, a program that lets you export your estimates to QuickBooks Pro for actual job costing. A 40-minute interactive video teaches you how to use this CD-ROM to estimate construction costs. And to top it off, to help you create professional-looking estimates, the disk includes over 40 construction estimating and bidding forms in a format that's perfect for nearly any *Windows*™ word processing or spreadsheet program.
CD Estimator is $73.50

Getting Financing & Developing Land

Developing land is a major leap for most builders — yet that's where the big money is made. This book gives you the practical knowledge you need to make that leap. Learn how to prepare a market study, select a building site, obtain financing, guide your plans through approval, then control your building costs so you can ensure yourself a good profit. Includes a CD-ROM with forms, checklists, and a sample business plan you can customize and use to help you sell your idea to lenders and investors. **232 pages, 8½ x 11, $39.00**

Construction Forms & Contracts

125 forms you can copy and use — or load into your computer (from the FREE disk enclosed). Then you can customize the forms to fit your company, fill them out, and print. Loads into *Word* for *Windows*™, *Lotus 1-2-3*, *WordPerfect*, *Works*, or *Excel* programs. You'll find forms covering accounting, estimating, fieldwork, contracts, and general office. Each form comes with complete instructions on when to use it and how to fill it out. These forms were designed, tested and used by contractors, and will help keep your business organized, profitable and out of legal, accounting and collection troubles. Includes a CD-ROM for *Windows*™ and Mac.
400 pages, 8½ x 11, $41.75

Basic Lumber Engineering for Builders

Beam and lumber requirements for many jobs aren't always clear, especially with changing building codes and lumber products. Most of the time you rely on your own "rules of thumb" when figuring spans or lumber engineering. This book can help you fill the gap between what you can find in the building code span tables and what you need to pay a certified engineer to do. With its large, clear illustrations and examples, this book shows you how to figure stresses for pre-engineered wood or wood structural members, how to calculate loads, and how to design your own girders, joists and beams. Included FREE with the book — an easy-to-use limited version of NorthBridge Software's *Wood Beam Sizing* program.
272 pages, 8½ x 11, $38.00

Moving to Commercial Construction

In commercial work, a single job can keep you and your crews busy for a year or more. The profit percentages are higher, but so is the risk involved. This book takes you step-by-step through the process of setting up a successful commercial business; finding work, estimating and bidding, value engineering, getting through the submittal and shop drawing process, keeping a stable work force, controlling costs, and promoting your business. Explains the design/build and partnering business concepts and their advantage over the competitive bid process. Includes sample letters, contracts, checklists and forms that you can use in your business, plus a CD-ROM with blank copies in several word-processing formats for both Mac and PC computers. **256 pages, 8½ x 11, $42.00**

National Repair & Remodeling Estimator

The complete pricing guide for dwelling reconstruction costs. Reliable, specific data you can apply on every repair and remodeling job. Up-to-date material costs and labor figures based on thousands of jobs across the country. Provides recommended crew sizes; average production rates; exact material, equipment, and labor costs; a total unit cost and a total price including overhead and profit. Separate listings for high- and low-volume builders, so prices shown are specific for any size business. Estimating tips specific to repair and remodeling work to make your bids complete, realistic, and profitable. Includes a CD-ROM with an electronic version of the book with *National Estimator*, a stand-alone *Windows*™ estimating program, plus an interactive multimedia video that shows how to use the disk to compile construction cost estimates.
296 pages, 8½ x 11, $53.50. Revised annually

Contractor's Survival Manual

How to survive hard times and succeed during the up cycles. Shows what to do when the bills can't be paid, finding money and buying time, transferring debt, and all the alternatives to bankruptcy. Explains how to build profits, avoid problems in zoning and permits, taxes, time-keeping, and payroll. Unconventional advice on how to invest in inflation, get high appraisals, trade and postpone income, and stay hip-deep in profitable work.
160 pages, 8½ x 11, $25.00

Working Alone

This unique book shows you how to become a dynamic one-man team as you handle nearly every aspect of house construction, including foundation layout, setting up scaffolding, framing floors, building and erecting walls, squaring up walls, installing sheathing, laying out rafters, raising the ridge, getting the roof square, installing rafters, subfascia, sheathing, finishing eaves, installing windows, hanging drywall, measuring trim, installing cabinets, and building decks. **152 pages, 5½ x 8½, $17.95**

Estimating With Microsoft Excel

Most builders estimate with Excel because it's easy to learn, quick to use, and can be customized to your style of estimating. Here you'll find step-by-step instructions on how to create your own customized automated spreadsheet estimating program for use with Excel. You'll learn how to use the magic of Excel to create detail sheets, cost breakdown summaries, and links. You'll put this all to use in estimating concrete, rebar, permit fees, and roofing. You can even create your own macros. Includes a CD-ROM that illustrates examples in the book and provides you with templates you can use to set up your own estimating system. **148 pages, 7 x 9, $39.95**

Professional Kitchen Design

Remodeling kitchens requires a "special" touch -- one that blends artistic flair with function to create a kitchen with charm and personality as well as one that is easy to work in. Here you'll find how to make the best use of the space available in any kitchen design job, as well as tips and lessons on how to design one-wall, two-wall, L-shaped, U-shaped, peninsula and island kitchens. Also includes what you need to know to run a profitable kitchen design business. **176 pages, 8½ x 11, $24.50**

Contractor's Plain-English Legal Guide

For today's contractors, legal problems are like snakes in the swamp — you might not see them, but you know they're there. This book tells you where the snakes are hiding and directs you to the safe path. With the directions in this easy-to-read handbook you're less likely to need a $200-an-hour lawyer. Includes simple directions for starting your business, writing contracts that cover just about any eventuality, collecting what's owed you, filing liens, protecting yourself from unethical subcontractors, and more. For about the price of 15 minutes in a lawyer's office, you'll have a guide that will make many of those visits unnecessary. **272 pages, 8½ x 11, $49.50**

How to Succeed With Your Own Construction Business

Everything you need to start your own construction business: setting up the paperwork, finding the work, advertising, using contracts, dealing with lenders, estimating, scheduling, finding and keeping good employees, keeping the books, and coping with success. If you're considering starting your own construction business, all the knowledge, tips, and blank forms you need are here. **336 pages, 8¹/₂ x 11, $28.50**

Markup & Profit: A Contractor's Guide

 In order to succeed in a construction business, you have to be able to price your jobs to cover all labor, material and overhead expenses, and make a decent profit. The problem is knowing what markup to use. You don't want to lose jobs because you charge too much, and you don't want to work for free because you've charged too little. If you know how to calculate markup, you can apply it to your job costs to find the right sales price for your work. This book gives you tried and tested formulas, with step-by-step instructions and easy-to-follow examples, so you can easily figure the markup that's right for your business. Includes a CD-ROM with forms and checklists for your use. **320 pages, 8½ x 11, $32.50**

Roofing Construction & Estimating

Installation, repair and estimating for nearly every type of roof covering available today in residential and commercial structures: asphalt shingles, roll roofing, wood shingles and shakes, clay tile, slate, metal, built-up, and elastomeric. Covers sheathing and underlayment techniques, as well as secrets for installing leakproof valleys. Many estimating tips help you minimize waste, as well as insure a profit on every job. Troubleshooting techniques help you identify the true source of most leaks. Over 300 large, clear illustrations help you find the answer to just about all your roofing questions. **432 pages, 8¹/₂ x 11, $38.00**

Steel-Frame House Construction

Framing with steel has obvious advantages over wood, yet building with steel requires new skills that can present challenges to the wood builder. This new book explains the secrets of steel framing techniques for building homes, whether pre-engineered or built stick by stick. It shows you the techniques, the tools, the materials, and how you can make it happen. Includes hundreds of photos and illustrations, plus a CD-ROM with steel framing details, a database of steel materials and manhours, with an estimating program. **320 pages, 8¹/₂ x 11, $39.75**

Basic Engineering for Builders

 If you've ever been stumped by an engineering problem on the job, yet wanted to avoid the expense of hiring a qualified engineer, you should have this book. Here you'll find engineering principles explained in non-technical language and practical methods for applying them on the job. With the help of this book you'll be able to understand engineering functions in the plans and how to meet the requirements, how to get permits issued without the help of an engineer, and anticipate requirements for concrete, steel, wood and masonry. See why you sometimes have to hire an engineer and what you can undertake yourself: surveying, concrete, lumber loads and stresses, steel, masonry, plumbing, and HVAC systems. This book is designed to help the builder save money by understanding engineering principles that you can incorporate into the jobs you bid. **400 pages, 8¹/₂ x 11, $36.50**

Handbook of Construction Contracting, Volume 1

Everything you need to know to start and run your construction business; the pros and cons of each type of contracting, the records you'll need to keep, and how to read and understand house plans and specs so you find any problems before the actual work begins. All aspects of construction are covered in detail, including all-weather wood foundations, practical math for the job site, and elementary surveying.
416 pages, 8¹/₂ x 11, $32.75

Plumber's Handbook Revised

 This new edition shows what will and won't pass inspection in drainage, vent, and waste piping, septic tanks, water supply, graywater recycling systems, pools and spas, fire protection, and gas piping systems. All tables, standards, and specifications are completely up-to-date with recent plumbing code changes. Covers common layouts for residential work, how to size piping, select and hang fixtures, practical recommendations, and trade tips. It's the approved reference for the plumbing contractor's exam in many states. Includes an extensive set of multiple choice questions after each chapter, and in the back of the book, the answers and explanations. Also in the back of the book, a full sample plumber's exam. **352 pages, 8¹/₂ x 11, $32.00**

Blueprint Reading for the Building Trades

How to read and understand construction documents, blueprints, and schedules. Includes layouts of structural, mechanical, HVAC and electrical drawings. Shows how to interpret sectional views, follow diagrams and schematics, and covers common problems with construction specifications. **192 pages, 5¹/₂ x 8¹/₂, $14.75**

Contractor's Guide to QuickBooks Pro 2004

This user-friendly manual walks you through QuickBooks Pro's detailed setup procedure and explains step-by-step how to create a first-rate accounting system. You'll learn in days, rather than weeks, how to use QuickBooks Pro to get your contracting business organized, with simple, fast accounting procedures. On the CD included with the book you'll find a QuickBooks Pro file preconfigured for a construction company (you drag it over onto your computer and plug in your own company's  data). You'll also get a complete estimating program, including a database, and a job costing program that lets you export your estimates to QuickBooks Pro. It even includes many useful construction forms to use in your business. **344 pages, 8½ x 11, $48.50**
Also available: **Contractor's Guide to QuickBooks Pro 2001, $45.25**
Contractor's Guide to QuickBooks Pro 2003, $47.75

Estimating Home Building Costs

Estimate every phase of residential construction from site costs to the profit margin you include in your bid. Shows how to keep track of manhours and make accurate labor cost estimates for footings, foundations, framing and sheathing finishes, electrical, plumbing, and more. Provides and explains sample cost estimate worksheets with complete instructions for each job phase. **320 pages, 5¹/₂ x 8¹/₂, $17.00**

Contractor's Guide to the Building Code Revised

This new edition was written in collaboration with the International Conference of Building Officials, writers of the code. It explains in plain English exactly what the latest edition of the *Uniform Building Code* requires. Based on the 1997 code, it explains the changes and what they mean for the builder. Also covers the *Uniform Mechanical Code* and the *Uniform Plumbing Code*. Shows how to design and construct residential and light commercial buildings that'll pass inspection the first time. Suggests how to work with an inspector to minimize construction costs, what common building shortcuts are likely to be cited, and where exceptions may be granted. **320 pages, 8¹/₂ x 11, $39.00**

National Renovation & Insurance Repair Estimator

 Current prices in dollars and cents for hard-to-find items needed on most insurance, repair, remodeling, and renovation jobs. All price items include labor, material, and equipment breakouts, plus special charts that tell you exactly how these costs are calculated. Includes a CD-ROM with an electronic version of the book with *National Estimator*, a stand-alone *Windows*™ estimating program, plus an interactive multimedia video that shows how to use the disk to compile construction cost estimates. **568 pages, 8¹/₂ x 11, $54.50. Revised annually**

Finish Carpenter's Manual

Everything you need to know to be a finish carpenter: assessing a job before you begin, and tricks of the trade from a master finish carpenter. Easy-to-follow instructions for installing doors and windows, ceiling treatments (including fancy beams, corbels, cornices and moldings), wall treatments (including wainscoting and sheet paneling), and the finishing touches of chair, picture, and plate rails. Specialized interior work includes cabinetry and built-ins, stair finish work, and closets. Also covers exterior trims and porches. Includes manhour tables for finish work, and hundreds of illustrations and photos. **208 pages, 8¹/₂ x 11, $22.50**

Wood-Frame House Construction

Step-by-step construction details, from the layout of the outer walls, excavation and formwork, to finish carpentry and painting. Contains all new, clear illustrations and explanations updated for construction in the '90s. Everything you need to know about framing, roofing, siding, interior finishings, floor covering and stairs — your complete book of wood-frame homebuilding. **320 pages, 8¹/₂ x 11, $25.50. Revised edition**

Finish Carpentry: Efficient Techniques for Custom Interiors

Professional finish carpentry demands expert skills, precise tools, and a solid understanding of how to do the work. This new book explains how to install moldings, paneled walls and ceilings, and just about every aspect of interior trim — including doors and windows. Covers built-in bookshelves, coffered ceilings, skylight wells and soffits, including paneled ceilings with decorative beams. **288 pages, 8¹/₂ x 11, $34.95**

National Construction Estimator

Current building costs for residential, commercial, and industrial construction. Estimated prices for every common building material. Provides manhours, recommended crew, and gives the labor cost for installation. Includes a CD-ROM with an electronic version of the book with *National Estimator*, a stand-alone *Windows*™ estimating program, plus an interactive multimedia video that shows how to use the disk to compile construction cost estimates. **616 pages, 8¹/₂ x 11, $52.50. Revised annually**

Roof Framing

Shows how to frame any type of roof in common use today, even if you've never framed a roof before. Includes using a pocket calculator to figure any common, hip, valley, or jack rafter length in seconds. Over 400 illustrations cover every measurement and every cut on each type of roof: gable, hip, Dutch, Tudor, gambrel, shed, gazebo, and more. **480 pages, 5¹/₂ x 8¹/₂, $24.50**

Construction Estimating Reference Data

Provides the 300 most useful manhour tables for practically every item of construction. Labor requirements are listed for sitework, concrete work, masonry, steel, carpentry, thermal and moisture protection, doors and windows, finishes, mechanical and electrical. Each section details the work being estimated and gives appropriate crew size and equipment needed. Includes a CD-ROM with an electronic version of the book with *National Estimator*, a stand-alone *Windows*™ estimating program, plus an interactive multimedia video that shows how to use the disk to compile construction cost estimates. **432 pages, 11 x 8¹/₂, $39.50**

Commercial Metal Stud Framing

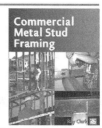

Framing commercial jobs can be more lucrative than residential work. But most commercial jobs require some form of metal stud framing. This book teaches step-by-step, with hundreds of job site photos, high-speed metal stud framing in commercial construction. It describes the special tools you'll need and how to use them effectively, and the material and equipment you'll be working with. You'll find the shortcuts, tips and tricks-of-the-trade that take most steel frames years on the job to discover. Shows how to set up a crew to maintain a rhythm that will speed progress faster than any wood framing job. If you've framed with wood, this book will teach you how to be one of the few top-notch metal stud framers. **208 pages, 8¹/₂ x 11, $45.00**

Building Layout

Shows how to use a transit to locate a building correctly on the lot, plan proper grades with minimum excavation, find utility lines and easements, establish correct elevations, lay out accurate foundations, and set correct floor heights. Explains how to plan sewer connections, level a foundation that's out of level, use a story pole and batterboards, work on steep sites, and minimize excavation costs. **240 pages, 5¹/₂ x 8¹/₂, $19.00**

Mail This Card Today
For a Free Full Color Catalog

Over 100 books, annual cost guides and estimating software packages at your fingertips with information that can save you time and money. Here you'll find information on carpentry, contracting, estimating, remodeling, electrical work, and plumbing.

All items come with an unconditional 10-day money-back guarantee. If they don't save you money, mail them back for a full refund.

Name_____

e-mail address (for special offers)

Company_____

Address_____

City/State/Zip

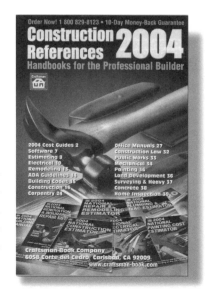

Craftsman Book Company / 6058 Corte del Cedro / P.O. Box 6500 / Carlsbad, CA 92018

BUSINESS REPLY MAIL

FIRST CLASS MAIL PERMIT NO. 271 CARLSBAD, CA

POSTAGE WILL BE PAID BY ADDRESSEE

 Craftsman Book Company
6058 Corte del Cedro
P.O. Box 6500
Carlsbad, CA 92018-9974

BUSINESS REPLY MAIL

FIRST CLASS MAIL PERMIT NO. 271 CARLSBAD, CA

POSTAGE WILL BE PAID BY ADDRESSEE

Craftsman Book Company
6058 Corte del Cedro
P.O. Box 6500
Carlsbad, CA 92018-9974

BUSINESS REPLY MAIL

FIRST CLASS MAIL PERMIT NO. 271 CARLSBAD, CA

POSTAGE WILL BE PAID BY ADDRESSEE

Craftsman Book Company
6058 Corte del Cedro
P.O. Box 6500
Carlsbad, CA 92018-9974